SYMMETRIES IN QUANTUM MECHANICS

T0362243

Graduate Student Series in Physics
Other books in the series

GRADUATE STUDENT SERIES IN PHYSICS

Series Editor:
Professor Douglas F Brewer, MA, DPhil
Emeritus Professor of Experimental Physics, University of Sussex

SYMMETRIES IN QUANTUM MECHANICS

FROM ANGULAR MOMENTUM TO SUPERSYMMETRY

MASUD CHAICHIAN

Department of Physics, University of Helsinki
and
Helsinki Institute of Physics

ROLF HAGEDORN

Retired Staff Member of Theory Division, CERN, Geneva

Taylor & Francis
Taylor & Francis Group
New York London

Taylor & Francis is an imprint of the
Taylor & Francis Group, an informa business

Published in 1998 by
Taylor & Francis Group
270 Madison Avenue
New York, NY 10016

Published in Great Britain by
Taylor & Francis Group
2 Park Square
Milton Park, Abingdon
Oxon OX14 4RN

No claim to original U.S. Government works
Printed in the United States of America on acid-free paper
10 9 8 7 6 5 4 3 2

International Standard Book Number-10: 0-7503-0408-1 (Softcover)
International Standard Book Number-13: 978-0-7503-0408-5 (Softcover)

Library of Congress Cataloging-in-Publication Data

Catalog record is available from the Library of Congress

**Visit the Taylor & Francis Web site at
http://www.taylorandfrancis.com**

Symmetry, as wide or as narrow as you may define its meaning, is one idea by which man through the ages has tried to comprehend and create order, beauty and perfection.

Hermann Weyl

CONTENTS

PREFACE

Symmetries are of primary importance in physics, particularly in quantum theory. Among the first symmetries which were remarked upon historically and conceptually were those of space and time. While Galilei invariance was later to be generalized into Lorentz invariance, the invariance under spatial rotations (though also a subgroup of the Lorentz group) has survived as such and has become an important subject of quantum theory. It is well known, and will also be discussed extensively in this book, that invariance generates conservation laws; in the case of rotational invariance the conserved quantity is the angular momentum. Therefore there is a dense interlacing between

- the description of rotations, geometrically and group theoretically
- their representations by unitary transformations in the Hilbert space of quantum mechanical states, and
- the quantum theory of angular momentum.

Somehow these are only three different facets of one and the same thing. According to circumstances one or the other aspect will naturally be pushed to the foreground.

The main *raison d'être*, however, of this book is that we have tried to exhibit the wholeness of this 'one and the same thing', namely *symmetry* as the basic concept underlying all the (sometimes tedious) formalisms. This implies, for instance, that the 'Wigner theorem'—stating that to every symmetry group there corresponds a representation (unitary or anti-unitary) in the quantum mechanical Hilbert space—is, in our opinion, of such fundamental importance that it should not simply be dealt with by saying 'Wigner has proved that . . .'. Therefore we give an explicit proof of it. There, as on many other occasions, our aim has been to stress the underlying ideas and motivations: the 'why is' and the 'how is'; in short, to make things plausible rather than overburden the reader with a formal and condensed proof. Lovers of rigour and compactness may be irritated by our often pedestrian length, as well as by some repetitions in which earlier arguments come up again and are discussed anew in another context.

Thus, in spite of its mathematical appearance, this is a didactic text written by physicists for physicists. We took our time in writing it and we hope our readers will take their time in reading it.

This, then, has been our philosophy in writing the book. It is necessarily biased and incomplete (for instance we did not include the graphical representation of formulae). Fortunately there are a sufficient number of other books on this subject with different aims:

- some putting the formal mathematical aspects in the foreground
- some trying to be as short and concentrated as possible
- some emphasizing practical applications and examples, and
- some serving as inexhaustible collections of formulae.

We will mention in appropriate chapters examples of such books where some different aspects of the subject are emphasized. Knowing of the existence of these other books and making the reader aware of them makes us feel that our text might be a useful complement to the existing literature. Thus we hope to help our readers more towards a feeling and an understanding of what it is all about rather than just a superficial knowledge. As three-dimensional rotations are conceptually simple and accessible to intuition and since it also happens that the three-dimensional rotation group is the simplest non-Abelian continuous symmetry in physics, this naturally occupies the largest part of the book.

In writing this book we have benefited from discussing various questions with many of our colleagues. It is a pleasure to express our gratitude to all of them, especially Khosrow Chadan, Wen-Feng Chen, Andrei Demichev, Petr Kulish, Kazuhiko Nishijima, Claus Montonen and Peter Prešnajder, and to acknowledge the stimulating discussions and their useful advice. In particular, our appreciation is extended to Matti Pitkänen for all his contributions, improvements and suggestions during the preparation of the manuscript.

We would also like to thank the commissioning editor, Jim Revill, for his patience, and the (unknown to us) referees for valuable complements, corrections and suggestions including even a brushing up of our English. Our special thanks go to Jeanne Rostand from CERN for having improved considerably the language of the book, although we asked her to leave some non-native English flavour in the text.

Masud Chaichian, Rolf Hagedorn
Helsinki, Geneva
July 1997

1

INTRODUCTION

1.1 Notation

We shall employ here the following notation and basic formulae of quantum mechanics; our units are such that $\hbar = c = 1$.

In quantum mechanics observables are represented by Hermitian operators in a Hilbert space. The physical fact that not all measurements (preparations) are compatible with each other corresponds to the fact that not all operators commute. The 'state' of a physical system is fully determined in one way or another if a complete set of compatible measurements has been carried through such that no further measurement exists which is compatible with, and which is not simply a calculable function of, those measurements already made. Since not all independent measurements are compatible, there exist several non-compatible 'complete sets of measurements' for a given physical system, which lead to different aspects of that system. Each complete set of commuting observables leads to a complete basis in the Hilbert space, the basis vectors being the simultaneous eigenvectors of all operators of the set. We label these basis vectors—which are assumed to be normalized to unit length—by the eigenvalues: if $A, B, C, D, \ldots, X$ is a complete set of commuting observables, then the physical states are represented by state vectors: $|a, b, c, d, \ldots, x\rangle$ and we have

$$
\begin{aligned}
A \,|a, b, c, d, \ldots, x\rangle &= a \,|a, b, c, d, \ldots, x\rangle \quad \text{etc} \\
\langle a', b', c', d', \ldots, x' \,|a, b, c, d, \ldots, x\rangle &= \delta_{aa'} \delta_{bb'} \ldots \delta_{xx'}
\end{aligned}
\tag{1.1.1}
$$

where $\delta_{aa'}$ is the Kronecker symbol if a is discrete and the Dirac δ function if continuous. We may conveniently compress all quantum numbers and all commuting operators into one symbol each:

$$
\begin{aligned}
\gamma &= \{a, b, c, d, \ldots, x\} & \Gamma|\gamma\rangle &= \gamma|\gamma\rangle \\
\Gamma &= \{A, B, C, D, \ldots, x\} & \langle\gamma'|\gamma\rangle &= \delta_{\gamma\gamma'}.
\end{aligned}
\tag{1.1.2}
$$

If there is another complete set of commuting observables, Λ, then the system may be described by states $|\lambda\rangle$, exhibiting another aspect of it. Since both sets, Γ and Λ, furnish a complete orthonormal basis in the Hilbert space, these two bases are connected by a unitary transformation U.

Any arbitrary state ψ may be expanded with respect to any basis, e.g.

$$|\psi\rangle = \sum_{\gamma} |\gamma\rangle\langle\gamma|\psi\rangle \quad \text{or}$$

$$|\psi\rangle = \sum_{\lambda} |\lambda\rangle\langle\lambda|\psi\rangle \quad \text{etc.} \tag{1.1.3}$$

From that follows the important 'spectral resolution of the unit operator', namely

$$1 \equiv \sum_{\gamma} |\gamma\rangle\langle\gamma| = \sum_{\lambda} |\lambda\rangle\langle\lambda| = \dots \tag{1.1.4}$$

where γ and λ must be summed over the whole range[1]. If the sum is carried only over a part of the whole range, then this is the unit operator in that part of the Hilbert space which is spanned by the vectors whose labels lie within the range of the summation; it is the null operator for all the rest, i.e. in the orthogonal subspace. Hence, if $\sum'$ indicates summation over a subset $\{\gamma\}'$ of the total range, then

$$\sideset{}{'}\sum_{\gamma} |\gamma\rangle\langle\gamma| = P_{\{\gamma\}'}$$

$$\sideset{}{'}\sum_{\lambda} |\lambda\rangle\langle\lambda| = P_{\{\lambda\}'} \tag{1.1.5}$$

are projection operators with the—necessary and sufficient—properties

$$P^2 = P \quad \text{and} \quad P^{\dagger} = P \tag{1.1.6}$$

(† means Hermitian conjugate). If the subset $\{\gamma\}'$ reduces to one single element γ, then

$$|\gamma\rangle\langle\gamma| = P_{\gamma} \tag{1.1.7}$$

is the projection onto the single state $|\gamma\rangle$. Since the operator Γ will multiply this single state by the eigenvalue, we obtain the total effect of Γ by decomposing the Hilbert space into subspaces $|\gamma\rangle = P_{\gamma}\mathcal{H}$, multiplying each by its eigenvalue γ and summing up over γ; hence

$$\Gamma = \sum_{\gamma} \gamma P_{\gamma} = \sum_{\gamma} |\gamma\rangle\gamma\langle\gamma| \tag{1.1.8}$$

which we call the 'spectral resolution of Γ', from which it follows at once that any (reasonable) function of Γ, $f(\Gamma)$, may be defined by its spectral resolution

$$f(\Gamma) = \sum_{\gamma} f(\gamma) P_{\gamma} = \sum_{\gamma} |\gamma\rangle f(\gamma)\langle\gamma| \tag{1.1.9}$$

which is a useful formula.

[1] This perhaps too schematic notation means, of course, integration wherever the spectrum is continuous. We consistently use this simplified notation.

Note that formulae (1.1.5), (1.1.7), (1.1.8) and (1.1.9) are invariant under the operation $|\gamma\rangle \rightarrow e^{iF(\gamma)}|\gamma\rangle$ implying $\langle\gamma| \rightarrow e^{-iF(\gamma)}\langle\gamma|$, where $F(\gamma)$ is any real function of γ. This invariance lead us to the expectation that phase factors of states may have little significance. This is indeed the case, as we shall see now.

If the quantum mechanical system is described by a certain state $|\psi(t)\rangle$, then the 'probability amplitude' A for finding in the next measurement the state $|\phi\rangle$ is given by

$$A(\psi(t) \rightarrow \phi) = \langle\phi|\psi(t)\rangle. \qquad (1.1.10)$$

The observable thing is, however, not A but its absolute square; $|A|^2$ is a probability (or, if $|\phi\rangle$ belongs to a continuum, a probability density which has to be multiplied by the corresponding interval $d\phi$)

$$\text{Prob}\,(|\psi(t)\rangle \rightarrow |\phi\rangle) = |\langle\phi|\psi(t)\rangle|^2. \qquad (1.1.11)$$

This leads to the important observation that the state vectors $|\psi\rangle$, etc, contain more information than can possibly be inferred from experiment. Indeed if we replace the above state vectors $|\phi\rangle$ and $|\psi\rangle$ by $e^{i\alpha}|\phi\rangle$ and $e^{i\beta}|\psi\rangle$, with α and β real, then the left-hand side (l.h.s.) of (1.1.11) would not change. As the maximal knowledge about a physical system is equivalent to specifying simultaneously all quantum numbers of one complete set, a state vector $|\gamma\rangle$ in the sense of (1.1.2) is a complete description of the actual physical state, but then $e^{i\alpha}|\gamma\rangle$ describes the same situation: (1.1.1) gives

$$\text{Prob}\,(|\gamma\rangle \rightarrow e^{i\alpha}|\gamma\rangle) = 1 \qquad (1.1.12)$$

which means that $e^{i\alpha}|\gamma\rangle$ and $|\gamma\rangle$ do not differ in any observable respect. We are thus forced to introduce the concept of 'rays' and 'physical states'. The system is said to be in a definite 'physical state' if a complete set of quantum numbers γ is specified. This does not yet define a state vector in Hilbert space, because all state vectors $e^{i\alpha}|\gamma\rangle$ with arbitrary real α represent the same physical state. The set of these state vectors

$$\hat{\gamma} \equiv \{e^{i\alpha}|\gamma\rangle;\ \alpha\ \text{real}\} \qquad (1.1.13)$$

is called a 'unit ray' $\hat{\gamma}$. (Correspondingly, 'rays' are defined; they differ from unit rays only by not being normalized.) We then state the postulate

There is a one-to-one correspondence between physical states and rays. $\qquad (1.1.14)$

As any element $e^{i\alpha}|\gamma\rangle$ of the unit ray $\hat{\gamma}$ represents the same physical state, we may always loosely say: 'the system is in the state $|\gamma\rangle$'. In other words: while a state vector uniquely specifies a physical state, the physical state does not specify a state vector but only the unit ray $\hat{\gamma}$. One point should be clear: although in the correspondence between physical states and state vectors the phase of the latter is entirely arbitrary, this arbitrariness is absent from the *relative* phases entering a linear superposition such as $a|\psi\rangle + b|\phi\rangle$. If this superposition is to represent a physical state, then its overall phase is arbitrary but not the angle between a and b in the complex plane.

Therefore the freedom in the phase factors must not be misinterpreted as implying their physical insignificance. We shall encounter enough examples where the phase factors are important.

The remark that the relative phase of a and b in $a|\psi\rangle + b|\phi\rangle$ is significant immediately rules out one possibility one might have thought of in view of the postulate (1.1.14); namely the possibility to formulate quantum mechanics with rays rather than state vectors. As one sees from the above example, rays cannot be added. The superposition principle (which holds for state vectors but not for rays) is, however, a basic requirement of quantum mechanics.

We are thus forced to formulate quantum mechanics using Hilbert vectors. This has some consequences regarding symmetries of the physical system, which will be investigated in more detail in chapter 2. The point is: assume the physical system is invariant under a certain group G—called a symmetry of the system. If a physical state S is mapped by $g \in G$ onto the physical state $S' = gS$, then the corresponding unit ray $\hat{\gamma}$ is mapped onto its image $\hat{\gamma}' = \hat{g}\hat{\gamma}$. (We have adopted the rule that Greek letters $\hat{\gamma}$, etc stand for unit rays, Latin letters $\hat{g}$, etc for ray operators which will be defined more pedantically later on.) As between S and $\hat{g}$ there is a one-to-one correspondence, it follows that between g and $\hat{g}(g)$ there is also a one-to-one correspondence. What we really need to know, however, is not so much how unit rays transform but how Hilbert vectors do. Obviously, there is an infinity of mappings of Hilbert vectors corresponding to one given ray transformation. The problem is to pick out the most suitable mappings. It will turn out that these are either unitary or anti-unitary.

We denote the unit matrix and unit operator simply by 1.

1.2 Some basic concepts in quantum mechanics

Observables are represented in quantum mechanics (QM) by Hermitian operators in Hilbert space. These operators are analogues of finite-dimensional Hermitian operators, but the infinite dimensionality requires some mathematical explanation. Appendices A and B at the end of the book provide additional details for the following discussion. For general mathematical aspects of QM we recommend the reader to consult the fundamental books of Dirac (1981) and von Neumann (1955).

If A is an $n \times n$ Hermitian matrix such that

$$\langle Au|v\rangle = \langle u|Av\rangle \tag{1.2.1}$$

for all $|u\rangle$, $|v\rangle$ in V, then A has a complete orthonormal set of eigenvectors:

$$A|u_i\rangle = \lambda_i|u_i\rangle, \quad \lambda_i \text{ real}, \quad \langle u_i|u_j\rangle = \delta_{ij}.$$

That means that any $|v\rangle \in V$ can be decomposed as

$$|v\rangle = \sum c_i|u_i\rangle. \tag{1.2.2}$$

In the infinite-dimensional Hilbert space $\mathcal{H}$ (1.2.1) and (1.2.2) remain valid provided the operator A is bounded (i.e. there exists a positive constant C, such that for any $|\varphi\rangle \in \mathcal{H}$ it holds that $\|A\varphi\| \leq C\|\varphi\|$).

An unbounded operator A in $\mathcal{H}$ (for example the Hamiltonian) can be defined in some subset D_A of $\mathcal{H}$. The analogue of condition (1.2.1)

$$\langle A\varphi|\psi\rangle = \langle\varphi|A\psi\rangle \text{ for } |\varphi\rangle, |\psi\rangle \in D_A$$

defines a *symmetric* operator. More restrictive is the notion of a *self-adjoint* operator possessing a complete set of eigenvectors:

(i) corresponding to a discrete spectrum of eigenvalues

$$A|\varphi_i\rangle = \lambda_i|\varphi_i\rangle, \quad \lambda_i \text{ real}, \quad \langle\varphi_i|\varphi_j\rangle = \delta_{ij} \tag{1.2.3}$$

(ii) and/or (generalized) eigenvectors

$$A|\varphi_\lambda\rangle = \lambda|\varphi_\lambda\rangle, \quad \lambda \text{ real}, \quad \langle\varphi_\lambda|\varphi_{\lambda'}\rangle = \delta(\lambda - \lambda') \tag{1.2.4}$$

for the continuous spectrum. The completeness means that any $|\varphi\rangle \in \mathcal{H}$ can be expanded as

$$|\varphi\rangle = \sum_i c_i|\varphi_i\rangle + \int d\lambda\, c(\lambda)|\varphi(\lambda)\rangle. \tag{1.2.5}$$

Obviously, the self-adjointness implies the symmetry of the operator (but the opposite is not true).

The existence of a complete set of eigenvectors corresponding to real eigenvalues (self-adjointness) is necessary for any operator assigned to a quantum observable. We shall call such operators Hermitian, as is usual in physics, but in what follows, the term 'Hermitian operator' will always mean 'self-adjoint operator' in the mathematical sense (and not the 'symmetric operator'). Namely the probability that in the state $|\varphi\rangle \in \mathcal{H}$ the measurement of the variable A will give a result lying in the interval I is given by the formula

$$P_\varphi(A, I) = \sum_{\lambda_i \in I} |c_i|^2 + \int |c(\lambda)|^2\, d\lambda \tag{1.2.6}$$

where c_i and $c(\lambda)$ are coefficients in the expansion (1.2.5). The existence of such an expansion, i.e. the self-adjointness of A, is inevitably connected with the interpretation of quantum mechanics.

Examples

(i) The momentum operator $A = (1/i)\partial_x$ acting on sufficiently smooth functions over a real line R is self-adjoint. This operator has a continuous spectrum for each $k \in R$; there is a generalized eigenfunction

$$\psi_k(x) = \frac{1}{\sqrt{2\pi}} e^{ikx} \tag{1.2.7}$$

satisfying

$$A\psi_k(x) = k\psi_k(x). \tag{1.2.8}$$

This is a generalized function (distribution) since it is non-normalizable

$$\langle\psi_k|\psi_k\rangle = \frac{1}{2\pi} \int_{-\infty}^{+\infty} dx\, |\psi_k|^2 = \infty. \tag{1.2.9}$$

However, for any $\varphi(x)$ from the Hilbert space $\mathcal{H} = \mathcal{L}^2(R^N, dx)$ the generalized (inner) product

$$\langle\psi_k|\varphi\rangle = \frac{1}{\sqrt{2\pi}} \int_{-\infty}^{+\infty} dx\, e^{ikx}\varphi(x) = \tilde{\varphi}(k) \tag{1.2.10}$$

exists and is related directly to the Fourier transform of $\varphi(x)$. Instead of (1.2.9) one should write

$$\langle\psi_k|\psi_{k'}\rangle = \frac{1}{2\pi} \int_{-\infty}^{+\infty} dx\, e^{i(k'-k)x} = \delta(k' - k). \tag{1.2.11}$$

This formula is frequently used in quantum mechanics.

(ii) The operator $A = (1/i)\partial_\varphi$ acting on smooth functions on the circle S^1 (where we introduce the angle variable φ instead of x):

$$A\psi(\varphi) = \frac{1}{i}\partial_\varphi\psi(\varphi) \tag{1.2.12}$$

where $\psi(2\pi) = \psi(0)$ (which is the same as to lie on a circle). This operator is self-adjoint and has only a discrete spectrum. The corresponding eigenfunctions are

$$\psi_n(\varphi) = \frac{1}{\sqrt{2\pi}} e^{in\varphi} \qquad n \text{ integer.} \tag{1.2.13}$$

They are orthonormal in the Hilbert space $\mathcal{H} = \mathcal{L}^2(S^1, d\varphi)$:

$$\langle\psi_n|\psi_{n'}\rangle = \frac{1}{2\pi} \int_0^{2\pi} e^{i(n'-n)\varphi}\, d\varphi = \delta_{nn'}. \tag{1.2.14}$$

Note. This is a very important example for this book. Such operators appear, using proper variables, in the description of rotations. It is known from elementary courses on QM that the operator $(1/i)\partial_\varphi$ is just the operator L_z of rotations around the z-axis (expressed in terms of spherical coordinates).

(iii) The operator $A = (1/i)\partial_x$ defined on sufficiently smooth functions on an interval $x \in (0, 2\pi)$ satisfying boundary conditions $\varphi(0) = \varphi(2\pi) = 0$ is symmetric, but not self-adjoint, since it has no eigenfunctions. However, as we have seen in example (ii), this operator with (weaker) periodic boundary conditions is self-adjoint.

(iv) The momentum operator $A = (1/i)\partial_x$ on the half-line $[0, \infty]$ is even worse, since it cannot be made self-adjoint by imposing any boundary condition.

Note. One can say that for a particle confined to an interval (with vanishing wave function in the end points) the momentum $(1/i)\partial_x$ does not represent a proper physical observable. However, the Hamiltonian $H = (1/2m)\partial_x^2$ still represents a physical observable since it is self-adjoint on the set of wave functions in question.

1.3 Some basic objects of group theory

In this subsection we recall some basic definitions from the theory of Lie groups and algebras which will be essentially used in the main text. Of course, rigorous definitions can be given only in the appropriate context of a complete exposition of the theory and we refer the reader to the books of Barut and Raczka (1977), Elliot and Dawber (1985), Humphrey (1972), Jacobson (1961), Jones (1990), Tung (1985), Vilenkin (1968), van der Waerden (1974), Weyl (1931), Wigner (1959), Wybourn (1974) and Zhelobenko (1973) for further details and clarifications.

1.3.1 Groups: finite, infinite, continuous, Abelian, non-Abelian; subgroup of a group, cosets

A set G of elements $g_1, g_2, g_3, \ldots$ is said to form a group if a law of *multiplication* of the elements is defined which satisfies certain conditions. The result of multiplying two elements g_a and g_b is called the *product* and is written $g_a g_b$. The conditions to be satisfied are the following.

(i)   The product $g_a g_b$ of any two elements is itself an element of the group, i.e.

$$g_a g_b = g_d \qquad \text{for some } g_d \in G.$$

(ii) In multiplying three elements g_a, g_b and g_c together, it does not matter which product is made first:

$$g_a(g_b g_c) = (g_a g_b)g_c$$

where the product inside the brackets is carried out first. This implies that the use of such brackets is unnecessary and we may simply write $g_a g_b g_c$ for the triple product. This property is called *associativity* of the group multiplication.

(iii) One element of the group, usually denoted by e and called *identity* or *unity*, must have the property

$$eg_a = g_a e = g_a.$$

(iv) To each element g_a of the group there corresponds another element of the group, denoted by g_a^{-1} and called the *inverse* of g_a, which has the properties

$$g_a g_a^{-1} = g_a^{-1} g_a = e.$$

In general, $g_a g_b$ is not the same element as $g_b g_a$. A group for which $g_a g_b = g_b g_a$ for all elements g_a and g_b is called an *Abelian* group. Its elements are said to *commute*. If at least one pair of elements do not commute, i.e. one has $g_a g_b \neq g_b g_a$, then the group is called *non-Abelian*. The simplest non-Abelian group is the rotation group in the three-dimensional space. The rotations in a two-dimensional space (on a plane), however, form an Abelian group.

The number of elements in a group may be finite, in which case this number is called the *order* of the group, or it may be infinite. The groups are correspondingly called finite or infinite groups. Among the latter the most important for physics are *continuous* ones, for which the group elements, instead of being distinguished by a discrete label, are labelled by a set of continuous parameters.

Given a set of elements forming a group G it is often possible to select a smaller number of these elements which satisfy the group definitions among themselves. They are said to form a *subgroup* of G. A normal subgroup is a subgroup H of G with the property that $gHg^{-1} = H$ for any $g \in G$. For example, translations and rotations of three-dimensional space generate a group which has translations as a normal subgroup.

For a given subgroup $H \subset G$ and $g \in G$, one can define right (left) coset Hg (gH) as the set consisting of elements hg (gh), $h \in H$. For finite groups the number of elements in each coset is clearly given by the order of H. One can define the set G/H as the set of right (left) cosets. For a normal subgoup H, left and right cosets are identical and G/H is a group. Cosets define decomposition of G into disjoint subsets and for finite groups this implies Lagrange's theorem stating that the order g of G is divisible by the order h of H (g/h gives the number of elements in G/H). For a Lie group G of dimension $d(G)^2$ and $H \subset G$ of dimension $d(H)$, an analogous result holds: $d(G/H) = d(G) - d(H)$. For example, if G is the three-dimensional rotation group and H is the one-dimensional group of rotations around a given axis, G/H is the two-dimensional sphere.

1.3.2 Isomorphism, automorphism, homomorphism

Let $\mathcal{X}$ and $\mathcal{X}'$ be two sets with some relations among elements within each set.

---

[2] For the definition of the dimension of a Lie group see section 1.3.3.

For example, $\mathcal{X}$ and $\mathcal{X}'$ can be groups, and the corresponding relations can be the group multiplications : $g_a g_b = g_c$ for $g_a, g_b, g_c \in \mathcal{X}$ and $g_a' g_b' = g_c'$ for $g_a', g_b', g_c' \in \mathcal{X}'$. Another example is ordered sets with defined inequalities $a > b$, $a, b \in \mathcal{X}$ and $a' > b'$, $a', b' \in \mathcal{X}'$.

Let there exist a one-to-one correspondence (map) $\rho : \mathcal{X} \leftrightarrow \mathcal{X}'$ preserving the relations among elements of $\mathcal{X}$ and $\mathcal{X}'$, i.e. if some relation is fulfilled for $a, b \in \mathcal{X}$ then the corresponding relation is fulfilled for $\rho(a), \rho(b) \in \mathcal{X}'$ and vice versa. In this case the sets $\mathcal{X}$ and $\mathcal{X}'$ are called *isomorphic* ones: $\mathcal{X} \cong \mathcal{X}'$, and the correspondence ρ is called *isomorphism*.

In particular, if the sets coincide $\mathcal{X} = \mathcal{X}'$, a one-to-one correspondence ρ, preserving structure relations, is called *automorphism*.

If each element $a \in \mathcal{X}$ is mapped into a unique image, a single element $a' \in \mathcal{X}'$, but the reverse is not in general true (e.g. a' may be the image of several elements of $\mathcal{X}$ or not be the image of any element of $\mathcal{X}$) and the map preserves structure relations in $\mathcal{X}$ and $\mathcal{X}'$, then this map is called *homomorphism*.

1.3.3 Lie groups and Lie algebras

The group elements $g(a_1, a_2, \ldots, a_r)$ of a continuous group depend on real parameters a_i which are all essential in the sense that the group elements cannot be distinguished by any smaller number. The number r is called the *dimension* of the group. Each parameter has a well defined range of values. For the elements to satisfy the group postulate, a multiplication law must be defined and the product of two elements

$$g(a_1, a_2, \ldots, a_r) g(b_1, b_2, \ldots, b_r) = g(c_1, c_2, \ldots, c_r)$$

must be another group element. Thus the new parameters c_i must be expressible as functions of the parameters a_i and b_i:

$$c_i = \phi_i(a_1, a_2, \ldots, a_r, b_1, b_2, \ldots, b_r) \qquad i = 1, \ldots, r.$$

It is customary to define the parameters in such a way that the identity element has all the parameters equal to zero. The r functions ϕ_i must satisfy several conditions in order for the group postulates to be satisfied. The groups with differentiable functions ϕ_i are called *Lie groups*.

An abstract *Lie algebra* is a vector space L together with a bilinear operation $[\cdot, \cdot]$ from $L \times L$ into L satisfying

$$[x_1 + x_2, y] = [x_1, y] + [x_2, y] \qquad x_1, x_2, y \in L$$

$$[\alpha x, y] = \alpha [x, y] \qquad \alpha \in \mathbb{C} \text{ or } \mathbb{R}, \ x, y \in L$$

$$[x, y] = -[y, x] \qquad x, y \in L$$

$$[x, [y, z]] + [y, [z, x]] + [z, [x, y]] = 0 \qquad x, y, z \in L \quad \text{(Jacobi identity)}.$$

In all cases of interest to us the bilinear operation $[\cdot, \cdot]$ can be understood as the commutator in the corresponding associative algebra

$$[x, y] = xy - yx.$$

There exists a tight interrelation between Lie groups and Lie algebras, which we shall consider after the introduction of group and algebra representations.

1.3.4 Representations: faithful, irreducible, reducible, completely reducible (decomposable), indecomposable, adjoint, fundamental

A *representation* of an algebra L (group G) is a homomorphism of L (or G) into an algebra (group) of linear transformations of some vector space V. If the dimension of the space V is d then the representation is said to be d dimensional.

A representation is said to be *faithful* if the homomorphism is an isomorphism.

A subspace $V_1 \subset V$ of a representation space V is called *invariant subspace* with respect to an algebra L (group G) if $Tv \in V_1$ for all $v \in V_1$ and all $T \in L$ (or $T \in G$).

A representation is called *irreducible* if the representation space V has no invariant subspaces (except the whole space V and zero space $\{0\}$). Otherwise, the representation is called *reducible*.

A representation is called *completely reducible* or *decomposable* if all linear transformations of the representation can be presented in the form of *block-diagonal* matrices, each block acting in the corresponding invariant subspace. Otherwise, the representation is called *indecomposable*.

Example. The simplest example of an indecomposable representation is provided by the two-dimensional representation of the Abelian group $G = \mathbb{R}$ (the group multiplication corresponds to the addition in $\mathbb{R}$, where $\mathbb{R}$ is the set of all real numbers)

$$x \longrightarrow T_x = \begin{pmatrix} 1 & x \\ 0 & 1 \end{pmatrix} \qquad x \in \mathbb{R}$$

which acts in the linear space $V^{(2)}$, i.e.

$$T_x \begin{pmatrix} u_1 \\ u_2 \end{pmatrix} = \begin{pmatrix} u_1 + x u_2 \\ u_2 \end{pmatrix}.$$

The subspace $V_{(1)}^{(2)}$ of vectors $u = \begin{pmatrix} u_1 \\ 0 \end{pmatrix}$ is invariant with respect to T_x, $\forall x \in \mathbb{R}$, while the orthogonal subspace $V_{(2)}^{(2)}$ consisting of vectors $u = \begin{pmatrix} 0 \\ u_2 \end{pmatrix}$ is not invariant. It is not possible to achieve the decomposition into invariant subspaces by any (linear) transformations of bases in $V^{(2)}$.

The representation of a Lie group G (Lie algebra L) in the vector space of the Lie algebra L itself is called the *adjoint representation* and the corresponding transformations are denoted by Ad_g, $g \in G$ (ad_X, $X \in L$). In the case of Lie algebra, the adjoint representation is defined by the commutator in L

$$\text{ad}_X Y = [X, Y] \qquad X, Y \in L.$$

Thus the dimension of the adjoint representation coincides with the dimension of the group.

The Lie group or Lie algebra (nontrivial) representation of the lowest dimension is said to be the *fundamental representation*.

1.3.5 Relation between Lie algebras and Lie groups, Casimir operators, rank of a group

Consider a representation $T(a_1, \ldots, a_r)$ of the group G in a space V. By convention the parameters are chosen such that the identity element has all $a_i = 0$, so that

$$T(0, \ldots, 0) = 1.$$

If all parameters a_i are small then, to first order in these parameters

$$T(a_i) \simeq 1 + \sum_i a_i X_i$$

where the X_i are some linear operators, independent of the parameters a_i. These operators are called the *infinitesimal operators* or *generators* of the group in a given representation and are expressed explicitly as partial derivatives

$$X_i = \left. \frac{\partial T(a_1, \ldots, a_r)}{\partial a_i} \right|_{a_1 = \cdots = a_r = 0}.$$

For any representation T of group G, the set of infinitesimal operators X_i satisfy the commutation relations

$$[X_i, X_j] = \sum_k c_{ij}^k X_k$$

where the numbers c_{ij}^k, called *structure constants*, are the same for all representations T of G. Thus the infinitesimal operators (generators) of a Lie group form the Lie algebra with the commutator playing the role of the bilinear operation in the abstract Lie algebra.

One can prove that there is the *exponential map* $L \to G$, which assigns to any element X of the Lie algebra L the element $\exp(X)$ of the group G. Choosing some basis I_i ($i = 1, \ldots, r$) in the vector space L, so that any element of L reads as $X = \sum_{i=1}^r a_i I_i$, one can write the exponential map in the form

$\exp(X) = \exp(\sum_i a_i I_i)$, where the group parameters a_i $(i = 1, \dots, r)$ appear explicitly.

A certain combination of elements of a Lie algebra which commutes with all the generators is called an *invariant* or *Casimir operator* of the group[3]. The maximal number of such independent operators is equal to the *rank* of the group, the latter being defined as the maximal number of mutually commuting elements of the Lie algebra.

1.3.6 Schur's lemmas

Schur's first lemma. Let $T(g)$ be an irreducible representation of a group G in a space V and let A be a given operator in V. Schur's first lemma states that if $T(g)A = AT(g)$ for all $g \in G$ then $A = \lambda 1$, where 1 is the identity (or unity) operator. In other words, any given operator which commutes with all the operators $T(g)$ of an irreducible representation of the group G is a constant multiple of the unit operator.

Schur's second lemma. Let $T_1(g)$ and $T_2(g)$ be two irreducible representations of a group G in two spaces V_1 and V_2 respectively, of dimensions s_1 and s_2, and let A be an operator which transforms vectors from V_1 into V_2. Schur's second lemma states that if $T_1(g)$ and $T_2(g)$ are inequivalent and $T_1(g)A = AT_2(g)$ for all $g \in G$, then $A = 0$, i.e. it is the null (or zero) operator.

1.3.7 Semidirect sum of Lie algebras and semidirect product of Lie groups (inhomogeneous Lie algebras and groups)

Let M and T be Lie algebras and $D : X \to D(X)$, $X \in M$ be a homomorphism of M into the set of linear operators in the vector space T, such that every D is a differentiation of T, i.e. D satisfies the Leibniz rule:

$$D(X)YZ = \big(D(X)Y\big)Z + Y\big(D(X)Z\big).$$

We define the Lie algebra structure in the whole set of elements of both vector spaces M and T in the following way:

(i)   for the pairs of elements from M, the commutators are defined as in the Lie algebra M;
(ii) for the pairs of elements from T, the commutators are defined as in the Lie algebra T;
(iii) for the mixed pairs, the commutators are defined by the operators D:

$$[X, Y] = D(X)Y \qquad X \in M, \ Y \in T.$$

---

[3] The Casimir operator is not an element of the Lie algebra, unless (as happens for some groups) it is linear in the Lie algbra elements.

One can check that for the above construction all the Lie algebra axioms are satisfied. The obtained Lie algebra L is called the *semidirect sum* of M and T

$$L = T \uplus M.$$

Such an algebra generates the *semidirect product* of the Lie groups G_M and G_T. The semidirect product $G = G_T \otimes G_M$ is the group of all ordered pairs (g, Λ), where $g \in G_T$ and $\Lambda \in G_M$, with the group multiplication

$$(g, \Lambda)(g', \Lambda') = (g\Lambda(g'), \Lambda\Lambda').$$

Here $\Lambda(g)$ defines an automorphism of the group G_T. It is easy to see that the unit element of the semidirect product has the form (e, id), e being the unity in G_T, id being the unity in G_M, and the inverse elements read as

$$(g, \Lambda)^{-1} = (\Lambda^{-1}(g^{-1}), \Lambda^{-1}).$$

Physically, the most important example of such a construction is the *Poincaré group* of relativistic space-time symmetry, which is the semidirect product of the Lorentz Lie group and the (Abelian) group of translations. The corresponding Lie algebra (semidirect sum) is called the *Poincaré Lie algebra*.

1.3.8 The Haar measure

The so-called Haar measure (see, e.g. Barut and Raczka (1977)) defines the invariant integration measure for Lie groups. This means that one can identify a volume element $d\mu(g)$ defining the integral of a function f over G as $\int_G f(g) \, d\mu(g)$ with the property that the integral is both left and right invariant

$$\int_G f(g^{-1}x) \, d\mu(x) = \int_G f(xg^{-1}) \, d\mu(x) = \int_G f(x) \, d\mu(x).$$

The invariance follows from the invariance of the volume element $d\mu(g)$:

$$d\mu(x) = d\mu(gx) = d\mu(xg)$$

which implies that the expression for $d\mu(g)$ in the neighbourhood of point g can be found by fixing the value of $d\mu(g)$ at $g = e$ (unit element) and by performing a left or right translation by g: $d\mu(g) = d\mu(e)$. Let the action of a map $x \to g(x)$ (left translation) be given by $x^i \to y^i(x^j)$, with x^i being the coordinates in the neighbourhood of the unit element e and denote by $dx^1 \ldots dx^n$ the volume element spanned by the coordinate differentials $dx^1, dx^2, \ldots, dx^n$ at point e. Then the volume element spanned by the same coordinate differentials at point g is given by

$$d\mu(g) = |J|^{-1} \, dx^1 \, dx^2 \ldots dx^n$$

where J is the Jacobian for the map $x \to g(x)$ evaluated at the unit element e:

$$J = \frac{\partial(y^1 \dots y^n)}{\partial(x_1 \dots x_n)}.$$

In a right or a left translation $dx^1 dx^2 \dots dx^n$ and $|J|$ are multiplied by the same Jacobian determinant so that $d\mu(g)$ is indeed right and left invariant. A straightforward manner to derive the Haar measure is to consider a faithful matrix representation of the group and take some subset of matrix elements as coordinates x^i. The Lie groups also allow an invariant metric and $d\mu(g)$ is just the volume element $\sqrt{g}\, dx^1 \dots dx^n$ associated with this metric.

Example. The volume element of the group $SU(2)$. The elements of $SU(2)$ can be represented as 2×2 unitary matrices of the form

$$x = \sum_\mu x^\mu \tilde\sigma^\mu \qquad \sum_\mu x^\mu x^\mu = 1$$

where the $\tilde\sigma$-matrices are defined by $\sigma_0 = 1$ and $\tilde\sigma_i = -i\sigma_i$, with σ_i the Pauli spin-matrices, and thus we have $\tilde\sigma_i \tilde\sigma_j = -\delta_{ij}\tilde\sigma_0 + \epsilon_{ijk}\tilde\sigma_k$. The coordinates of $SU(2)$ can be taken as the coordinates x^i, $i = 1, 2, 3$, so that one has $x^0 = \pm\sqrt{1 - r^2}$, $r \equiv \sqrt{\sum_i x^i x^i}$. Clearly, the $SU(2)$ group manifold can be regarded as a three-dimensional sphere of unit radius ($\sum_{\mu=1}^4 x^\mu x^\mu = 1$) in Euclidian space E^4. The unit element e corresponds to the origin: $x^i = 0$, $x^0 = 1$. The left action of the element x on y can be written as $z = xy = \sum_\mu z^\mu \sigma^\mu$, where the coordinates z^μ are given by

$$z^i = (x^0 y^i + x^i y^0) + \epsilon_{ijk}x^j y^k$$
$$z^0 = \sqrt{1 - \sum_i z^i z^i}.$$

From this the Jacobian matrix at $y^i = 0$ can be deduced: $\partial z^i / \partial y^j = x^0 \delta_{ij} + \epsilon_{ijk}x^k$ and its determinant is $J = \pm\sqrt{1 - r^2}$ depending on the sign of x^0. The invariant integration measure reads as

$$d\mu = \frac{1}{\sqrt{1 - r^2}}\, dx^1 dx^2 dx^3.$$

Note that the metric of $SU(2)$ can be deduced as the metric induced from the Euclidian space E^4 into which $SU(2)$ is embedded as a sphere.

1.4   Remark about the introduction of angular momentum

There are essentially two ways to introduce angular momentum into quantum mechanics. The one used in most elementary books starts with the classical definition (we use bold-italic letters for 3-vectors)

$$L = r \times p \tag{1.4.1}$$

and considers this as an operator equation defining the operator L. From this the commutation relations follow and from them the whole algebra is built up.

The other way starts with the consideration of the rotation group in the physical three-dimensional space: it is found that the operators of infinitesimal rotations follow the same commutation relation as, and physically correspond to, the angular momentum. One then builds up the algebra from the commutation relations as in the other procedure.

We shall follow the second alternative, because it leads to a deeper insight; in fact, even in classical mechanics, the true essence of angular momentum is that it is the quantity which is conserved in systems with invariance under rotations. Thus, starting from the classical definition $L = r \times p$ means beginning the story with its second chapter. But since we prefer to begin with its first, i.e. with the discussion of the rotations, we naturally apply it directly to quantum mechanics. This has the advantage that we also circumvent the uneasy feelings which arise when it turns out that the quantum mechanical algebra yields angular momentum states with half-integer total angular momentum and corresponding operators which cannot be represented by $L = r \times p$, the very equation we would have started from. If, on the other hand, we start from the true basis, namely the rotational invariance, then we do not (or should not) have any prejudice as to what form of the conserved operators we are to expect. We will be satisfied, of course, if we find that there are some which can be written $L = r \times p$—but we do not expect that this is necessarily so.

The plan of the book is then roughly as follows:

- we first consider quite generally the quantum mechanical symmetry and establish the existence of unitary or anti-unitary transformations corresponding to the elements of the symmetry group;
- we work out the physical significance of the generators of the symmetry group;
- we specialize these considerations to the rotation group in three dimensions and to the proper orthochronous Lorentz group;
- we discuss some related subjects which are of interest in physics: two-dimensional rotations and supersymmetry.

2

SYMMETRY IN QUANTUM MECHANICS

2.1 Definition of symmetry

2.1.1 General considerations

Consider any given structure—whether a physical system or a geometrical figure, a set of rules, an equation or a set of points in abstract space—whatever it may be, we only require the possibility of its mathematical description. Such a structure can always be cast in the following abstract form: it contains elements with names such as 'electron', 'angle', 'sequence', 'five' or 'vector' and it contains relations between these elements. Obviously the relations are the important things, for if two structures happen to contain different elements but the same relations, then knowing these relations we know everything about both structures. It might well happen that the relations of one structure are considered as elements of another structure when things are described on a higher level. In any case, the elements of a given structure can be represented by points in an abstract space. We then consider transformations in that space. Among all possible transformations in that space there might be a group of transformations which leave the relations between the points unaffected. We say the relations are invariant under this particular group of transformations and we call this group the symmetry group of our structure. Very different structures may have the same symmetry group; therefore, a symmetry group is itself a new independent structure with elements (the transformations considered above) and relations (namely the group properties and the particular law of multiplication of that group). This leads to the consideration of the various symmetry groups as abstract mathematical objects (e.g. the permutation group, Lorentz group, gauge group etc); individual realizations of these groups (e.g. the hydrogen molecule, the free particle of spin $\frac{1}{2}$, the Lagrangian of charged particles—to mention only examples corresponding to the groups just mentioned)—such realizations are not considered in the study of abstract groups.

Here we are, in contrast, not so much concerned with the symmetry groups themselves as with symmetry groups being properties of physical systems. That is, we take the symmetry group for granted and assume it to be known and studied. The question is then: given such and such a physical system S with such and such a symmetry group G what are the consequences for the description of that system?

Our structures are physical systems and/or their mathematical descriptions consisting of elements such as particles, fields, observables, operators etc and of relations between these elements: the equations of motion and the rules of quantum mechanics. 'The physical system S has a symmetry group G' means, then, that there is a group of transformations leaving the equations of motion and the rules of quantum mechanics invariant. In particular no transformation of G is allowed to produce an observable effect.

Let then $g \in G$ be one of the symmetry operations. By applying g to the system S, the latter is transformed into S'. If S can be described in the framework of quantum mechanics by

S: observables A; states $|\psi\rangle$, $|\phi\rangle$, ..., then S' will be described by
S': observables A'; states $|\psi'\rangle$, $|\varphi'\rangle$, ...

If the transformation $S \leftrightarrow S'$ is to be a symmetry of the system, then no observable effect can result from the transformation. Thus if in S and S' corresponding states and corresponding operators are taken, then for arbitrary ψ, φ and A

$$|\langle\psi|A|\varphi\rangle|^2 = |\langle\psi'|A'|\varphi'\rangle|^2 \qquad (2.1.1)$$

must hold. Notice however that, conversely, the condition (2.1.1) alone does not necessarily imply that the transformation $S \rightarrow S'$ is a symmetry of the system (since (2.1.1) holds for any unitary transformation as will be found later).

There are two ways of interpreting symmetries: active and passive. The active interpretation means changing the material system S into another material system S'. The passive interpretation means leaving the system S untouched but changing the environment such that S, described with respect to the new environment, behaves as S' would have behaved in the old one.

Example. The two interpretations of one and the same rotation read

(i) active: the material system (sometimes we say 'the space') is bodily rotated by an angle α, say;
(ii) passive: the coordinate system is rotated by the angle $-\alpha$, whereas the material system (or the space) S is left as it was.

We shall take the active point of view, since it is the more natural one: we may be able, for instance, to prepare a system S' in which all motions are inverse to those in S, but we certainly are not able to force time to run backwards.

With this convention, our above definition of symmetry amounts to saying

G is a symmetry group of S, if for any $g \in G$ there exists another material system $S' = gS$ (symbolically) and also a uniquely defined operator function F_g for observables A such that

$\qquad (2.1.2)$

● $A' = F_g(A)$ is again an observable of S and
● measuring A' in S' leads to the same eigenvalues with the same probability distributions as measuring A in S—as expressed in equation (2.1.1).

This definition of symmetry needs comments.

(i)   Saying that S is carried materially into $gS = S'$ means that there is something by which one can distinguish S and S'—for instance the orientations or the velocities of S and S' relative to an observer. Now, if any difference between S and S' can be observed, we do not have the symmetry we speak of. The observers—or more generally the surroundings of the system—indeed violate the symmetry by their very presence, but it is assumed that the effect of the surroundings on the system can be neglected (except for measurements). If there were no surroundings (characterized, e.g., by a coordinate system fixed to the laboratory), then S and S' would be just the same. That is, if the system S is considered as isolated, then there is nothing inherent in the system which would allow us to distinguish S' and S.

Therefore one should perhaps rather say that S' and S do not stand for different systems (whose differences are only defined by a symmetry-breaking relation to the observer) but rather for two different states of one system (different states with respect to the symmetry-breaking presence of the observer). That we nevertheless use the term 'another system S''', is because S' might, for instance, consist of antimatter where S is made of matter: everybody would call them different systems—although there would be nothing *in the system itself* which would enable one to distinguish two systems S and $\overline{S}$, where S and $\overline{S}$ are mirror images of each other and $\overline{S}$ is made of antimatter. It is the observer, who, being made of matter, can distinguish both by touching them: if he does not survive the touch, it was $\overline{S}$.

(ii) Since—except with respect to a symmetry-breaking environment—the systems S and S' are the same and indistinguishable (if $gS = S'$ is a symmetry), the operation g must map the set of all states of S onto itself and it also must map the set of all observables of S onto itself. Therefore all states $|\psi\rangle'$ of S' are also possible states of S: if $|\psi\rangle'$ is a state of S', then there is always a state $|\varphi\rangle$ of S such that $|\psi\rangle' \equiv |\varphi\rangle$. A similar interpretation holds for the 'new' observables A'.

This has then an important consequence. Let us consider the evolution of S and S' in time, using firstly the Schrödinger picture and then the Heisenberg picture.

In S all states develop as

$$|t\rangle = e^{-iHt}|0\rangle$$

and in S', they evolve as

$$|t\rangle' = e^{-iH't}|0\rangle'.$$

Let $|\psi\rangle'$ be a certain state in S'; then $|\varphi\rangle = |\psi\rangle'$ is a possible state in S (to quote an example: if S and S' are rotated against each other by $90°$, then $|\psi\rangle' \equiv$ plane wave in the x' direction, $|\varphi\rangle \equiv$ plane wave in the y ($\equiv x'$) direction).

We now take a state $|\psi\rangle$ and its image $|\psi\rangle'$ and integrate from zero to t. We find in S

$$|\psi(t)\rangle = e^{-iHt}|\psi(0)\rangle$$

and in S'

$$|\psi(t)\rangle' = e^{-iH't}|\psi(0)\rangle'.$$

$|\psi(0)\rangle'$ is, however, a possible state of the system S too. We thus can write in S

$$|\psi(t)\rangle'' = e^{-iHt}|\psi(0)\rangle'.$$

If now the states $|\psi(t)\rangle''$ and $|\psi(t)\rangle'$ differed by more than a single phase factor, then the probability to find any given state $|\varphi\rangle$ would be different in the two cases:

$$|\langle\varphi|\psi(t)\rangle'|^2 \neq |\langle\varphi|\psi(t)\rangle''|^2.$$

That would mean that S and S' could be distinguished from each other by observing their internal history. That contradicts the assumption that $S' = gS$ was a symmetry operation. The conclusion is, then, that if g is a symmetry, $H' = H + \lambda\cdot 1$ must hold and λ may be any real number, which we can put equal to zero.

In the Heisenberg picture, the argument goes like this: consider the set of all observables A of S and call it $\{A\}_S$. If $S' = gS$ is a symmetry, then $\{A'\}_{S'} \equiv \{A\}_S$, i.e. each observable A' of S' is also a possible observable of S. We now take the equations of motion

in S: $\dot{A} = i[H, A]$
in S': $\dot{A}' = i[H', A']$
in S: $\dot{A}' = i[H, A']$ (because A' is also an observable in S).

Hence $H - H'$ commutes with all the observables and is therefore a multiple of the unit operator. We thus arrive at the same result as above:

among all observables, the Hamiltonian is distinguishable by
being invariant under the symmetry transformations of the   (2.1.3)
physical system.

Remark. We have used here the Hamiltonian as expressing the time development of the system. We did this for convenience only. What our result implies is that a symmetry operation must leave the equations of motion invariant whether or not they are written in Hamiltonian form. Therefore Lorentz transformations under which H transforms like the time component of a four-vector are by no means excluded from our consideration. In fact one can almost literally carry over all our arguments into a language where the 'Hamiltonian' is replaced by 'scattering operator' (or 'S-matrix').

2.1.2   Formal definition of symmetry; ray correspondence

So far we have argued in a rather intuitive, physical way. In what follows, we shall be a little more formal. This will not much deepen our understanding of symmetry; it will, however, clear up its mathematical content and open the gate to an aspect of quantum mechanics which is possibly *the* aspect: namely, that quantum mechanics is the theory of representations of the symmetry groups of the physical systems considered; or at least that the Hilbert space of a given system is nothing other than a representation space of the symmetries.

We go back to our statement (see (1.1.14)) that the state vectors $|\psi\rangle$, $|\varphi\rangle$ etc do not correspond in a one-to-one way to physical states, but that unit rays $\hat{\psi}, \hat{\varphi}$, etc do. Therefore to any group G of transformations of a given physical system S—no matter whether G is a symmetry group of S or not—there exists a 'ray representation' $\hat{G}$ isomorphic to G, which transforms the rays in the same way as G transforms the physical states.

(It has to be noted that in the one-to-one correspondence between physical states and rays only a certain class of rays is admitted and only transformations between these rays are contained in $\hat{G}$: if $|\varphi\rangle \in \hat{\varphi}$, then $\hat{\varphi}$ is admitted as representing a physical state if and only if $|\varphi\rangle$ lies entirely within one super-selection subspace of $\mathcal{H}$. In what follows we restrict ourselves always to one particular super-selection subspace which we do not give a new symbol but simply call $\mathcal{H}$. The reader who is not yet familiar with the concept of super-selection rules will appreciate these remarks only a little later, when at the end of the chapter the full meaning of symmetry has become clear and the opportunity of discussing super-selection rules arises quite naturally. For the time being this reader may simply ignore all remarks referring to the particular type of symmetry called super-selection symmetry.) We now fix up some notation and then define symmetry.

In the present chapter Hilbert vectors are written with Greek letters: $|\psi\rangle$, $|\varphi\rangle$, etc. The norm of a vector is defined by

$$\|\varphi\| \equiv \||\varphi\rangle\| = \sqrt{\langle\varphi|\varphi\rangle}. \tag{2.1.4}$$

Any vector in Hilbert space can then be written as a multiple of a corresponding unit vector: $|\varphi\rangle = a|\varphi_0\rangle$ with $a \geq 0$

$$a = \|\varphi\| \qquad |\varphi_0\rangle = \frac{1}{a}|\varphi\rangle. \tag{2.1.5}$$

Rays are defined by (1.1.13); we repeat

$$\hat{\varphi} = \{\omega|\varphi\rangle \qquad |\omega| = 1\}. \tag{2.1.6}$$

Any $|\varphi\rangle \in \hat{\varphi}$ can serve as 'representative'. We define the scalar product of rays:

$$\hat{\varphi} \cdot \hat{\psi} = |\langle\varphi|\psi\rangle|. \tag{2.1.7}$$

$|\varphi\rangle \in \hat{\varphi}$ and $|\psi\rangle \in \hat{\psi}$ may be any representatives; the definition of (2.1.7) does not depend on the choice. The sum of rays is not defined.

From (2.1.4) and (2.1.7) it follows that

$$\sqrt{\hat{\varphi} \cdot \hat{\varphi}} = \|\varphi\| \qquad (2.1.8)$$

is independent of the choice of the representative and may serve to define the norm of a ray:

$$\|\hat{\varphi}\| = \sqrt{\langle\varphi|\varphi\rangle} = \|\varphi\|. \qquad (2.1.9)$$

Then any ray can be written as $\hat{\varphi} = \|\varphi\| \cdot \hat{\varphi}_0$, where $\hat{\varphi}_0$ is a unit ray. As physical states correspond to the set of unit rays, the ray representations $\hat{G}$ are defined only on the set of unit rays; we extend this definition to all rays by *defining* $\hat{g}\hat{\varphi} = \hat{g}\|\hat{\varphi}\|\hat{\varphi}_0 = \|\hat{\varphi}\|\hat{g}\hat{\varphi}_0$.

In what follows we shall omit the subscript 0 for unit rays, although in most cases the rays we have in mind will in fact be unit rays.

We now give the formal definition of symmetry:

> G is a symmetry group of the quantum mechanics system S if the ray representation $\hat{G}$ of G leaves the ray products (for S) invariant:
>
> G: $\quad g \in G$ transforms $\qquad S \to S' = gS$ $\qquad\qquad$ (2.1.10)
> $\hat{G}$: $\quad \hat{g} \in \hat{G}$ transforms $\qquad \hat{\psi} \to \hat{\psi}' = \hat{g}\hat{\psi}$
> $\qquad$ such that $\hat{\varphi}\cdot\hat{\psi} = \hat{\varphi}'\cdot\hat{\psi}'$.
>
> The transformations $\hat{g}$ are then called 'ray correspondence'.

(From the remarks above it should be clear that $\hat{G}$ maps admissible rays belonging to one super-selection subspace into rays belonging to the same super-selection subspace. This definition guarantees the invariance of all probabilities and eigenvalues; in other words: of all observable data.)

2.2 Wigner's theorem: the existence of unitary or anti-unitary representations

The formal definition which we have just given is all that can be inferred from the operational sense of the word *symmetry*. Dealing with quantum mechanics, we are forced to transform not only rays but also Hilbert vectors. It seems that there is an infinity of choices of vector transformations emerging from a given ray transformation; we visualize this in figure 2.1, where we have drawn the rays $\hat{\varphi}$ and $\hat{\psi}$ and their images $\hat{\varphi}'$ and $\hat{\psi}'$.

The transformation of the vectors $|\varphi\rangle \in \hat{\varphi}$ is completely arbitrary, subject to the condition that $|\varphi'\rangle \in \hat{\varphi}'$. We have symbolically indicated the transformation of three vectors $|\cdot\rangle$, $|\circ\rangle$, $|\otimes\rangle$. Any other choice will be as good as the one we took. Having chosen the image of $|\varphi\rangle$ nothing is said about the image of $e^{i\alpha}|\varphi\rangle$ whatever α is! This means that the infinite set $\{T\}$ of vector transformations,

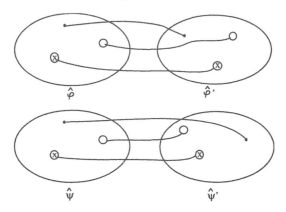

Figure 2.1. Ray transformation and vector transformation.

which is compatible with $\hat{\varphi} \rightarrow \hat{\varphi}'$, contains all kinds of discontinuous and non-linear transformation. Having fixed a vector transformation which maps $\hat{\varphi}$ onto $\hat{\varphi}'$, again nothing is defined for the mapping $\hat{\psi}$ onto $\hat{\psi}'$; the whole story is repeated.

We shall now show that things are not really so bad. The infinite set $\{T\}$ of vector transformations T described above (compatible with a ray transformation induced by a symmetry) always contains a subset of either unitary or anti-unitary vector transformations—but not both simultaneously—from which we can choose the ones representing our symmetry. This important theorem was first demonstrated by Wigner and many other proofs have been given since[1], all of them based on the same idea: to fix the phase of the vectors $|\varphi\rangle$ and $T|\varphi\rangle$ such that T becomes unitary (or anti-unitary).

Consider then the set $\{T\}$ of vector transformations T corresponding to a symmetry. We shall narrow down this infinite set by making explicit the constraints imposed by the symmetry and by using all available freedom in choosing T until we remain with either a unitary or an anti-unitary vector transformation. In other words, we show by construction that Wigner's theorem holds.

We first observe that every $T \in \{T\}$ transforms a complete orthonormal basis $\{|\varphi_n\rangle\}$ of $\mathcal{H}$ into another complete orthonormal basis $\{|\varphi_n'\rangle\}$. This follows from $\hat{\varphi} \cdot \hat{\psi} = \hat{\varphi}' \cdot \hat{\psi}'$. The norm is preserved too. The new basis must be complete, because if the transformed basis $\{|\varphi_n'\rangle\}$ were incomplete, then a vector $|\psi'\rangle$, orthogonal to all $|\varphi_n'\rangle$, would exist; consequently also a $|\psi\rangle$ would exist orthogonal to all $\{|\varphi_n\rangle\}$ contrary to the supposition of $\{|\varphi_n\rangle\}$ being complete.

---

[1] The original proof by Wigner (1931) was not complete (because—not being interested in time reversal—he ruled out the anti-unitary transformations); a complete proof was given, for instance, by Bargmann (1964). Another mathematically rigorous proof can be found in Weinberg's book *Quantum Theory of Fields* (Weinberg (1995)).

Let us now choose a basis $\{|\varphi_n\rangle\}$. To this basis corresponds a set of unit rays $\{\hat{\varphi}_n\}$ and by the ray transformation $\hat{g}$ its image $\{\hat{\varphi}_n'\}$. We choose arbitrarily a representative $|\overset{\circ}{\varphi}_n'\rangle$ from each $\hat{\varphi}_n'$. This choice of a basis $\{|\overset{\circ}{\varphi}_n'\rangle\}$ is only temporary; we shall construct a more suitable one.

We shall now reach the main conclusion in several steps.

(i)   Consider an arbitrary ray $\hat{\psi}$ and its image $\hat{\psi}'$. Choose a representative from each and expand $|\psi\rangle$ in $\{|\varphi_n\rangle\}$ and $|\psi'\rangle$ in $\{|\overset{\circ}{\varphi}_n'\rangle\}$: $|\psi\rangle = \sum A_n |\varphi_n\rangle$; $|\psi'\rangle = \sum A_n' |\overset{\circ}{\varphi}_n'\rangle$. Then, because of the symmetry and by (2.1.7)

$$\hat{\varphi}_n \cdot \hat{\psi} = |A_n| = \hat{\varphi}_n' \cdot \hat{\psi}' = |A_n'|.$$

This result is independent of the choice of the basis, of $\hat{\psi}$ and of the representatives $\hat{\psi}$ and $\hat{\varphi}'$. From now on we write

$$A_n = a_n \, e^{i\alpha_n} \qquad a_n \geq 0 \qquad \alpha_n \text{ real}$$

then

$$a_n = a_n'. \tag{2.2.1}$$

(ii) Consider a vector $|\psi\rangle$ with real non-negative coefficients, but otherwise arbitrary:

$$|\psi\rangle = \sum a_n |\varphi_n\rangle \qquad a_n \geq 0.$$

$|\psi\rangle$ is the representative of some ray $\hat{\psi}$. Now choose an arbitrary representative $|\psi'\rangle \in \hat{\psi}'$ and expand it in the basis $\{|\overset{\circ}{\varphi}_n'\rangle\}$:

$$|\psi'\rangle = \sum a_n \, e^{i\alpha_n'} |\overset{\circ}{\varphi}_n'\rangle \tag{2.2.2}$$

where (2.2.1) has been used. We now use our freedom to *define* T such that $|\psi'\rangle$ shall be the image of $|\psi\rangle$:

$$T|\psi\rangle = |\psi'\rangle \tag{2.2.3}$$

and then we use our freedom again to correct our choice of the basis $\{|\overset{\circ}{\varphi}_n'\rangle\}$: we put $e^{i\alpha_n'}|\overset{\circ}{\varphi}_n'\rangle = |\varphi_n'\rangle$ and adopt $\{|\varphi_n'\rangle\}$ as a new basis. All this is compatible with the given ray correspondence. We have thus achieved that a particular vector $|\psi\rangle$ with real non-negative coefficients (in the old basis) is transformed into another one, $|\psi'\rangle$, *with the same coefficients* (in the new basis).

(iii) We now show that we can achieve the same for any arbitrary $|\varphi\rangle$ with real non-negative coefficients—contrary to the suspicion one might have that it could perhaps hold only for $|\psi\rangle$, because the phases α_n' entered into the definition of the basis $\{|\varphi_n'\rangle\}$. Let

$$|\varphi\rangle = \sum b_n |\varphi_n\rangle \qquad b_n \geq 0$$

and choose a representative $|\varphi\rangle' \in \hat{\varphi}'$ such that

$$|\varphi'\rangle = \sum b_n \, e^{i\beta'_n} |\varphi'_n\rangle \qquad \text{with} \quad \beta'_1 = 0.$$

This is always possible. Again (2.2.1) was used.

From $\hat{\varphi} \cdot \hat{\psi} = \hat{\varphi}' \cdot \hat{\psi}'$ it now follows that (we indicate by $\sum'$ the sum without its first term)

$$a_1 b_1 + \sum{}' a_n b_n = \left| a_1 b_1 + \sum{}' a_n b_n \, e^{i\beta'_n} \right|.$$

On the left-hand side each term is zero or greater; this implies that all β'_n belonging to non-vanishing b_n must be zero. (Visualize

$$a_1 b_1 + \sum{}' a_n b_n \, e^{i\beta'_n}$$

as a sum of vectors in the complex plane: on the left-hand side they are all stretched out along the real axis; if the right-hand side is to yield the same length, they also must all be stretched out—hence $\beta'_n = 0$.) We *define* the above choice to be the image of $|\varphi\rangle$, whereby we have constructed T, such that all vectors with real non-negative coefficients are transformed into vectors with *the same* real non-negative coefficients:

$$|\varphi\rangle = \sum b_n |\varphi_n\rangle \rightarrow T|\varphi\rangle = |\varphi'\rangle = \sum b_n |\varphi'_n\rangle. \qquad (2.2.4)$$

This holds in particular for all those vectors having only one non-vanishing component $a_n = 1$; in other words for the basis vectors $|\varphi_n\rangle$ themselves. Hence the transformation T is such that now the image of the basis $\{|\varphi_n\rangle\}$ is $\{|\varphi'_n\rangle\}$

$$T|\varphi_n\rangle = |\varphi'_n\rangle. \qquad (2.2.5)$$

(iv) Next we choose two arbitrary vectors

$$|\psi\rangle = \sum a_n \, e^{i\alpha_n} |\varphi_n\rangle \qquad |\varphi\rangle = \sum b_n \, e^{i\beta_n} |\varphi_n\rangle.$$

Their (not yet fully defined) images must be of the form

$$|\psi'\rangle = \sum a_n \, e^{i\alpha'_n} |\varphi'_n\rangle \qquad |\varphi'\rangle = \sum b_n \, e^{i\beta'_n} |\varphi'_n\rangle.$$

From $\hat{\varphi} \cdot \hat{\psi} = \hat{\varphi}' \cdot \hat{\psi}'$, it follows that

$$\left| \sum a_n b_n \, e^{i(\beta_n - \alpha_n)} \right| = \left| \sum a_n b_n \, e^{i(\beta'_n - \alpha'_n)} \right|.$$

This must hold for whatever a_n, b_n, α_n and β_n we choose. We now specialize $|\varphi\rangle$ such that all $\beta_n = 0$ and all $b_n = 0$ except b_l and b_k. Then all $\beta'_n = 0$ and $b'_n = 0$ except $b_l = b'_l$ and $b_k = b'_k$. Hence

$$\left| a_l b_l \, e^{-i\alpha_l} + a_k b_k \, e^{-i\alpha_k} \right| = \left| a_l b_l \, e^{-i\alpha'_l} + a_k b_k \, e^{-i\alpha'_k} \right|$$

or

$$\left| a_l b_l + a_k b_k\, e^{-i(\alpha_k - \alpha_l)} \right| = \left| a_l b_l + a_k b_k\, e^{-i(\alpha_k' - \alpha_k')} \right|.$$

Thus for arbitrary l and k

$$\alpha_k - \alpha_l = \pm(\alpha_k' - \alpha_l'). \tag{2.2.6}$$

One easily sees that either the plus sign holds for all l and k or the minus sign holds for all l and k; suppose it were not so:

$$\alpha_k - \alpha_l = \alpha_k' - \alpha_l'$$
$$\alpha_k - \alpha_j = -\alpha_k' + \alpha_j'$$

and add: the result is $2\alpha_k - \alpha_j - \alpha_l = \alpha_j' - \alpha_l'$, contrary to (2.2.6). Now we put

$$\alpha_k' = \pm\alpha_k - \lambda_k$$

and find

$$\alpha_k' - \alpha_l' = \pm(\alpha_k - \alpha_l) + (\lambda_k - \lambda_l).$$

Then (2.2.6) implies that $\lambda_k - \lambda_l = 0$ for all k and l; hence $\lambda_1 = \lambda_2 = \ldots = \lambda$ is a simple constant. Then

$$|\psi\rangle = \sum a_n\, e^{i\alpha_n}|\varphi_n\rangle \rightarrow |\psi'\rangle = e^{i\lambda}\sum a_n\, e^{\pm i\alpha_n}|\varphi_n'\rangle.$$

We still have the freedom to choose the phase $e^{i\lambda}$ of $|\psi'\rangle$ arbitrarily. The simplest choice is $\lambda = 0$; this is also consistent with the fact that if $|\psi\rangle$ has only real non-negative coefficients, the same must hold for $|\psi'\rangle$.

(v) We have seen that for each particular vector $|\psi\rangle$ either all $\alpha_k \rightarrow \alpha_k$ or $\alpha_k \rightarrow -\alpha_k$.

Suppose now that there exist in $\mathcal{H}$ both kinds of vector simultaneously and let

$$|\psi_+\rangle = \sum a_n\, e^{i\alpha_n}|\varphi_n\rangle \rightarrow |\psi_+'\rangle = \sum a_n\, e^{i\alpha_n}|\varphi_n'\rangle$$
$$|\varphi_-\rangle = \sum b_n\, e^{i\beta_n}|\varphi_n\rangle \rightarrow |\varphi'\rangle = \sum b_n\, e^{-i\beta_n}|\varphi_n'\rangle.$$

Then the vector

$$|\varphi\rangle = |\psi_+\rangle + |\varphi_-\rangle$$

would be neither of the plus type nor of the minus type—contrary to the above-established fact that each vector definitely belongs to one of these classes. We thus see that the $+$ or the $-$ sign is a property neither of the individual phases nor of the vector as a whole; *it is a property of the transformation considered.* Obviously, if the ray correspondence considered is continuously connected with the identity, then only the $+$ sign is possible.

(vi) If the transformation is such that $\alpha_n \rightarrow \alpha_n$, then with (2.2.5)

$$|\psi\rangle = \sum A_n |\varphi_n\rangle \rightarrow T|\psi\rangle = \sum A_n |\varphi'_n\rangle = \sum A_n T|\varphi_n\rangle.$$

Hence in this case T is linear; (2.1.10) implies then that it is unitary: $T = U$, because now

$$\langle U\varphi | U\psi \rangle = \langle \varphi | \psi \rangle.$$

If, on the other hand, $\alpha_n \rightarrow -\alpha_n$, then similarly

$$|\psi\rangle = \sum A_n \, e^{i\alpha_n} |\varphi_n\rangle \rightarrow |\psi'\rangle = \sum A_n^* |\varphi'_n\rangle.$$

In this case T is antilinear and therefore an anti-unitary transformation $\bar{U}$, so that

$$\langle \bar{U}\varphi | \bar{U}\psi \rangle = \langle \psi | \varphi \rangle = \langle \varphi | \psi \rangle^*.$$

Whether T is unitary or anti-unitary depends only on the symmetry group G and not on our choice.

We thus have the result that, to a given ray correspondence $\hat{g}$, one can always construct a vector transformation T, either unitary or anti-unitary and compatible with $\hat{g}$.

From now on we will forget about all the other vector transformations and consider only unitary (or anti-unitary) transformations T.

(vii) The next question is: what is the class of all unitary (anti-unitary) transformations compatible with a given $\hat{g}$? Suppose T and R both are compatible with $\hat{g}$, then

$$T|\varphi\rangle = \omega(\varphi)R|\varphi\rangle \qquad |\omega(\varphi)| = 1$$
$$T|\psi\rangle = \omega(\psi)R|\psi\rangle \qquad |\omega(\psi)| = 1$$

and if $|\psi\rangle$ and $|\varphi\rangle$ are linearly independent

$$T(|\varphi\rangle + |\psi\rangle) = T|\varphi\rangle + T|\psi\rangle = \omega(\varphi)R|\varphi\rangle + \omega(\psi)R|\psi\rangle$$
$$T(|\varphi\rangle + |\psi\rangle) = \omega(\varphi, \psi) \cdot R(|\varphi\rangle + |\psi\rangle)$$
$$= \omega(\varphi, \psi)R|\varphi\rangle + \omega(\varphi, \psi)R|\psi\rangle$$

which shows that $\omega(\varphi) = \omega(\varphi, \psi) = \omega(\psi) = \omega$ independent of the transformed vector. If, however, $|\psi\rangle$ and $|\varphi\rangle$ are not linearly independent, then a third vector $|\varphi'\rangle$ exists which is linearly independent of both $|\psi\rangle$ and $|\varphi\rangle$. Then $\omega(\varphi') = \omega(\psi)$ and $\omega(\varphi') = \omega(\varphi', \psi)$ and all these are equal to the one ω we had before.

The result is that

the unitary (anti-unitary) transformations which are compatible with a given ray correspondence (symmetry transformation) differ from each other at most by complex factors ω of modulus one.　　(2.2.7)

(viii) We may therefore introduce the notion of an 'operator ray', namely the set of all unitary (anti-unitary) vector transformations compatible with a given ray correspondence $\hat{g}$

$$\hat{U}(\hat{g}) = \{\omega(\hat{g})U(\hat{g}); \ |\omega| = 1\} \tag{2.2.8}$$

by which we have then established a one-to-one relationship between the ray correspondences $\hat{g}$ and unitary (anti-unitary) operator-rays $\hat{U}^2$. Since the ray correspondences $\hat{g}$ form a group, the same is automatically true for $\hat{U}$ and the two groups are isomorphic. Not so, however, for the U themselves. They do not form even a group representation. This is our next point.

(ix) Assume we consider the set of all elements $\hat{g}$ of a certain symmetry group $\hat{G}$ and together with it the corresponding unitary (anti-unitary) operator-rays $\hat{U}$. We are free to choose a representative $U \in \hat{U}$ from each, and there is no indication as to which one we should choose. Here the difficulty arises: let $\hat{g}$ and $\hat{h}$ be two ray correspondences and $\hat{g} \cdot \hat{h} = \hat{j}$ their product. We choose

$$U_g \in \hat{U}_g$$
$$U_h \in \hat{U}_h$$
$$U_j \in \hat{U}_j = \hat{U}_{g \cdot h}.$$

Then $U_g \cdot U_h \in \hat{U}_j$, but in general $U_g \cdot U_h \neq U_j$. They can differ, however, only by a factor of modulus one, hence in general

$$U_g \cdot U_h = \omega(g; h)U_{g \cdot h}. \tag{2.2.9}$$

The result is that

the unitary (anti-unitary) operators compatible with the ray correspondences form in general only a 'representation up to a factor of modulus one'. (2.2.10)

It depends on the symmetry group considered whether and to what extent the factors $\omega(g; h)$ can be simplified by a suitable choice of the $U \in \hat{U}$. We shall not go into this detail here. As we shall see when we study the rotation group, there $\omega = \pm 1$ is the best we can achieve.

---

[2] Note that up to here we have only established the existence of a unitary (anti-unitary) U corresponding to one arbitrarily given ray correspondence $\hat{g}$. The same construction done for another ray correspondence $\hat{g}'$ might lead to a very different U' even if $\hat{g}$ and $\hat{g}'$ are near to each other in parameter space. The following considerations deal with this circumstance.

(x) Let us now collect all results in the theorem (Wigner's theorem):

> For every symmetry group G of a given physical system there exists
>
> (a)   a group $\hat{G}$ of ray correspondences $\hat{g}$;
> (b)   a group $\hat{U}_G$ of unitary (anti-unitary) operator-rays $\hat{U}$;
> (c)   a set of unitary (anti-unitary) operators $U \in \hat{U}$;
>
> such that the groups G, $\hat{G}$ and $\hat{U}_G$ are isomorphic. If one chooses (arbitrarily) one representative U from each $\hat{U}$, then the unitary (anti-unitary) transformations $U(g)$ form a representation of G up to a factor of modulus one[a].   (2.2.11)

[a] Although we mentioned explicitly the possibility that anti-unitary representations might come up, they do so only when time reversal is considered. As we do not discuss this case within the general framework, we will encounter *in this book* only unitary representations.

2.3   Continuous matrix groups and their generators

2.3.1   General considerations

In this section we shall consider a type of group which makes up most of the physical symmetry groups: namely those depending continuously on certain parameters, for instance, the rotation group. In their most general form they are considered as continuous and even differentiable transformation groups in some n-dimensional vector space. We shall be modest and restrict ourselves to linear transformation groups in n dimensions, i.e. to $n \times n$ matrix groups. The reasons will be given below. This restriction makes possible derivations which, for more general 'Lie groups', would not hold in this form and would become more complicated.

We shall not elaborate the abstract group theoretical machinery in any detail. Our aim is, however, to say enough about the common group theoretical framework of symmetry problems in physics that the reader may see that angular momentum is but another example of a general and beautiful theory, and that he may forget the somewhat uneasy feelings he might have had when he first encountered angular momentum in quantum mechanics at a too elementary level.

If we deal with groups in physics, then they are not as abstractly defined as in purely mathematical considerations. It would be of no interest for us to make things artificially more complicated than they are: the groups we have to consider in this book are the Poincaré group (= inhomogeneous Lorentz group), i.e. a group of transformations in space-time which can be written as a 5×5 matrix group, and the three-dimensional rotations in space (a subgroup of the Poincaré group). That means that we have to do with a group which—as far as its physical definition goes—is a matrix group of 5×5 matrices (which are not unitary) and with a particular subgroup of it, which is—taken by itself—a matrix group of 3×3 matrices. In other words our groups are defined as groups of linear transformations in a (at most) five-dimensional space. That by introducing a quantum mechanical description we are forced to consider representations of these groups in any, even infinite, dimensions—is quite another story. Basically, practically all non-finite symmetry groups in physics are very simple: namely groups of linear transformations (matrices) in vector spaces of a finite (and small!) number of dimensions and depending on only a few parameters. That makes their theory clear and elementary; the infinite-dimensional unitary (anti-unitary) representations in Hilbert space are a difficulty not inherent in the groups but added from the outside. This basic simplicity holds in particular also for the symmetry groups called *unitary symmetry* of elementary particles. The analysis in all these cases is similar and the theory of angular momentum is a good model for the others, not to mention its own importance.

2.3.2 Continuous matrix groups; decomposition into pieces

According to what we just mentioned, we shall consider continuous groups of finite-dimensional matrices; continuous means the group elements depend continuously on parameters. The number of parameters depends on the group considered; the number of *real* parameters $a_1, a_2, \ldots, a_i$ is called the *dimension* of the group; we denote it by r. Indeed the word dimension is the most natural, because the group elements, if written as $g(a_1, a_2, \ldots, a_i)$, can be taken as points of an r-dimensional manifold. If the group is a matrix group of $n \times n$ complex matrices—the full linear group in n dimensions—then $r = 2n^2$; when further conditions are added, for instance unitarity, reality or unimodularity, then $r < 2n^2$; in fact r is equal to $2n^2$ minus the number of real equations expressing the supplementary conditions.

The r parameters may be chosen such that the unit transformation is represented by

$$g(0, 0, 0, \ldots) = 1. \tag{2.3.1}$$

The group may or may not *decompose into several pieces*—namely the set of parameters may or may not be connected. If space reflections are included, then the rotation group decomposes into two pieces (namely the rotations with determinant $+1$ and the products of one reflection with all rotations: determinant -1) and if space reflection and time reflection are included, then the Lorentz group decomposes into four pieces (determinant $+1$ or -1; forward or backward cone interchanged or not interchanged; these two choices are independent and give rise to four possibilities). If a group decomposes, then only one of the pieces is continuously connected to the unit element; this piece is in fact a normal subgroup. The other pieces taken separately are, of course, not subgroups. We prove that the piece continuously connected to the identity is a normal subgroup; we say that a group element $g(a_1, a_2, \ldots, a_r)$ is continuously connected with the identity, if we can introduce a suitable set of continuous functions $a_i(\lambda)$ such that $a_i(0) = 0$ and $a_i(1) = a_i$; then $g(\lambda)$ for $0 \leq \lambda \leq 1$ is a *path* leading from unity to $g(a_1, a_2, \ldots, a_r)$. The set of all group elements which can be connected to unity by any path is the piece of the group we are considering. Let $g(\lambda)$ be one path and $h(\lambda)$ another one $(h(1) \neq g(1))$, then $g(\lambda)h^{-1}(\lambda)$ is again such a path; this proves that the set of all elements lying on these paths is a subgroup. Next let f be a group element not belonging to the piece containing the identity; then $fg(\lambda)f^{-1}$ is an element of the group which lies on a path starting from the unit element; hence it belongs also to the piece containing the identity; this proves that this piece is a normal subgroup. (In the Lorentz group and the rotation group these normal subgroups are the proper orthochronous Lorentz group (determinant $+1$; forward cone $\rightarrow$ forward cone) and the proper rotation group (determinant $+1$) respectively.) The remaining pieces are its cosets.

2.3.3 The Lie algebra (Lie ring, infinitesimal ring)

In what follows we shall restrict ourselves to the piece containing the unit element and in that piece to a suitable neighbourhood N of the unit. Let the group be really r dimensional in that whole neighbourhood; furthermore assume in the following that all functions considered are at least twice differentiable in that neighbourhood, and that all products of all group elements which we might consider, again lie in that neighbourhood.

We then define the 'Lie algebra' (also Lie ring, infinitesimal ring) as follows.

Take any group element $g(a_1, a_2, \ldots, a_r) \in N$ and connect it by a path $g(\lambda)$ to the identity (i.e. choose arbitrary functions $a_1(\lambda), \ldots, a_r(\lambda)$ such that $g(0) = 1$ and $g(1) = g(a_1, a_2, \ldots, a_r)$). The derivative of $g(\lambda)$ at $\lambda = 0$ is called $\overset{\circ}{g}$:

$$\overset{\circ}{g} = \frac{\mathrm{d}}{\mathrm{d}\lambda} g(\lambda)_{|\lambda=0}. \qquad (2.3.2)$$

Then $\overset{\circ}{g}$ is an element of the infinitesimal ring $\overset{\circ}{G}$ of the group G and the whole ring consists of all elements $\overset{\circ}{g}$ which can be obtained in this way.

Comments are as follows.

(i)   We might have taken another path from $g(a_1, a_2, \ldots, a_r)$ to unity and obtained another $\overset{\circ}{g}$. Indeed $\overset{\circ}{g}$ is the tangent to the path at unity. Strictly speaking we should have written $g(\Lambda; \lambda)$ where Λ denotes the path. Then $\overset{\circ}{g}_\Lambda = (\partial/\partial\lambda)g(\Lambda; \lambda)$ is an element of $\overset{\circ}{g}$. We shall, however, not employ this pedantic notation but simply write $\overset{\circ}{g}, \overset{\circ}{h}, \ldots$ for the elements of $\overset{\circ}{G}$.

(ii) As derivatives of $n \times n$ matrices, the $\overset{\circ}{g}$ are again $n \times n$ matrices; *they are, however, not group elements.*

We now show that the set $\overset{\circ}{G}$ is indeed a ring, i.e. an r-dimensional vector space equipped with a multiplication law.

(i) $\overset{\circ}{G}$ is an r-dimensional (real) vector space.

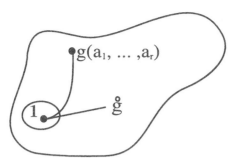

Figure 2.2. The elements of the Lie algebra.

(a) Take a path $g(\lambda)$; it yields a certain $\overset{\circ}{g}$. Take the same path but replace λ by $\alpha \cdot \lambda$ (α real; $0 \le \alpha \cdot \lambda \le 1$); it yields the element $\alpha \cdot \overset{\circ}{g}$, which therefore also belongs to $\overset{\circ}{G}$.

(b) Take $g(\lambda)h(\lambda) \in N$; then $g(\lambda)$ yields $\overset{\circ}{g}$, $h(\lambda)$ yields $\overset{\circ}{h}$ and $g(\lambda) \cdot h(\lambda)$ yields $\overset{\circ}{g} + \overset{\circ}{h}$; hence with $\overset{\circ}{g}$ and $\overset{\circ}{h}$ also $\overset{\circ}{g} + \overset{\circ}{h} \in \overset{\circ}{G}$.

(c) $\overset{\circ}{G}$ is an r-dimensional real space, because the r elements $\overset{\circ}{g}_i = (\partial g / \partial a_i)(0, 0, \ldots, 0)$ are linearly independent (we have assumed that the group is really r dimensional) and they span, with real coefficients, the whole of $\overset{\circ}{G}$. Namely every $\overset{\circ}{g} \in \overset{\circ}{G}$ can be written

$$\overset{\circ}{g} = \frac{\mathrm{d}}{\mathrm{d}\lambda} g(\lambda)|_{\lambda=0} = \sum \frac{\partial g}{\partial a_i}(0, 0, \ldots, 0) \frac{\mathrm{d}a_i}{\mathrm{d}\lambda}|_{\lambda=0} = \sum \alpha_i \overset{\circ}{g}_i$$

with real α_i: conversely, whatever α_i we choose, this sum belongs to $\overset{\circ}{G}$.

(ii) The real r-dimensional vector space $\overset{\circ}{G}$ is equipped with a multiplication law.

(a) Let $g(\lambda)$ and h (fixed) $\in N$; then $hg(\lambda)h^{-1}$ is a path and it yields $h\overset{\circ}{G}h^{-1} \in \overset{\circ}{G}$. Thus if $h \in N$ and $\overset{\circ}{g} \in \overset{\circ}{G}$, then $h\overset{\circ}{g}h^{-1} \in \overset{\circ}{G}$.

(b) Now let $\overset{\circ}{g}$ be fixed but take $h(\lambda)$ to be a path. Then for each fixed value of λ, we have $h(\lambda)\overset{\circ}{g}h^{-1}(\lambda) \in \overset{\circ}{G}$. Then also $(1/\lambda)[h(\lambda)\overset{\circ}{g}h^{-1}(\lambda) - \overset{\circ}{g}] \in \overset{\circ}{G}$ and with it also its limit for $\lambda \to 0$. This limit is the derivative of $h(\lambda)\overset{\circ}{g}h^{-1}(\lambda)$ and equals $\overset{\circ}{h}\overset{\circ}{g} - \overset{\circ}{g}\overset{\circ}{h} \equiv [\overset{\circ}{h}, \overset{\circ}{g}]$. Thus with any two elements, $\overset{\circ}{h}$ and $\overset{\circ}{g}$, also their commutator $[\overset{\circ}{h}, \overset{\circ}{g}]$ belongs to $\overset{\circ}{G}$. We define $[\overset{\circ}{h}, \overset{\circ}{g}]$ as 'the product' of $\overset{\circ}{h}$ and $\overset{\circ}{g}$.

We have thus shown that

the r-dimensional real vector space $\overset{\circ}{G}$ (associated with an r-dimensional continuous group G) is equipped with the multiplication law $[\overset{\circ}{h}, \overset{\circ}{g}] = \overset{\circ}{h}\overset{\circ}{g} - \overset{\circ}{g}\overset{\circ}{h}$ and this 'product', called a 'commutator', belongs to $\overset{\circ}{G}$. Hence $\overset{\circ}{G}$ is a ring (called a 'Lie algebra', 'Lie ring' or 'infinitesimal ring').   (2.3.3)

Comments are as follows.

(i)   If $\overset{\circ}{g} \in \overset{\circ}{G}$, then $\alpha\overset{\circ}{g} \in \overset{\circ}{G}$ for real α; but in general not for complex α because if $g(\lambda) \in G$ then $g(\alpha \cdot \lambda)$ with complex α need not belong to the group G (think of the rotation group: replace the rotation angle λ by $i\lambda$; this is no longer a rotation). Therefore $\overset{\circ}{G}$ is a real vector space (although the matrices $\overset{\circ}{g}$ may have complex elements $\overset{\circ}{g}_{ik}$).

(ii) Note the difference between the group and its Lie algebra: for the group addition is *not* defined (of course one can add two matrices $g \in G$ and $h \in G$ but in general the matrix $g + h$ does not belong to the group G); multiplication in G is just matrix multiplication. For the Lie algebra $\overset{\circ}{G}$ addition *is* defined; multiplication, however, is *not* simple matrix multiplication (of course one can multiply two matrices $\overset{\circ}{g} \in \overset{\circ}{G}$ and $\overset{\circ}{h} \in \overset{\circ}{G}$, but in general the matrix $\overset{\circ}{h}\overset{\circ}{g}$ does *not* belong to the Lie algebra $\overset{\circ}{G}$, whereas $[\overset{\circ}{h}, \overset{\circ}{g}]$ does). We shall illustrate these statements by an explicit calculation in subsection 3.3.2.

(iii) The term 'Lie algebra' is used because the $n \times n$ matrix groups considered here are a particular example of the much more general Lie groups, which also possess a Lie algebra.

2.3.4   Canonical coordinates

We have an infinity of choices when it comes to adopting a definite set $a_1, a_2, \ldots, a_r$ of parameters; indeed any set of functions $b_1(a_1, a_2, \ldots, a_r)$, $b_2(a_1, a_2, \ldots, a_r)$, $\ldots$, $b_r(a_1, a_2, \ldots, a_r)$ will do as well as long as the correspondence $a \leftrightarrow b$ is continuous and one-to-one (at least in the neighbourhood of zero) and $a \to 0 \leftrightarrow b \to 0$. There is, however, a particular choice which is very advantageous. We shall explain it now.

Let us start from any path, defined by a set of functions $a_1, a_2, \ldots, a_r$ and write down

$$\overset{\circ}{g} = \sum_{i=1}^{r} \frac{\partial g}{\partial a_i}(0, 0, \ldots, 0)\frac{\partial a_i}{\partial \lambda}(0) = \sum_{i=1}^{r} \alpha_i \overset{\circ}{g}_i \qquad (2.3.4)$$

where $\alpha_i = (\partial a_i/\partial\lambda)(0)$. As we have discussed already, every element of $\overset{\circ}{G}$ can be written that way. Thus the $\overset{\circ}{g}_i \equiv (\partial g/\partial a_i)(0, 0, \ldots, 0)$ serve as basis elements and the α_i are then nothing other than the 'coordinates' of the point $\overset{\circ}{g}$ in the vector space $\overset{\circ}{G}$. In other words: the α_i can be considered as the components of the vector $\overset{\circ}{g}$. These α_i are called 'canonical coordinates' or 'canonical parameters'. Before we investigate their relevance to the group, we note what the result would have been had we started from another parametrization $b_i(\mu)$ instead of $a_i(\lambda)$; describing the same path once in terms of $a_i(\lambda)$ and once in terms of $b_i(\mu)$, we obtain (without changing the symbol g)

$$\overset{\circ}{g} = \sum_{i=1}^{r} \frac{\partial g(a_1, \ldots, a_r)}{\partial a_i}\Big|_0 \cdot \frac{\partial a_i(0)}{\partial\lambda} = \sum_{i=1}^{r} \frac{\partial g(b_1, \ldots, b_r)}{\partial b_i}\Big|_0 \cdot \frac{\partial b_i(0)}{\partial\mu}$$

but with $b_i = b_i(a_1, \ldots, a_r)$ we have

$$\frac{\partial g(a_1, \ldots, a_r)}{\partial a_i}\Big|_0 = \sum_{j=1}^{r} \frac{\partial g(b_1, \ldots, b_r)}{\partial b_j}\Big|_0 \cdot \frac{\partial b_j(a_1, \ldots, a_r)}{\partial a_i}\Big|_0. \qquad (2.3.5)$$

Having supposed a one-to-one correspondence $a \leftrightarrow b$, we see that the determinant $|\partial b_j/\partial a_i| \neq 0$. Thus the change of the parameters corresponds to a non-singular linear transformation of the basis $\overset{\circ}{g}$. This, as we shall see, leaves invariant all those properties of the canonical coordinates which really matter.

Let us now see what are the advantages of canonical coordinates.

If the set $\{\overset{\circ}{g}_i\}$ is a basis defined by an arbitrary choice of parameters, then, as we saw, any $\overset{\circ}{g}$ can be written by means of the canonical coordinates α_i:

$$\overset{\circ}{g} = \overset{\circ}{g}(\alpha) = \sum_{i=1}^{r} \alpha_i \overset{\circ}{g}_i. \qquad (2.3.6)$$

Consider the elements (λ real)

$$\lambda\overset{\circ}{g} = \sum (\lambda \cdot \alpha_i) \cdot \overset{\circ}{g}_i.$$

This lies on the same straight line in the r-dimensional space $\overset{\circ}{G}$ as does $\overset{\circ}{g}$. Obviously two such elements $\lambda\overset{\circ}{g}$ and $\mu\overset{\circ}{g}$ (λ, μ real) commute:

$$[\lambda\overset{\circ}{g}, \mu\overset{\circ}{g}] = \lambda\mu[\overset{\circ}{g}, \overset{\circ}{g}] = 0. \qquad (2.3.7)$$

Let us write the canonical parameters as follows:

$$(\alpha_1, \ldots, \alpha_r) = \alpha = \alpha \cdot n_\alpha$$

$$\alpha = \left(\sum \alpha_i^2\right)^{1/2} \qquad n_\alpha = \alpha/\alpha = \text{unit vector in direction } \alpha. \qquad (2.3.8)$$

So far the canonical parameters have been used only as coordinates in the Lie algebra. Any arbitrary choice of a path $a_1, \ldots, a_r(\lambda)$ leads to a set of canonical coordinates. Consider now the group elements $g(\alpha_1, \ldots, \alpha_r)$, where simply $a_k(\lambda)$ has been replaced by $\alpha_k = (\partial a_k/\partial \lambda)|_{\lambda=0}$. We may, with n_α kept fixed, assume that the α_k are variable, because this can be achieved by varying the scale of λ.

We now study the correspondence between group elements $g(\alpha)$ and elements $\overset{\circ}{g}(\alpha)$ of the infinitesimal ring, keeping the direction n_α fixed. For the time being, we suppress the subscript α on n_α.

Consider the element $\overset{\circ}{g}$ associated with $g(\alpha n)$; it is defined by (2.3.2):

$$\overset{\circ}{g}(n) = \frac{\mathrm{d}}{\mathrm{d}\alpha} g(\alpha n)|_{\alpha=0}.$$

Replacing α by $\mu\alpha$ (μ real) we find $\overset{\circ}{g}(\mu\alpha) = \mu\overset{\circ}{g}(n)$. We obtain thus the following correspondences:

$$\overset{\circ}{g}(n) \leftrightarrow g(\alpha)$$
$$\mu\overset{\circ}{g}(n) \leftrightarrow g(\mu\alpha)$$
$$(\mu + \lambda)\overset{\circ}{g}(n) \leftrightarrow g[(\mu + \lambda)\alpha]$$

but, on the other hand, by the definition (2.3.2)

$$(\mu + \lambda)\overset{\circ}{g}(n) \leftrightarrow g(\mu\alpha) \cdot g(\lambda\alpha).$$

From the last two correspondences it follows that $g[(\mu+\lambda)\alpha]$ and $g(\mu\alpha) \cdot g(\lambda\alpha)$ lead to the same element of the Lie algebra; in other words for $\alpha \longrightarrow 0$ the elements $g((\mu + \lambda)\alpha)$ and $g(\mu\alpha) \cdot g(\lambda\alpha)$ become equal.

The following question arises. Can we introduce, instead of $g(\alpha)$, a new functional dependence $\bar{g}(\alpha)$ of the group element upon the parameters α, such that

(i) if α varies in the whole neighbourhood of zero, the group element $\bar{g}(\alpha)$ varies in the whole neighbourhood of unity and

(ii) in that whole neighbourhood (i.e. even for finite α) it holds true that $\bar{g}(\mu\alpha) \cdot \bar{g}(\lambda\alpha) = \bar{g}[(\mu + \lambda)\alpha]$?

We shall soon show by construction that this is possible. In order not to complicate the notation, we shall omit the bar over $\bar{g}$ and designate the new function by $g(\alpha)$ again. But we shall now *postulate that $g(\alpha)$ depends in such a way on α that*

$$g(\mu\alpha) \cdot g(\lambda\alpha) = g[(\mu + \lambda)\alpha] \qquad \alpha \text{ fixed} \qquad \mu, \lambda \text{ real.} \qquad (2.3.9)$$

We then say that the group elements are labelled by *canonical parameters*: for a fixed direction n in the space of canonical parameters, group multiplication is simply addition of canonical parameters. Next, we observe that

$$g(\mu\alpha)g(\alpha)g^{-1}(\mu\alpha) = g(\alpha) \qquad (2.3.10)$$

on account of the just proved additivity; in other words: group elements with the same direction n in the space of canonical parameters commute. This corresponds to (2.3.7). So far we have postulated properties without proving that objects with these properties really exist: in order to show that they exist, we have to prove that by varying α in the neighbourhood of zero, the whole neighbourhood of unity can be obtained from group elements $g(\alpha)$ with the postulated properties. We now construct these functions $g(\alpha)$.

The properties (2.3.9) and (2.3.10) enable us to write down an exponential form for the group elements belonging to a fixed direction n:

$$g(\alpha) = g\left(m \cdot \frac{\alpha}{m}\right) = \left[g\left(\frac{\alpha}{m}\right)\right]^m.$$

For large m we find

$$g\left(\frac{\alpha}{m}\right) = g\left(\frac{\alpha}{m} \cdot n\right) = g(0) + \frac{\alpha}{m} \cdot \frac{d}{d(\alpha/m)} g\left(\frac{\alpha}{m}n\right)\Big|_{\alpha/m=0} + \dots$$

$$= 1 + \frac{\alpha}{m}\overset{\circ}{g}(n) + \dots.$$

If we raise this to the power m and let $m \to \infty$, then the result is

$$g(\alpha) = \exp[\alpha\overset{\circ}{g}(n)] \equiv \exp[\overset{\circ}{g}(\alpha)]. \tag{2.3.11}$$

The exponential of a matrix is defined by the power series, which always converges. Thus (2.3.11) exists for all values of α (we consider here only real ones). Keeping the direction n still fixed, it follows again that

$$g(\mu\alpha) \cdot g(\lambda\alpha) = g(\lambda\alpha) \cdot g(\mu\alpha) = g[(\lambda + \mu)\alpha]$$

and the product $g[(\lambda + \mu)\alpha] = \exp[(\lambda + \mu) \cdot \overset{\circ}{g}(\alpha)]$ always exists. For $\alpha = 0$ we obtain the unit element; hence the group elements $g(\alpha) = \exp[\overset{\circ}{g}(\alpha)]$ with fixed direction n_α constitute a one-parameter Abelian subgroup $G(n_\alpha)$ of G.

Conversely, given any element $g(\beta)$ in the neighbourhood of unity, there exists an Abelian subgroup $G(n_\beta)$ to which $g(\beta)$ belongs: namely the one with $n_\beta = \beta/\beta$.

It remains to mention (without proof) that the mapping $A = \exp[B]$ of the neighbourhood of zero (matrices B) onto the neighbourhood of unity (matrices A) is one-to-one. We leave the proof as an exercise for the reader.

This fact shows that

$$g(\alpha) = \exp[\alpha\overset{\circ}{g}(n)] = \exp\left[\sum_i \alpha_i \overset{\circ}{g}_i\right] = \exp[\overset{\circ}{g}(\alpha)]$$

indeed fulfils all requirements contained in the definition of canonical parameters and in particular constitutes a one-to-one mapping of the neighbourhood of the zero of the infinitesimal ring onto the neighbourhood of the unity of the group. This enables one to study the local group properties not on the group directly, but rather by looking at the Lie algebra and its structure. For obvious reasons the basis elements $\overset{\circ}{g}_i$ of the Lie algebra are often called 'generators of the group'.

All this is true if one uses canonical coordinates. Clearly, linear transformations in the space of canonical coordinates again yield canonical coordinates. From now on we assume tacitly that canonical coordinates are used.

Remark. In the literature two types of canonical coordinate are encountered (see subsection 3.3.4):

- canonical coordinates of the first kind

$$g(\alpha_1, \alpha_2, \ldots, \alpha_r) = \exp[\alpha_1 \overset{\circ}{g}_1 + \alpha_2 \overset{\circ}{g}_2 + \ldots + \alpha_r \overset{\circ}{g}_r]$$

- canonical coordinates of the second kind

$$g(\gamma_1, \gamma_2, \ldots, \gamma_r) = \exp(\gamma_1 \overset{\circ}{g}_1) \exp(\gamma_2 \overset{\circ}{g}_2) \ldots \exp(\gamma_r \overset{\circ}{g}_r).$$

In the neighbourhood of zero these two kinds of canonical coordinate are analytic functions of each other. So far, we have dealt with the canonical coordinates of the first kind. We shall employ in practice both kinds of coordinate. In the parametrization of the rotation group the angle α of rotation and the direction n of the axis of rotation are a set of three canonical coordinates of the first kind; the Euler angles α, β, γ are of the second kind.

2.3.5 The structure of the group and its infinitesimal ring

We have seen that there is—at least in the neighbourhood N of the unit element of the group—a one-to-one correspondence between the group elements and the elements of the Lie algebra; in canonical coordinates this takes the form

$$g(\alpha) = \exp[\overset{\circ}{g}(\alpha)].$$

This suggests that the whole structure of the group in the neighbourhood of the unit element is determined by the structure of its Lie algebra. Group multiplication associates with two vectors α and β in the space of canonical coordinates a third one, γ. If α and β lie sufficiently near to zero, γ is uniquely determined. Then

$$g(\alpha) \cdot g(\beta) = \exp[\overset{\circ}{g}(\alpha)] \exp[\overset{\circ}{g}(\beta)] = \exp[\overset{\circ}{g}(\gamma)]$$

where the two exponentials do not, in general, commute. In the space $\overset{\circ}{G}$ we have then an element $\overset{\circ}{g}(\gamma)$, which is a unique function of the two elements $\overset{\circ}{g}(\alpha)$ and $\overset{\circ}{g}(\beta)$. This function should, of course, be computable by means of operations (addition, $\lambda \overset{\circ}{g} + \mu \overset{\circ}{h}$ with λ, μ real; multiplication, $[\overset{\circ}{g}, \overset{\circ}{h}]$) defined in the ring $\overset{\circ}{G}$; i.e. $\overset{\circ}{g}(\gamma)$ should be calculable by adding elements, commutators and commutators of commutators etc with real coefficients. This is indeed the case, as the Baker–Campbell–Hausdorff theorem asserts:

$$e^A e^B = e^C$$
$$C = A + B + \tfrac{1}{2}[A, B] + \tfrac{1}{12}\{[A, [A, B]] + [B, [B, A]]\} + \dots . \tag{2.3.12}$$

In this expansion all terms consist of iterated commutators; the coefficients are real. The proof of this theorem will not be given here; the reader should calculate as an exercise the first terms written in (2.3.12).

With the help of the Baker–Campbell–Hausdorff theorem we can immediately show that indeed the structure of the Lie algebra determines the structure of the group in the neighbourhood N of the unity.

The structure of the Lie algebra is completely determined if for any two of its elements, $\overset{\circ}{g}$ and $\overset{\circ}{h}$ say, the result of their product $[\overset{\circ}{g}, \overset{\circ}{h}]$ is specified. We may expand

$$\overset{\circ}{g} = \sum \alpha_i \overset{\circ}{g}_i \qquad \overset{\circ}{h} = \sum \beta_i \overset{\circ}{g}_i$$

and have

$$[\overset{\circ}{g}, \overset{\circ}{h}] = \sum_{i,k} \alpha_i \beta_k [\overset{\circ}{g}_i, \overset{\circ}{g}_k]$$

which indicates that it suffices to know all possible products of the basis elements $\overset{\circ}{g}_i$. Since the commutator of any two elements is again an element of the Lie algebra, it follows that it can be expanded with respect to the given basis:

$$[\overset{\circ}{g}_i, \overset{\circ}{g}_k] = \sum_{j=1}^{r} c_{ik}^j \overset{\circ}{g}_j. \tag{2.3.13}$$

The 'structure constants' c_{ik}^j therefore determine completely the structure of the Lie algebra. Not only that, in fact they also determine completely the structure of the group in the neighbourhood of the unity. Namely, in that neighbourhood

$$g(\alpha)g(\beta) = g(\gamma)$$

reads

$$\exp[\overset{\circ}{g}(\alpha)] \exp[\overset{\circ}{g}(\beta)] = \exp[\overset{\circ}{g}(\gamma)]$$

and the Baker–Campbell–Hausdorff theorem yields

$$\overset{\circ}{g}(\gamma) = \overset{\circ}{g}(\alpha) + \overset{\circ}{g}(\beta) + 1/2[\overset{\circ}{g}(\alpha), \overset{\circ}{g}(\beta)] + \dots$$

where the first and all following commutators are completely determined by the structure constants c_{ik}^j (and, of course, by $\overset{\circ}{g}(\alpha)$ and $\overset{\circ}{g}(\beta)$).

As the commutator is already a quadratic function of α and β, one can write $\overset{\circ}{g}(\gamma) = \overset{\circ}{g}(\alpha + \beta)$ + terms of second and higher orders in α and β.

In other words: in an *infinitesimal* neighbourhood of unity, group multiplication is (almost) commutative and is in a first approximation equivalent to vector addition in the space of canonical parameters. This is the meaning of the somewhat misleading saying that 'infinitesimal transformations commute'.

Although the Lie algebra generates by exponentiation a group which is isomorphic in the neighbourhood of the unity to the group from which the ring was derived, it need not be that these two groups are isomorphic outside the neighbourhood of the unity. The group generated by

$$g(\alpha) = \exp[\overset{\circ}{g}(\alpha)]$$

is called the 'local Lie group' of the group G, because it is locally isomorphic to G but not necessarily globally.

A simple example for a local Lie group, which is not globally isomorphic to the group from which it was derived, is obtained if the original group decomposes into two pieces, as for instance does the rotation group (in three dimensions):

$$R = \{R_+, SR_+\}$$

where R_+ is the proper rotation group (determinant $+ 1$), S a reflection and SR_+ the coset (determinant $- 1$) of R_+. Then near to the unity the group R and its normal subgroup R_+ are locally isomorphic, because no element of the coset SR_+ lies near to unity. If one then constructs the Lie algebra of either R or R_+, it is the same for the two cases but the group generated by its elements is R_+, not R.

2.3.6   Summary: continuous matrix groups and their Lie algebra

Let us summarize the results in the following statement (we do not say theorem, because we do not formulate it rigorously and also have made things only partly plausible instead of proving them): to every continuous r-parameter group G of $n \times n$ matrices one can choose particular (real!) parameters (called canonical parameters) $\alpha_1, \ldots, \alpha_r$ such that the group elements $g(\alpha_1, \ldots, \alpha_r)$ have the following properties.

(i)

$$g(0, 0, \ldots, 0) = 1. \tag{2.3.14}$$

(ii)   The derivatives

$$\overset{\circ}{g}(n_\alpha) = \frac{d}{d\alpha} g(\alpha)|_{\alpha=0} \qquad \alpha = |\alpha| \qquad n_\alpha = \alpha/|\alpha| \tag{2.3.15}$$

(again $n \times n$ matrices) constitute a real linear vector space of dimension r, which becomes a ring by defining the product of any two elements

$$[\overset{\circ}{g}, \overset{\circ}{h}] = \overset{\circ}{g}\overset{\circ}{h} - \overset{\circ}{h}\overset{\circ}{g}. \tag{2.3.16}$$

This is the Lie algebra $\overset{\circ}{G}$ (also Lie ring, infinitesimal ring) of the group G.

Its basis is given by the elements $\overset{\circ}{g}_i = (\partial/\partial\alpha_i)g(0, 0, \ldots, 0)$; $i = 1, \ldots, r$.

(iii) The neighbourhood of the unity of the group and of the zero of its Lie algebra are mapped one-to-one onto each other by

$$g(\alpha) = \exp\left[\sum \alpha_i \overset{\circ}{g}_i\right] = \exp[\overset{\circ}{g}(\alpha)]. \qquad (2.3.17)$$

The $\overset{\circ}{g}_i$ are often called generators of the group.

(iv) The local structure of the group in the neighbourhood of the identity (and with some precaution one can say in the neighbourhood of any of its elements) is completely determined by the structure of its Lie algebra, i.e. by the structure constants c_{ik}^j in

$$[\overset{\circ}{g}_i, \overset{\circ}{g}_k] = \sum_{j=1}^{r} c_{ik}^j \overset{\circ}{g}_j . \qquad (2.3.18)$$

(v) If one starts from a group G and constructs its Lie algebra $\overset{\circ}{G}$, then by means of (2.3.17) a new group G_L is generated which is locally but not necessarily globally isomorphic to G. One calls G_L the local Lie group of G.

(vi) The group elements lying on the same straight line α in the space of canonical parameters constitute an Abelian, one-parameter subgroup of G_L and each element of G_L belongs to such a subgroup.

2.3.7 Group representations

In section 2.2 we proved the Wigner theorem, which states that for every symmetry group G in physics there exists a unitary (anti-unitary) representation in Hilbert space. This representation is no longer an $n \times n$ matrix group—it is in fact a group of matrices in an infinite-dimensional space $\mathcal{H}$. For the moment we shall ignore that it is, in general, only a representation up to a factor.

The set of unitary (anti-unitary) $\infty \times \infty$ matrices constituting this representation may or may not decompose into a direct sum[3] of finite-dimensional matrices.

In the case of the rotation group (as for all compact groups) it in fact does so. We have then the following situation.

We start from an $n \times n$ continuous matrix group which is the group with a direct physical meaning; we find that there exist unitary (forget about the anti-unitary, which do not come up in this book) representations $U_k(g)$ in a space of k dimensions (where k may be any number, including zero and infinity); this k-dimensional space is a subspace (called an 'invariant subspace') of $\mathcal{H}$. The direct

[3] See section 4.3 for the definition of the direct sum.

sum of all these representations U_k makes up the unitary representation U. We must mention here that the invariant subspace may be labelled by continuous quantum numbers (for instance momentum) when the direct sum becomes a direct integral; we ignore this complication.

We also do not go into any detail here about such questions as when invariant subspaces exist and how they are found; we only need for the following consideration the fact that there may be representations of the $n \times n$ matrix group G by $k \times k$ (or infinite) unitary matrices U where k in general differs from n.

If such a set of unitary matrices U is a representation of the group G, then we may describe them in exactly the same way as the group elements $g(\alpha_1, \ldots, \alpha_r)$ as a function of the canonical parameters: $U(\alpha_1, \ldots, \alpha_r)$. These matrices, taken by themselves, constitute now a $k \times k$ (possibly infinite) continuous matrix group. The dimension of this group may be smaller than r, because we have not specified k. Indeed the number of independent parameters of the unitary group in k dimensions is $2k^2$ minus the number of real equations expressing unitarity. These equations are $\frac{1}{2}k(k-1)$ complex equations expressing orthogonality of different rows (or of columns—this gives nothing new), hence $k^2 - k$ real equations and k equations expressing the normalization of the rows— these equations are real. Hence there are k^2 real equations between the $2k^2$ real parameters. The dimension of the full unitary group in k-dimensional space is therefore k^2. As k may be any integer, it follows that the representation $U_k(\alpha)$, taken by itself, need not be of dimension r although it may be described by r parameters. Its dimension is equal to the smaller of the two numbers, k^2 or r. If $k^2 < r$, the group elements $U(\alpha)$ and $U(\alpha')$ need not be different for $\alpha \neq \alpha'$ (both near to zero).

Let us now consider the unitary matrix group $U_k(\alpha_1, \ldots, \alpha_r)$ with fixed k (suppressed) and forget for a moment that it is a unitary representation of G. Nothing can prevent us from going in this case once more through all the preceding arguments on continuous matrix groups, their Lie algebra and the generation of the piece of the group connected to the identity by simply exponentiating the elements of the Lie algebra. In particular, we define the elements of the Lie algebra $\overset{\circ}{U}$ of U by

$$I(n_\alpha) = \frac{\mathrm{d}}{\mathrm{d}\alpha} U(\alpha n_\alpha)|_{\alpha=0} \tag{2.3.19}$$

in complete analogy to $\overset{\circ}{g}(n_\alpha)$ defined by (2.3.2). The only difference to be kept in mind is that the r 'basis elements'

$$I_i = \frac{\partial}{\partial \alpha_i} U(0, \ldots, 0) \tag{2.3.20}$$

need not be linearly independent and thus need not span an r-dimensional Lie algebra. They will do that, however, if $k^2 \geq r$. Let us assume this to be the case from now on.

The local Lie group of U will then simply consist of the elements

$$U(\alpha) = \exp[I(\alpha)] = \exp\left[\sum_{i=1}^{r} \alpha_i I_i\right]. \qquad (2.3.21)$$

Now we remember that U is a representation of the group G, and that the structure of the local Lie group and that of the Lie algebra determine each other uniquely (2.3.18). It follows that (under the restriction $k^2 \geq r$) the elements of $\overset{\circ}{g}$ and those of $\overset{\circ}{U}$ are mapped one-to-one onto each other, thereby preserving the commutation relations; we call the ring $\overset{\circ}{U}$ a representation of the ring $\overset{\circ}{G}$.

Thus if U is a representation of G then the Lie algebra $\overset{\circ}{U}$ is a representation of the Lie algebra $\overset{\circ}{G}$. What is so important here is the observation that one can go the other way round. If $\overset{\circ}{U}$ is a representation of the Lie algebra $\overset{\circ}{G}$ of G, then (in an obvious shorthand notation)

$$U = \exp[\overset{\circ}{U}]$$

is a representation of G (in the neighbourhood of the unity; i.e. U is locally isomorphic to G).

This enables one to construct group representations by working with the Lie algebra and finding its representations [4]; exponentiation then yields a (local) group representation. Exactly this is done in the standard theory of angular momentum: as we shall see in any desirable detail later on, the angular momentum operators J_1, J_2 and J_3 are a basis of a Lie algebra with elements

$$J(\alpha) = \sum \alpha_i J_i$$

yielding a representation

$$U(\alpha) = \exp[-iJ(\alpha)]$$

where $\alpha = \alpha n$ designates the axis and angle of rotation and where the factor i has been introduced to have Hermitian operators J (this is not in contradiction with our previous remark that the group parameters must be kept real: here the whole set of group parameters and all elements of the infinitesimal ring are multiplied simultaneously by i—which does not matter. In fact it simply means a redefinition of the elements of the Lie algebra).

Contrary to the general theory of Lie groups we have started from a definite group of *linear* transformations in an n-dimensional space (Lorentz group, rotation group). Among all $(k \times k)$-dimensional representations of the group

---

[4] In fact one obtains this way all representations which depend *analytically* on the group parameters; in general one does not, however, obtain all continuous representations (see Boerner (1963), chapter 5, section 8).

and of its Lie algebra, one particular representation is exhibited: namely that representation which consists of the $n \times n$ matrices by which the group is defined. This representation logically precedes all the other ones. We shall call it the *fundamental representation* (also *original representation, self-representation*).

2.4   The physical significance of symmetries

2.4.1   Continuous groups connected to the identity; Noether's theorem

Let G be a continuous symmetry group of our physical system. Then either the whole of G or at least a piece of it (this piece being a normal subgroup) is continuously connected to the identity. We consider this piece.

What do we know then? We know from the preceding analysis five facts which now will combine into one beautiful theorem[5]:

for every continuous symmetry (continuously connected to the identity) there exists a corresponding conservation law; i.e. there exists a conserved observable.

We shall explain this more precisely now. The five facts mentioned above are the following.

(i)   The Wigner theorem (2.2.11) asserts that for the given group G with elements $g(\alpha_1, \ldots, \alpha_r)$ there exists a unitary representation $U(\alpha_1, \ldots, \alpha_r)$ (it cannot be anti-unitary for a group which is continuously connected to the identity).

(ii)   The considerations on continuous matrix groups have shown that with $g(\alpha_1, \ldots, \alpha_r)$ as well as with $U(\alpha_1, \ldots, \alpha_r)$ there exists a set of matrices

$$\overset{\circ}{g}_i = \frac{\partial}{\partial \alpha_i} g(0, 0, \ldots, 0) \qquad \overset{\circ}{U}_i = \frac{\partial}{\partial \alpha_i} U(0, 0, \ldots, 0)$$

respectively; each set can serve as a basis of a representation of the Lie algebra $\overset{\circ}{G}$ of the group G; the $\overset{\circ}{g}_i$ span the original representation of the Lie algebra.

(iii)   In the neighbourhood of the identity the group elements can be written

$$g(\alpha_1, \ldots, \alpha_r) = \exp\left\{\sum_{k=1}^{r} \alpha_k \overset{\circ}{g}_k\right\} \tag{2.4.1}$$

and their unitary representations become

$$U(\alpha_1, \ldots, \alpha_r) \equiv \exp\left\{\sum_{k=1}^{r} \alpha_k \overset{\circ}{U}_k\right\} \equiv \exp\left\{\sum_{k=1}^{r} i\alpha_k \overset{\circ}{B}_k\right\}. \tag{2.4.2}$$

[5] 'Noether's theorem' (Noether (1918)). For an extensive discussion of Noether's theorem see Hill (1957).

The last member of this equation defines the Hermitian operators $\overset{\circ}{B}_k = -i\overset{\circ}{U}_k$.

(iv) By this procedure the α have become canonical parameters; putting $(\alpha_1, \ldots, \alpha_r) \equiv \alpha \equiv \alpha \cdot n_\alpha$ we obtain a one-parameter Abelian subgroup for n_α fixed and α variable. This holds in particular when we choose $n_\alpha = (0, 0, \ldots, 0, 1, 0, \ldots, 0) \equiv n_i$ with a one at the ith place and zeros elsewhere.

(v) The Hamiltonian and the scattering operator (S-matrix) are invariant under the transformations of the symmetry group (see (2.1.3)).

Let then (according to (i)) a unitary representation $U(\alpha_1, \ldots, \alpha_r)$ be given, and the $\overset{\circ}{U}$ be calculated (according to (ii)). Then the local representation reads

$$U(\alpha_1, \ldots, \alpha_r) = \exp\left\{i\sum \alpha_k \overset{\circ}{B}_k\right\} \qquad \overset{\circ}{B}_k = -i\overset{\circ}{U}_k.$$

We choose in the canonical parameter space the direction $n_k = (0, \ldots, 0, 1, 0, \ldots, 0)$ and obtain the Abelian subgroup (according to (iv))

$$U(\alpha) = \exp\left\{i\alpha \overset{\circ}{B}_k\right\} \ .$$

Under this unitary transformation the Hamiltonian and the S-matrix are invariant (according to (v))[6]:

$$H' = U(\alpha) H U^{-1}(\alpha) = H.$$

For $\alpha \to 0$ we obtain

$$(1 + i\alpha \overset{\circ}{B}_k) H (1 - i\alpha \overset{\circ}{B}_k) = H - i\alpha[H, \overset{\circ}{B}_k] = H$$

hence

$$[H, \overset{\circ}{B}_k] = 0$$

$$[S, \overset{\circ}{B}_k] = 0. \tag{2.4.3}$$

This implies that the Hermitian operator $\overset{\circ}{B}_k$ is a constant of the motion. Having assumed that we stay during all these considerations within one definite superselection subspace (we come back to this very soon and then explain it in detail), it follows that the operator $\overset{\circ}{B}_k$ can be considered to be an observable. The same holds for all other $\overset{\circ}{B}_i$, $i = 1, \ldots, r$.

---

[6] Do not confuse this S with the letter S used to designate a physical system; the context always makes clear which of the two is meant.

Every element of the Lie algebra[7] can be written as

$$B(\alpha) = \sum \alpha_k \overset{\circ}{B}_k = \alpha \sum \frac{\alpha_k \overset{\circ}{B}_k}{\alpha} = \alpha \overset{\circ}{B}(n_\alpha) \tag{2.4.4}$$

and the above consideration can be repeated for any given direction n_α. Thus the operators $\overset{\circ}{B}(n_\alpha)$ as well as their multiples $\alpha \cdot \overset{\circ}{B}(n_\alpha)$—in other words, all elements of the Lie algebra—are observables which are constant in time.

However, as the basis elements $\overset{\circ}{B}_k$ ($k = 1, \ldots, r$) span the whole Lie algebra, it is sufficient to consider this basis.

The $\overset{\circ}{B}_k$ do not all commute with each other; in fact the whole structure of the local group and of its Lie algebra is contained in their commutation relations (see subsection 2.3.5 above)

$$[\overset{\circ}{B}_i, \overset{\circ}{B}_k] = \sum_{l=1}^{r} C_{ik}^l \overset{\circ}{B}_l \tag{2.4.5}$$

where the structure constants C_{ik}^ℓ are the heart of the matter. (They are of course not invariant under a change of the basis $\overset{\circ}{B}_k \to \overset{\circ}{B}_k{}'$, but that is not essential; it permits the choice of a basis such that a certain normal form of the C_{ik}^ℓ is achieved. We do not go into this detail here.)

We now may take any set $\{B_1, \ldots, B_\ell\}$ of commuting operators out of the Lie algebra. Since they commute among each other and with the Hamiltonian, we shall label the quantum states by their quantum numbers $b_1, b_2, \ldots, b_\ell$. Such a set might, however, not be suitable for the description of quantum states. It might be that certain 'functions' $F(\overset{\circ}{B}_1, \ldots, \overset{\circ}{B}_r)$ from the set $\{\overset{\circ}{B}_1, \ldots, \overset{\circ}{B}_r\}$ are convenient constants of the motion (e.g. the Casimir operator). For details see section 2.4.4 below.

Resumé (Noether's theorem)

The generators $\overset{\circ}{B}_1, \ldots, \overset{\circ}{B}_r$ of a symmetry group (continuously connected to the identity) are conserved observables whose commutation relations are uniquely determined by the structure of the group.

Remark. This obviously holds only if the equations of motion (the dynamics of the system) can be formulated in Hamiltonian form and if then $[H, \overset{\circ}{B}] = 0$. The invariance of the equations of motion is not sufficient. As is well known, conservation of the linear momentum follows from translational invariance. In the case of the motion of the body in a homogeneous viscous fluid the linear

---

[7] As discussed above, the Lie algebra of the $\overset{\circ}{B}$ and that of the $\overset{\circ}{U}$ differ only trivially by an overall factor i.

momentum is *not* conserved in spite of the translation invariance. Why? Because the equations contain a dissipative term (for friction) and therefore are not of the Hamilton type.

2.4.2   Pieces not connected to the identity; discrete groups

The above arguments involved one essential step, namely the expansion of the relation

$$U(\alpha)HU^{-1}(\alpha) = H \text{ for } \alpha \rightarrow 0 \text{ where } U(\alpha) \approx 1 + i\alpha \overset{\circ}{B}.$$

From this it followed that $[H, \overset{\circ}{B}] = 0$ and that therefore $\overset{\circ}{B}$ is an operator constant in time. Such a conservation law cannot be derived from $U(\alpha)HU^{-1}(\alpha) = H$ if $U(\alpha)$ is not continuously connected to the identity, i.e. if (for $\alpha \rightarrow 0$) $U(\alpha) \neq 1 + i\alpha \overset{\circ}{B}$. In this case the representation may also be anti-unitary. Examples of such symmetries are reflections, permutations of particles, time reversal (represented by antilinear and anti-unitary tranformation!) and so on.

The discrete symmetries (which are responsible for the groups not continuously connected to the identity) have, however, in most cases the remarkable property that their square gives unity. In that case, if they are of the unitary type, it follows from

$$UU^{\dagger} = U^{\dagger}U = 1 = U^2$$

that

$$UUU^{\dagger} = U = U^{\dagger}.$$

Hence U is not only unitary, it is even Hermitian and therefore it represents itself an observable, which, by $[U, H] = 0$, is a constant of the motion. Note, however, that this argument is not valid if $U^2 \neq 1$ and/or U is anti-unitary.

Example. A parity quantum number exists, but not a time-reversal quantum number.

We shall not go further into these symmetries, as they only occasionally come up in this text; in these few cases we deal with them directly and intuitively without needing a general theory.

2.4.3   Super-selection rules

We finish this consideration of symmetry by discussing the selection rules. Evidently the quantum numbers associated with the generators of continuous symmetries have to be the same before and after a (symmetry-preserving) reaction and they thus characterize the selection rules of the (symmetry-preserving) interaction. Also the discrete symmetries lead to conservation laws frequently expressible in terms of quantum numbers, but sometimes requiring a more elaborate description.

There is one particular kind of selection rule, which are called 'super-selection rules'. They are characterized by the fact that their generators commute with *all* observables of the system under consideration. In other words, the observables corresponding to the super-selection quantum numbers can be measured without disturbing any other measurement; their values are always sharp. Examples are the electric charge, the baryonic charge and—as far as only strong interactions are concerned—the strangeness. The term 'super-selection rule' does not imply that these quantities are super-strictly conserved: they may in fact be conserved only with respect to certain interactions (like the strangeness), whereas the four-momentum is absolutely conserved, although it does not lead to a super-selection rule.

Why then, are they called 'super-selection rules'? They are so called because they split the Hilbert space, in which our system is described, into subspaces between which no observable can have matrix elements and between which no linear combinations are allowed in constructing state vectors of physical significance.

This is seen as follows.

Assume that an operator X' commutes with all observables of our system S. Then also the Hermitian conjugate $X'^\dagger$ commutes with all observables and therefore the Hermitian operator $X = 1/2(X' + X'^\dagger)$ commutes also with all of them. Therefore, whatever complete set of commuting observables we may select to define a basis in $\mathcal{H}$, our X will be among the observables of this set and consequently have a common basis with them in $\mathcal{H}$.

We write down the spectral representation of X

$$X = \sum_{\xi,\gamma} \xi |\xi, \gamma\rangle\langle\xi, \gamma| = \sum_{\xi} \xi P_\xi \qquad (2.4.6)$$

where the projection operators $P_\xi = \sum_\gamma |\xi, \gamma\rangle\langle\xi, \gamma|$ project into that subspace $\mathcal{H}_\xi$ of the Hilbert space $\mathcal{H}$ where X has the eigenvalue ξ; γ in $|\xi, \gamma\rangle$ indicates further quantum numbers—in fact $\mathcal{H}_\xi$ is made up of all states $|\xi, \gamma\rangle$ in which ξ is kept fixed and the other quantum numbers γ assume all possible values. Whatever basis one chooses, that is, whatever the quantum numbers represented by γ mean, the ξ will always occur, because X commutes with all observables and therefore any complete system of commuting observables contains X and has to share its basis with it. This induces the mentioned decomposition of the Hilbert space $\mathcal{H}$ into subspaces $\mathcal{H}_\xi$. This decomposition is by its very nature unaffected when we change the basis in $\mathcal{H}$ by going over to the eigenvectors of another complete set of commuting observables—just because X is also a member of the new set of observables. Hence all those unitary transformations which transform from one basis (defined by a complete set of commuting observables) to another one leave this decomposition of $\mathcal{H}$ invariant. This holds also for the unitary (or anti-unitary) representations of the symmetry groups;

otherwise one would have an observable effect, namely a change from ξ to another eigenvalue ξ'.

No observable A can have matrix elements between states of *different* ξ:

$$\langle \xi, \gamma | A | \xi', \gamma \rangle = \langle \xi, \gamma | P_\xi A P_{\xi'} | \xi', \gamma \rangle$$

$$= \langle \xi, \gamma | A P_\xi P_{\xi'} | \xi', \gamma \rangle = 0$$

because $[P_\xi, A] = 0$, $P_\xi = P_\xi^\dagger$ and $P_\xi P_{\xi'} = \delta_{\xi\xi'} P_\xi$.

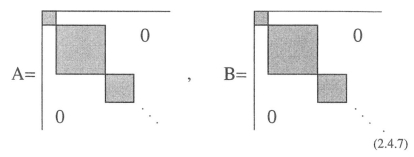

$$(2.4.7)$$

Thus all observables in S are split in the same way (whether they commute or not) and in each of these boxes (of in general infinite dimensionality) ξ has another value, constant throughout the box. The box structure is possibly different for two different physical systems. For instance, in strong interactions the strangeness S and the charge Q lead to independent box structures; if weak interactions are added, only the charge boxes remain.

By no physical process can a transition between different boxes be induced—as long as we do not enlarge the system under consideration by interaction terms not commuting with X. Therefore all probabilities involve only matrix elements inside one super-selection subspace: superpositions of state vectors belonging to different super-selection subspaces do not make sense.

Let us illustrate this by a simple example.

All physical systems—be they rotationally invariant or not—are transformed into themselves under a rotation by 2π about an arbitrary axis. Rotations have unitary representations, hence $U(2\pi)$ is unitary. One might think that $U(2\pi)$ must be equal to the unity operator; this is, however, not true: as we shall see later, a state with total angular momentum j is multiplied by $\exp(2\pi i j)$ under a rotation by 2π and this is ± 1 according to whether j is integer or half-integer. In any case we have $U^2(2\pi) = U(4\pi) = 1$.

We proved above that a unitary operator U whose square is unity is a Hermitian operator. Hence its eigenvalues will be $+1$ and -1. For obvious physical reasons the operator $U(2\pi)$ commutes with all observables and hence induces a decomposition of the total Hilbert space into super-selection subspaces: $\mathcal{H} = \mathcal{H}_+ + \mathcal{H}_-$. All states $\mathcal{H}_+$ are invariant under a rotation by 2π and all states of $\mathcal{H}_-$ are multiplied by -1 under such a rotation. Fermions have half-integer j and bosons as well as orbital angular momenta have integer j, hence $\mathcal{H}_+$

contains all states with the number of fermions plus antifermions being even, whereas $\mathcal{H}_-$ contains the states where this number is odd. This decomposition holds whether or not we have rotational invariance.

Assume now one tried to assign a physical significance to the superposition of one state belonging to $\mathcal{H}_+$ and one to $\mathcal{H}_-$:

$$|\psi\rangle = a|\varphi_+\rangle + b|\varphi_-\rangle \qquad |a|^2 + |b|^2 = 1.$$

Having prepared such a state, the probability to find it in a measurement (immediately following the preparation) must be unity. Indeed

$$|\langle\psi|\psi\rangle|^2 = (aa^* + bb^*)^2 = 1.$$

But then, as the *physical state (not the state vector)* rotated by 2π must be the same as the unrotated one, the probability to find the rotated state must also be one. The rotated state vector is, however, $|\psi'\rangle = a|\varphi_+\rangle - b|\varphi_-\rangle$ and the probability becomes

$$|\langle\psi'|\psi\rangle|^2 = (aa^* - bb^*)^2 \leq 1$$

which would mean that a rotation by 2π would lead to an observable effect— except if either $a = 0$ or $b = 0$. In other words a superposition of states belonging to different super-selection subspaces is not allowed. The super-selection rule forbidding the superposition of states with different statistics remains true also in the presence of super-symmetry (generalized symmetry transformations relating bosons and fermions to each other; see chapter 11).

What *is* allowed, however, is to combine such states of different super-selection subspaces in a density matrix, because the relative phases become irrelevant and the density matrix only expresses our knowledge that the system under the observation is, with certain probabilities, in such and such states (such a combination of states is called a mixture). Indeed, if we know that the system is with probability a in a certain state $|\varphi_+\rangle$ and with probability b in a state $|\varphi_-\rangle$, where a and b are real, $a + b = 1$, then the density matrix is

$$\rho = a|\varphi_+\rangle\langle\varphi_+| + b|\varphi_-\rangle\langle\varphi_-|$$

and this transforms into itself under a rotation by 2π:

$$\rho' = a|\varphi_+\rangle\langle\varphi_+| + b\cdot(-1)|\varphi_-\rangle\langle\varphi_-|\cdot(-1) = \rho.$$

The expectation value of any operator A is

$$\langle A\rangle = \mathrm{Tr}(A\varphi)$$

and the probability to find the mixture described by ρ, in an experiment immediately following its preparation, is given by

$$\langle 1\rangle = \mathrm{Tr}(\rho) = 1.$$

All this is not affected by a rotation by 2π, because ρ is invariant.

More generally, if $\{|\varphi_i\rangle\}$ is a complete orthogonal system spanning the whole of $\mathcal{H}$, and if we know the probabilities a_i for each $|\varphi_i\rangle$ (of course $\sum a_i = 1$; all a_i real) then

$$\rho = \sum_i a_i |\varphi_i\rangle\langle\varphi_i|$$

and this is obviously independent of the phases of the $|\varphi_k\rangle$, because if

$$|\varphi_k\rangle \to e^{i\alpha}|\varphi_k\rangle$$

then

$$\rho' = \sum_k a_k\, e^{i\alpha_k} |\varphi_k\rangle\langle\varphi_k|\, e^{-i\alpha_k} = \rho.$$

Our above considerations show that only Hermitian operators having a box structure (2.4.7), and therefore having vanishing matrix elements between different super-selection subspaces, can be admitted as observables; in fact they even must be—in other words any such operator corresponds to an observable quantity. We have seen that the unitary (or anti-unitary) representations of the symmetries leave the decomposition of $\mathcal{H}$ into super-selection subspaces invariant (otherwise one would have an observable effect); the generators of these groups cannot therefore have matrix elements between different subspaces; consequently they are indeed observables even in the presence of super-selection rules. One can also argue the other way round: in the definition (2.1.2) of symmetry we required $A' = UAU^\dagger$ to be an observable of S if A was one. Since all observables in S have to have the *same* box structure, A' must have it too. This is true for all observables if and only if U itself has this box structure. Hence $\overset{\circ}{B}$ in $U = \exp(i\alpha \overset{\circ}{B}) = 1 + i\alpha \overset{\circ}{B} + \ldots$ must be of the same box form and consequently is an observable.

The super-selection rules make quantum theory neither more complicated nor simpler: in discussing a given physical system with given super-selection rules we can restrict its description to one of the subspaces because no interaction (contained in the definition of the system) can lead out of it. That is, we can write on top of our calculation: 'We have eigenvalues $\xi', \eta', \ldots$ for the super-selection observables $X, Y, \ldots$' and then forget about it; having thus fixed all super-selection quantum numbers $\xi, \eta, \ldots$ we are back in an ordinary Hilbert space $\mathcal{H}(\xi, \eta, \ldots)$ in which no super-selection box structure remains.

2.4.4 Complete symmetry group, complete sets of commuting observables, complete sets of states

The following considerations will be only qualitative.

Let a definite physical system S be given. When we examine it carefully, we find a set of symmetry operations. If we look more carefully, we may find some more such symmetries. Assume that we have done our best and now believe we have found all of them; i.e. we believe we know *the* symmetry group G of the system S. Maybe later on we find further symmetries; then the arguments which will now follow must be repeated including the new symmetries. We should agree, however, on how we shall describe our system S with the best knowledge at hand; this present best knowledge contains a statement like this: G is *the* symmetry group of S, i.e. G is the complete set of symmetries of S. The group G will depend on parameters; continuously on some of them, $\alpha_1, \ldots, \alpha_r$, say, and discontinuously on some others, $\epsilon_1, \ldots, \epsilon_t$, say.

To the continuous (canonical) parameters corresponds a Lie algebra and to the unitary group representation there corresponds a representation of the Lie algebra. The basis of this representation, which is defined by the choice of the canonical parameters (changing the choice of the parameters implies only a linear transformation of the basis into another one—see section 2.3.4), consists of a set of observables $\overset{\circ}{B}_1, \overset{\circ}{B}_2, \ldots, \overset{\circ}{B}_r$ with well defined commutation relations. To this set is added the set of those unitary representations of discrete symmetries whose square is unity and which therefore are themselves Hermitian operators; let us call them $\overset{\circ}{C}_1, \overset{\circ}{C}_2, \ldots, \overset{\circ}{C}_s$. The set

$$\left\{ \overset{\circ}{B}_1, \overset{\circ}{B}_2, \ldots, \overset{\circ}{B}_r; \ \overset{\circ}{C}_1, \overset{\circ}{C}_2, \ldots, \overset{\circ}{C}_s \right\} \tag{2.4.8}$$

now contains only constant operators, which, however, do not all commute with each other. The operators $\overset{\circ}{B}_k$ of this set can be considered from two aspects: on one hand, they are basis elements of a Lie algebra and therefore only the 'product' $[\overset{\circ}{B}_i, \overset{\circ}{B}_k]$ is defined, and not the ordinary product $\overset{\circ}{B}_i \overset{\circ}{B}_k$; on the other hand, they—as well as the $\overset{\circ}{C}_k$—are operators in Hilbert space and then such products as $\overset{\circ}{B}_i \overset{\circ}{B}_k$, $\overset{\circ}{B}_i \overset{\circ}{C}_k$, $\overset{\circ}{C}_i \overset{\circ}{C}_k$ and further products are well defined. Such products—more generally, functions (power series) of these operators—no longer all belong to the Lie algebra, but they are nevertheless constant operators. We add these operators to the set (2.4.8) and obtain a new set, containing all functions of the $\overset{\circ}{B}_1, \overset{\circ}{B}_2, \ldots, \overset{\circ}{B}_r; \overset{\circ}{C}_1, \overset{\circ}{C}_2, \ldots, \overset{\circ}{C}_s$ (and among these functions of course the $\overset{\circ}{B}$ and $\overset{\circ}{C}$ themselves):

$$\left\{ F(\overset{\circ}{B}_1, \ldots, \overset{\circ}{C}_s) \right\}. \tag{2.4.9}$$

Example. In the theory of angular momentum L_x, L_y and L_z are conserved generators. As they do not commute among each other, we choose L_z and $L^2 = L_x^2 + L_y^2 + L_z^2$ (L^2 is a 'Casimir operator') as a commuting set.

The set (2.4.9), if derived from a complete symmetry group, now contains by definition all constant observables, because if a constant observable $\overset{\circ}{D}$ were not contained in it, it could not be a function of the $\overset{\circ}{B}$ and $\overset{\circ}{C}$. Hence it is an entirely new operator. By supposition it commutes with the Hamiltonian and so does $U_D(\alpha) = \exp(i\alpha\overset{\circ}{D})$.

As a unitary operator this U leaves all probabilities invariant and thus meets the definition of a symmetry. It therefore would have to be included in the symmetry group of our given physical system (although its physical interpretation might be difficult), which thereby would be enlarged. Starting again from this enlarged symmetry group we would now end up with an enlarged set (2.4.9) which this time would indeed contain the operator $\overset{\circ}{D}$. Having assumed, however, that the group was complete, no such additional symmetries leading to an enlargement exist—at least not with respect to the present experimental knowledge and the corresponding theoretical models.

Of course, in physics we are never sure that the group is really complete; but at any stage of our knowledge we build up a theory of the system, which tacitly assumes that we currently know the complete symmetry group—until one day new experimental material reveals something which has been overlooked so far. This is actually the way in which the known symmetry groups are enlarged; one finds experimentally new selection rules, formulates the rules in terms of observed operators and defines the corresponding symmetry. The history of the strong interactions, from the discovery of isospin invariance to the $SU(3)$ symmetry—and finally to QCD—is just an illustration of the above statement.

If we agree to label the basis vectors in $\mathcal{H}$ by 'good quantum numbers', i.e. by eigenvalues of conserved observables (we do not have to label them that way, but we *can*), then the usual complete set of commuting observables must be chosen from the set (2.4.9). As this set is determined by the complete symmetry group of the system, it follows that those complete sets of states which are labelled by good quantum numbers are determined by the complete symmetry group of the system; in other words, the complete symmetry group of the system determines the Hilbert space, which then appears to be nothing other than the representation space of the symmetries. Although this does not imply that the symmetries fix everything, yet they at least lay down the framework for the description.

In this respect even approximate symmetries are useful. These are symmetries which hold only for the 'strongest' part of the Hamiltonian but are violated for some 'weaker' part of it. The symmetries of elementary particles (isospin, strangeness, hypercharge, $SU(3)$ etc) are good examples. A similar example will be discussed in section 4.4, where we shall show that the formalism of angular momentum can remain useful even in situations where rotational symmetry is not valid.

2.4.5 Summary of the chapter

If one were to try to concentrate the result of this whole chapter into one single (necessarily not very precise) sentence, then it would read

> the Lie algebra of a physical symmetry group is so important because each of its elements is a conserved observable.   (2.4.10)

A more detailed summary, which also shows the logical flow of the arguments, is given in table 2.1; with this table we conclude the chapter.

Table 2.1. Summary of the chapter.

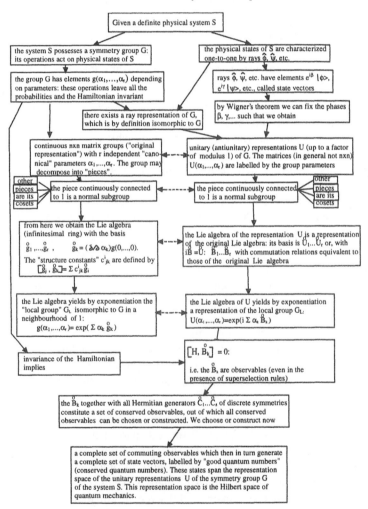

3

ROTATIONS IN THREE-DIMENSIONAL SPACE

3.1 General remarks on rotations

There are unfortunately a large variety of possibilities for writing down a rotation; also there are two different interpretations of the word 'rotation'. These points will be cleared up in this chapter.

3.1.1 Interpretation

Let a rotation R be given e.g. by fixing a directed axis of rotation and the angle of rotation (right screw with the direction of the axis). Suppose we have a coordinate frame x, y, z in which we describe our physical system. Then, if we speak of the rotation R, it still remains open *what* shall be rotated: the coordinate frame or the physical system. We shall call the rotation

> *active*, R_a, if the physical system under consideration is bodily rotated, whereas the coordinate frame remains fixed. As a shorthand notation, we will speak of 'rotating the space' against a fixed coordinate frame.
>
> *passive*, R_p, if the space remains fixed and the coordinate frame is rotated.

(3.1.1)

Obviously, if we rotate first the coordinate frame and then, by the same rotation, the space, all points will obtain again the same coordinate as before these rotations; the same is true in the reverse order. Hence in a somewhat symbolic notation (since strictly speaking this is true for the corresponding transformation matrices and representations)

$$R_a R_p = R_p R_a = 1 \qquad R_a = R_p^{-1}. \tag{3.1.2}$$

One may feel, perhaps, that this point is trivial and hardly worth mentioning. However, for reasons of intuition one may prefer sometimes the active and sometimes the passive interpretation—just as physical intuition suggests. Without carefully keeping track of what is rotated against what, one gets almost immediately lost in confusion about $+$ and $-$ signs in matrices and exponential operators, and even about the order in which non-commuting operators must be applied one after another. The subscripts a and p for active and passive rotations respectively are the cheapest and most efficient safeguard against any such confusion.

3.1.2 Parameters describing a rotation

We use two descriptions of a rotation. The first one fixes the axis by giving a unit vector n and the angle α of rotation so that the vector $\boldsymbol{\alpha} \equiv \alpha n$ defines uniquely a rotation with the convention that the positive rotation forms a right-hand screw with n. Conversely, a rotation does not uniquely define a vector $\boldsymbol{\alpha}$, since instead of rotating by α about n we may equally well rotate by $2\pi - \alpha$ about $-n$. We may improve the situation by restricting $0 \leq \alpha \leq \pi$. Then, if we look at the vectors $\boldsymbol{\alpha}$ we see that they fill a sphere of radius π and that the correspondence of the possible rotations and the points in the sphere is one to one, except for the surface (rotations by π) where any pair of opposite points represents the same rotation. We shall write $R(\boldsymbol{\alpha})$ if we have this parametrization in mind. It should be clear what $R_a(\boldsymbol{\alpha})$ and $R_p(\boldsymbol{\alpha})$ mean. Keeping n fixed and varying α, we obtain the rotations about a fixed axis, i.e. an Abelian subgroup:

$$R(\alpha_1 n) \cdot R(\alpha_2 n) = R((\alpha_1 + \alpha_2) \cdot n) \text{ and } R(0) = 1.$$

Hence this parametrization is one by canonical parameters (of the first kind), which automatically ensures an exponential form of the local group (see subsection 2.3.4).

The other description is by means of the so-called Euler angles α, β, γ and the prescription is

$R_p(\alpha, \beta, \gamma)$:

> (α) Rotate the coordinate frame K about the z-axis by the angle $0 \leq \alpha < 2\pi$. Call the new frame K' with coordinate axes x', y', z'.
>
> (β) Rotate the new coordinate frame K' about the new y'-axis by $0 \leq \beta \leq \pi$ into the position K'' with axes x'', y'', z''.
>
> (γ) Rotate the new coordinate K''-frame about the new z''-axis by the angle $0 \leq \gamma < 2\pi$ into the final position x''', y''', z''' making up the final coordinate frame K'''.

$R_a(\alpha, \beta, \gamma)$: Attach to the material system (to 'the space') three coordinate axes ξ, η, ζ coinciding with x, y, z (e.g. by three perpendicular pointers fixed on the system). Carry through the same operations R_p as described above, with the difference, however, that this time they operate on this body-fixed system of axes which on its way through the various positions $(\xi, \eta, \zeta) \rightarrow (\xi', \eta', \zeta') \rightarrow (\xi'', \eta'', \zeta'') \rightarrow (\xi''', \eta''', \zeta''')$ always carries the material system ('the space') with it; this time the main coordinate frame $x\,y\,z$ does not participate in the motions.

Suppose we have a rigid body ('the space') with material axes ξ, η, ζ coinciding with the coordinate axes x, y, z. Carry out first $R_a(\alpha, \beta, \gamma)$ and afterwards $R_p(\alpha, \beta, \gamma)$ or first $R_p(\alpha, \beta, \gamma)$ and then $R_a(\alpha, \beta, \gamma)$: in both cases the final positions of $\xi''', \eta''', \zeta'''$ and of x''', y''', z''' coincide again, just as if no rotation had been made at all. Hence, also here, although $R(\alpha, \beta, \gamma)$ is a product of three rotations

$$R_a(\alpha, \beta, \gamma) = R_p^{-1}(\alpha, \beta, \gamma). \tag{3.1.3}$$

That the Euler angles are canonical parameters of the second kind will become clear somewhat later.

3.1.3 Representation of a rotation

The rotations form a group and we have denoted the abstract group element by R with a subscript a, p referring to what is rotated and an argument α or α, β, γ indicating which parametrization we choose. We may sometimes leave these choices open, and simply write R.

A given rotation R_p, say, carries the frame of reference K into a new one K', and a point P fixed in space will have coordinates x, y, z in K and x', y', z' in K'. The old coordinates x, y, z and the new ones x', y', z' are related to each other by an orthogonal 3×3 matrix $M(R)$

$$\begin{pmatrix} x' \\ y' \\ z' \end{pmatrix} = M(R) \begin{pmatrix} x \\ y \\ z \end{pmatrix} \qquad M(R_p) = M^{-1}(R_a). \tag{3.1.4}$$

These matrices $M(R)$ furnish one particular representation of the rotation group, since the correspondence $M(R) \leftrightarrow R$ is one to one. This representation is called the 'original representation'.

There are, however, other representations of the group and they are the ones which will turn up in the theory of angular momentum: these are the unitary representations discussed quite generally in chapter 2. Applied to rotations the argument runs as follows.

Consider a physical system S. Its physical states are described in quantum mechanics by state vectors $|\gamma\rangle$, where $\gamma \equiv (a, b, \ldots, x)$ is the symbol for a complete set of quantum numbers. Suppose that the system is invariant under rotation, i.e. the Hamiltonian is invariant—not the actual states of the system; they may be as asymmetric as they like. Carry out a bodily rotation R_a of the physical system when it is in a definite state $|\gamma\rangle$, without disturbing it otherwise: the state should remain $|\gamma\rangle$ for an observer who rotates with the system. Rotational invariance means then that the result of this rotation, seen by an observer who did not participate in the rotation, is again a possible state, $|\gamma\rangle'$ say, of the unrotated system. The Wigner theorem asserts then that the states $|\gamma\rangle'$ and $|\gamma\rangle$ are connected by a unitary transformation $U(R)$

$$|\gamma\rangle' = U(R)|\gamma\rangle \tag{3.1.5}$$

and these unitary transformations form a group which is isomorphic up to a factor to the original group of rotations R. The matrices corresponding to the transformations $U(R)$ will then also furnish representations of the rotation group.

One says the rotations in the physical three-dimensional space *induce* unitary transformations in the Hilbert space, whose matrices give rise to (infinitely many independent) representations of the rotation group. We shall see this in detail. In particular it will turn out that the matrices of these unitary transformations can be chosen such that they split up into unconnected finite square boxes along the diagonal:

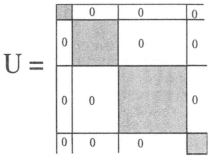

$$(3.1.6)$$

This will be seen to happen if one chooses as quantum numbers labelling the states those of the angular momentum (among others), i.e. if the basis vectors in the Hilbert space are eigenstates of the total angular momentum. If the total angular momentum of the system is j (remember $\hbar = 1$), then j will be unaffected by any rotation and the matrix representing U will transform a state $|j, \ldots\rangle$ into another state $|j, \ldots\rangle'$ with *the same* j. As we know from elementary quantum mechanics and as we shall see explicitly later on, there are (other quantum numbers held fixed) just $2j + 1$ linearly independent states for each value of j; consequently U will transform them among themselves. This part of U, acting in the subspace of total angular momentum j, is one of the $(2j+1)$-dimensional boxes of the structure indicated in (3.1.6). Keeping j fixed and letting R go through all the rotation group, this $(2j+1)$-dimensional matrix will become a $(2j + 1)$-dimensional representation of the rotation group. We shall introduce a particular symbol for these representations:

$$D^{(j)}(R) = (2j + 1)\text{-dimensional representation of } R \qquad (3.1.7)$$

(the letter D comes from the German word *Darstellung*, representation).
We collect in table 3.1 the results for ready reference.

3.2 Sequences of rotations

The use of the three Euler angles implies that we consider a sequence of three rotations. We shall now derive some simple statements concerning the product of three rotations. These statements will be true for the product of any number n of rotations; but as the generalization is obvious, we shall limit ourselves to the frequently encountered case $n = 3$.

Table 3.1. Symbols used for various rotations and their representations.

R	general symbol for the abstract operation 'rotation' to be specified by subscript and arguments
R_a	active rotation: the 'space' is rotated, the coordinate system remains fixed
R_p	passive rotation: the coordinate system is rotated, the 'space' remains fixed
$R(\alpha, \beta, \gamma)$	rotate (as specified by a subscript a or p)
	(α) first about the z-axis by α
	(β) then about the y'-axis by β
	(γ) finally about the z''-axis by γ
$R(\alpha)$	rotate (as specified by a subscript a or p) by α
$M(R)$	rotation matrix relating the old and new coordinates in the three-dimensional space: $x' = Mx$.
$U(R)$	unitary transformation induced by R in the Hilbert space
$D^{(j)}(R)$	$(2j + 1)$-dimensional representation (*Darstellung*) of the rotation group, connecting eigenstates (eigenfunctions) of total angular momentum j

Consider then a coordinate system K with axes x, y, z. In this system we arbitrarily fix three unit vectors n_1, n_2 and n_3 which are rotated with the coordinate axes (i.e. these unit vectors do not in general coincide with the three coordinate axes).

3.2.1 Considering the 'abstract' rotations R

Any three successive rotations of the coordinate axes will have the effect

$$K \rightarrow K' \rightarrow K'' \rightarrow K'''$$
$$n_1 n_2 n_3 \rightarrow n_1' n_2' n_3' \rightarrow n_1'' n_2'' n_3'' \rightarrow n_1''' n_2''' n_3'''.$$

To each of the unit vectors n_i we attach an angle α such that $\alpha_n = \alpha_i n_i$, and consider now the three particular successive rotations

$$\begin{aligned}
R_p(\alpha_1) &= \text{rotation } K \rightarrow K' \text{ by } \alpha_1 \text{ about } n_1 \\
R_p(\alpha_2') &= \text{rotation } K' \rightarrow K'' \text{ by } \alpha_2 \text{ about the new } n_2' \\
R_p(\alpha_3'') &= \text{rotation } K'' \rightarrow K''' \text{ by } \alpha_3 \text{ about the new } n_3''.
\end{aligned} \qquad (3.2.1)$$

The product of these three rotations can be written[1]

$$R_p(\alpha_1, \alpha_2', \alpha_3'') = R_p(\alpha_3'') R_p(\alpha_2') R_p(\alpha_1). \qquad (3.2.2)$$

---

[1] It might seem more suggestive to write the arguments in $R_p(\alpha_1, \alpha_2', \alpha_3'')$ in reverse order, namely in the same order as they appear in the r.h.s. product. However, the generally accepted convention in the current literature is as in our formulae above. The reason for this will become clear soon; see (3.2.4).

(The active interpretation is obtained if the three rotation axes are bodily fixed to the rotated physical system rather than to the coordinate frame.)

It must be possible to obtain the same resultant rotation by rotating by certain angles η_1, η_2, η_3 about the three *original* axes n_1, n_2, n_3 which then are thought to be kept fixed all the time: we now ask whether it is possible to give a simple relation between the above and this latter description? This relation is indeed surprisingly simple and also surprisingly simple is its proof.

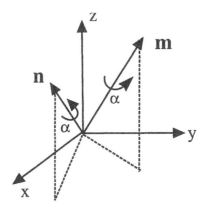

Figure 3.1. Rotation about an axis n.

Suppose a coordinate system is to be rotated by an angle α about an axis $\alpha = \alpha n$ but that we are requested to carry this out using the same angle α but another given axis m. As figure 3.1 shows, we can accomplish the task by first rotating such that n is turned into the position of m, then rotating by α and finally rotating n back into its old position. Let $R(n \to m)$ denote the rotation which brings n into m; then if nothing else is done

$$R(\alpha n) = R(m \to n)R(\alpha m)R(n \to m).$$

With this in mind we return to (3.2.2) and observe

$$R_p(\alpha_2') = R_p(\alpha_2 \to \alpha_2')R_p(\alpha_2)R_p(\alpha_2' \to \alpha_2).$$

However $R_p(\alpha_2 \to \alpha_2') = R_p(\alpha_1)$ so that

$$R_p(\alpha_2') = R_p(\alpha_1)R_p(\alpha_2)R_p^{-1}(\alpha_1).$$

Similarly

$$R_p(\alpha_3'') = R_p(\alpha_2')R_p(\alpha_3')R_p^{-1}(\alpha_2')$$

$$= R_p(\alpha_1)R_p(\alpha_2)R_p(\alpha_3)R_p^{-1}(\alpha_2)R_p^{-1}(\alpha_1).$$

This yields altogether, if inserted into (3.2.2)

$$R_p(\alpha_3'')R_p(\alpha_2')R_p(\alpha_1) = R_p(\alpha_1)R_p(\alpha_2)R_p(\alpha_3)R_p^{-1}(\alpha_2)R_p^{-1}(\alpha_1)$$
$$\times R_p(\alpha_1)R_p(\alpha_2)R_p^{-1}(\alpha_1)R_p(\alpha_1) = R_p(\alpha_1)R_p(\alpha_2)R_p(\alpha_3).$$

The general result is

> if a sequence $\alpha_1, \alpha_2, \ldots, \alpha_n$ of rotation axes is given then the resulting rotation is the same, whether we
>
> (i) let the system of rotation axes participate in the rotation and rotate in the order from 1 to n, using for each rotation its corresponding axis in the position where it was left after the preceding rotation, or
> (ii) keep the system of rotation axes fixed and rotate in the order from n to 1, using for each rotation its corresponding axis in its original position.

(3.2.3)

$$R(\alpha_1, \alpha_2', \alpha_3'', \ldots, \alpha_n^{(n-1)}) = R(\alpha_n, \alpha_{n-1}, \ldots, \alpha_3, \alpha_2, \alpha_1).$$

In particular, regarding the Euler parametrization, $e_{1,2,3}$ are the unit vectors along the x, y, z axes

$$R(\alpha, \beta, \gamma) = R(\gamma e''_3)R(\beta e'_2)R(\alpha e_3)$$

$$= R(\alpha e_3)R(\beta e_2)R(\gamma e_3). \qquad (3.2.4)$$

3.2.2 Considering the 3×3 rotation matrices $M(R)$

Consider any rotation matrix $M(\alpha)$, active or passive. If we rotate the coordinate system by $R_p(\beta)$ say, then, if α is kept fixed, the matrix $M(\alpha)$ will assume a new form $M'(\alpha)$ in the new coordinate system.

Let $M(\alpha)$ transform x into ξ (active or passive)

$$(i)\, \xi = M(\alpha)x$$

and let $M_p(\beta)$ transform x and ξ into x' and ξ' respectively:

$$x' = M_p(\beta)x.$$

Hence

$$(ii)\ x = M_p^{-1}(\beta)x'$$
$$(iii)\ \xi' = M_p(\beta)\xi.$$

We define now $M'(\alpha)$ by

$$\xi' = M'(\alpha)x'.$$

If we insert into (i) into (iii) and then replace x by means of (ii) we obtain

$$\xi' = M_p(\beta)M(\alpha)M_p^{-1}(\beta)x' = M'(\alpha)x'.$$

Hence

$$M'(\alpha) = M_p(\beta)M(\alpha)M_p^{-1}(\beta). \tag{3.2.5}$$

If, on the other hand, the axis α is also rotated by $R_p(\beta)$, i.e. if we consider it to be fixed to the rotated coordinate system, then the matrix expressing a rotation α in the old coordinates K must be the same as that expressing the rotation α' in the new coordinates K', because α' has the same position with respect to K' as α had with respect to K:

$$M'(\alpha') = M(\alpha). \tag{3.2.6}$$

Now let two orthogonal matrices M_1 and M_2 be given, in a definite numerical form, and consider the transformations (rotations in fact)

$$x' = M_1 x , \quad x'' = M_2 x' = M_2 M_1 x. \tag{3.2.7}$$

What does this mean?

Passive interpretation. x' are the new coordinates of a given point P, fixed in space, after the coordinate system has been turned into a new position. If e.g.

$$M_1 = \begin{pmatrix} \cos\varphi & \sin\varphi & 0 \\ -\sin\varphi & \cos\varphi & 0 \\ 0 & 0 & 1 \end{pmatrix}$$

then the coordinate system has been rotated by $+\varphi$ about the z-axis. If e.g.

$$M_2 = \begin{pmatrix} 1 & 0 & 0 \\ 1 & \cos\vartheta & \sin\vartheta \\ 1 & -\sin\vartheta & \cos\vartheta \end{pmatrix}$$

and we write

$$x'' = M_2 x'$$

then we may forget about the first rotation and look at this equation as stating that a point P has coordinates x' before and x'' after the rotation of the coordinate axes. What rotation? Obviously the rotation about *the first axis of the system in which P has coordinates x'*, that is, about the x'-axis. Thus, the matrix M_2 is the matrix representation, in the system with axes x', y', z', of a rotation about the x'-axis. Therefore, more pedantically, we should write

$$x'' = M'(\vartheta e_1')x' = M'(\vartheta e_1')M(\varphi e_3)x \tag{3.2.8}$$

and in the general case (with the help of (3.2.6))

$$x'' = M(\alpha_2)M(\alpha_1)x = M'(\alpha_2')M(\alpha_1)x. \tag{3.2.9}$$

This means first rotate the system K by α_1, then rotate the system K' by α_2 around the *new* axis α_2'.

Generally

> if a sequence $\alpha_1, \alpha_2, \ldots, \alpha_n$ of rotation axes and angles is given, then the product of the matrices $M_p(\alpha_1)$, $M_p(\alpha_2)$, $\ldots$, $M_p(\alpha_n)$ is the matrix which represents the following sequence of rotations of the coordinate system: the system of axes $\alpha_1, \ldots, \alpha_n$ participates in the rotations, which are carried out in the sequence from 1 to n, using for each rotation its corresponding axis in the position where it was left after the last rotation: (3.2.10)

$$M_p(\alpha_n)M_p(\alpha_{n-1})\ldots M_p(\alpha_2)M_p(\alpha_1)$$
$$= M_p^{(n-1)}(\alpha_n^{(n-1)})\ldots M_p''(\alpha_3'')M_p'(\alpha_2')M_p(\alpha_1)$$
$$\equiv M[R_p(\alpha_1\alpha_2'\alpha_3''\ldots\alpha_n^{(n-1)})]$$
$$\equiv M[R_p(\alpha_n^{(n-1)})\ldots R_p(\alpha_3'')R_p(\alpha_2')R_p(\alpha_1)].$$

Therefore matrix multiplication of rotation matrices in the passive interpretation always means that the next rotation is about the corresponding axis in its *new* position—in spite of the fact that the matrices are formally written down always as if everything applied to the originally given (and then fixed) axes of rotation and in the original coordinates. This is due to $M'(\alpha') = M(\alpha)$. The most adequate notation for the passive interpretation is, however, that of the r.h.s. of equation (3.2.10), $M_p^{(n-1)}(\alpha_n^{(n-1)})\ldots M_p''(\alpha_3'')M_p'(\alpha_2') M_p(\alpha_1)$.

Active interpretation. Consider now the corresponding product of matrices in the active interpretation:

$$x' = M_a(\alpha_1)x$$
$$x'' = M_a(\alpha_2)x'$$
$$\vdots$$
$$x^{(n)} = M_a(\alpha_n)x^{(n-1)} = M_a(\alpha_n)\ldots M_a(\alpha_3)M_a(\alpha_2)M_a(\alpha_1)x.$$

This means now that the point P has been transferred by a sequence of 'space' rotations from a position where its coordinates were x, to another position, where its coordinates are $x^{(n)}$—always with respect to the same fixed frame of reference. The matrices therefore refer to the axes $\alpha_1 \ldots \alpha_n$, fixed once and forever with respect to the fixed coordinate system and the matrices themselves are all meant as representatives of these rotations in terms of this same coordinate system:

> if a sequence $\alpha_1, \ldots, \alpha_n$ of rotation axes and angles is given, then the product of the matrices $M_a(\alpha_1)$, $M_a(\alpha_2)$, ..., $M_a(\alpha_n)$ is the matrix which represents the sequence of rotations (mappings) of the 'space' in the order from 1 to n; the coordinate frame and the axes $\alpha_1, \ldots, \alpha_n$ remain fixed once and forever: they do not participate in any rotation. Here the adequate notation is   (3.2.11)

$$M_a(\alpha_n)M_a(\alpha_{n-1})\ldots M_a(\alpha_2)M_a(\alpha_1)$$
$$\equiv M\left[R_a(\alpha_n)R_a(\alpha_{n-1})\ldots R_a(\alpha_2)R_a(\alpha_1)\right]$$
$$\equiv M[R_a(\alpha_1\alpha_2\ldots\alpha_n)].$$

In view of these statements we may now consider the relation between a given abstract rotation R and its matrix $M(R)$ in the two forms of parametrization.

If R is given in the α-parametrization, everything is trivial, because it is one single rotation:

$$M(R_p(\alpha)) \equiv M_p(\alpha)$$

as found in subsection 3.1.3. It is immediately obvious that

$$M_p(\alpha) = M_a^{-1}(\alpha) = M_a(-\alpha).$$

If, however, R is given in the Euler parametrization, then some care is required, since a product of rotations is involved. As we just have found, the active interpretation means carrying out the sequence of rotations using a fixed system of rotation axes; the passive interpretation means carrying out the rotations using a system of rotation axes which participates in the rotation. The identity (3.2.4)

$$R(\alpha, \beta, \gamma) = R(\gamma e_3'')R(\beta e_2')R(\alpha e_3)$$
$$= R(\alpha e_3)R(\beta e_2)R(\gamma e_3)$$

allows us to arrange the same total rotation to be carried out with moving or with fixed rotation axes, according to what we need in writing the matrices. We use the moving rotation axes in the passive and the fixed ones in the active interpretation:

$$M[R_p(\alpha, \beta, \gamma)] \equiv M[R_p(\gamma e_3'')R_p(\beta e_2')R_p(\alpha e_3)]$$
$$= M_p''(\gamma e_3'')M_p'(\beta e_2')M_p(\alpha e_3) \qquad (3.2.12)$$
$$= M_p(\gamma e_3)M_p(\beta e_2)M_p(\alpha e_3) \equiv M_p(\alpha, \beta, \gamma)$$

(see (3.2.10))

$$M[R_a(\alpha, \beta, \gamma)] = M[R_a(\alpha e_3) R_a(\beta e_2) R_a(\gamma e_3)]$$
$$= M_a(\alpha e_3) M_a(\beta e_2) M_a(\gamma e_3) \tag{3.2.13}$$
$$\equiv M_a(\alpha, \beta, \gamma)$$

(see (3.2.11)). Note the different order of operators in (3.2.12) and (3.2.13)! As one sees immediately

$$M_a(\alpha, \beta, \gamma) M_p(\alpha, \beta, \gamma) = M_p(\alpha, \beta, \gamma) M_a(\alpha, \beta, \gamma) = 1 \tag{3.2.14}$$

since for each single rotation $M_a(\eta) = M_p^{-1}(\eta)$ and the order in the products is just reversed. This is exactly as it should be according to the consideration leading to (3.1.3).

3.3 The Lie algebra and the local group

As an illustration we connect in this section the 3×3 matrices of rotations to the general considerations of chapter 2. We do this in several steps.

(i)   We work out the matrices $M_p(\eta)$, and check them by specializing to $M_p(\eta e_1)$, $M_p(\eta e_2)$, $M_p(\eta e_3)$, i.e. to the rotations about the three coordinate axes.

(ii) We derive from these matrices the Lie algebra of the rotation group and specify the commutation relations (structure constants).

(iii) We show that by exponentiation of the Lie algebra we obtain the local group which in this case is even globally isomorphic to the group we started from.

(iv) We show that the Euler angles are canonical parameters of the second kind and work out the Euler parameters as functions of $\eta = (\eta_1, \eta_2, \eta_3)$.

3.3.1 The rotation matrix $M_p(\eta)$

The rotation matrix for the passive rotation about a given axis $\eta = \eta n$ is complicated. Nevertheless it is not difficult to find it by using a trick which we already employed earlier: we first perform a rotation such that the z-axis is turned into the direction n; call this rotation $N(\vartheta, \varphi)$ where ϑ and φ are the angles defining n (see figure 3.2). Then we rotate by the angle η about the new axis z' ($\equiv n$) and finally by $N^{-1}(\vartheta, \varphi)$ we rotate the z'-axis into its final position z'' (i.e. the position into which $M_p(\eta)$ carries the z-axis). Hence

$$M_p(\eta) = N^{-1}(\vartheta, \varphi) M_p'(\eta) N(\vartheta, \varphi) \tag{3.3.1}$$

where $M_p'(\eta)$ is the matrix of the rotation $R(\eta)$ in the new coordinate system obtained by the rotation $N(\vartheta, \varphi)$—but this is now a simple rotation about the

z'-axis:

$$M_p'(\eta) = \begin{pmatrix} \cos\eta & \sin\eta & 0 \\ -\sin\eta & \cos\eta & 0 \\ 0 & 0 & 1 \end{pmatrix}. \tag{3.3.2}$$

Now we determine $N(\vartheta, \varphi)$ by decomposing it into, first, a rotation by φ about the z-axis and, second, a rotation by ϑ about the new y' axis (see figure 3.2); these two operations bring the z-axis into the position n

$$N(\vartheta, \varphi) = \begin{pmatrix} \cos\vartheta & 0 & -\sin\vartheta \\ 0 & 1 & 0 \\ \sin\vartheta & 0 & \cos\vartheta \end{pmatrix} \begin{pmatrix} \cos\varphi & \sin\varphi & 0 \\ -\sin\varphi & \cos\varphi & 0 \\ 0 & 0 & 1 \end{pmatrix}. \tag{3.3.3}$$

We leave it to the reader to verify that in the left-hand matrix the minus sign is *not* in the wrong place. $N^{-1}(\vartheta, \varphi)$ is obtained by inverting the order of the two matrices in (3.3.3) and changing $\vartheta \to -\vartheta$, $\varphi \to -\varphi$. Finally with figure 3.2

$$n = \eta/\eta = (\sin\vartheta \cos\varphi, \sin\vartheta \sin\varphi, \cos\vartheta) \equiv (n_1, n_2, n_3).$$

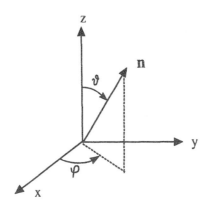

Figure 3.2. Rotation about an axis n.

Inserting (3.3.3) and (3.3.2) into (3.3.1), we obtain the final result

$$M_p(\eta) = \sin\eta \begin{pmatrix} 0 & n_3 & -n_2 \\ -n_3 & 0 & n_1 \\ n_2 & -n_1 & 0 \end{pmatrix}$$

$$+ (1 - \cos\eta) \begin{pmatrix} n_1^2 & n_1 n_2 & n_1 n_3 \\ n_2 n_1 & n_2^2 & n_2 n_3 \\ n_3 n_1 & n_3 n_2 & n_3^2 \end{pmatrix} \tag{3.3.4}$$

$$+ \cos\eta \cdot 1.$$

From this, by putting $n = e_1$, e_2 or e_3, we obtain respectively

$$M_p(\eta e_1) = \begin{pmatrix} 1 & 0 & 0 \\ 0 & \cos\eta & \sin\eta \\ 0 & -\sin\eta & \cos\eta \end{pmatrix}$$

$$M_p(\eta e_2) = \begin{pmatrix} \cos\eta & 0 & -\sin\eta \\ 0 & 1 & 0 \\ \sin\eta & 0 & \cos\eta \end{pmatrix} \qquad (3.3.5)$$

$$M_p(\eta e_3) = \begin{pmatrix} \cos\eta & \sin\eta & 0 \\ -\sin\eta & \cos\eta & 0 \\ 0 & 0 & 1 \end{pmatrix}.$$

(Of course, (3.3.5) is nothing new, as we have already used these three matrices above; it only checks our calculation.)

3.3.2 The generators of the rotation group

According to (2.3.15) a basis of the Lie algebra is given by the three matrices

$$\overset{\circ}{M}_k = \frac{d}{d\eta} M_p(\eta e_k)|_{\eta=0} \qquad (3.3.6)$$

from which we obtain the generators (basis of the Lie algebra)

$$\overset{\circ}{M}_1 = \begin{pmatrix} 0 & 0 & 0 \\ 0 & 0 & 1 \\ 0 & -1 & 0 \end{pmatrix} \qquad \overset{\circ}{M}_2 = \begin{pmatrix} 0 & 0 & -1 \\ 0 & 0 & 0 \\ 1 & 0 & 0 \end{pmatrix} \qquad \overset{\circ}{M}_3 = \begin{pmatrix} 0 & 1 & 0 \\ -1 & 0 & 0 \\ 0 & 0 & 0 \end{pmatrix}.$$

$$(3.3.7)$$

The general element of the Lie algebra becomes

$$\overset{\circ}{M} = a\overset{\circ}{M}_1 + b\overset{\circ}{M}_2 + c\overset{\circ}{M}_3 = \begin{pmatrix} 0 & c & -b \\ -c & 0 & a \\ b & -a & 0 \end{pmatrix} \qquad a, b, c \text{ real.} \qquad (3.3.8)$$

We observe that no element of the Lie algebra can be an element of the group, since all elements of the Lie algebra (3.3.8) have determinant zero, while all group elements (3.3.4) have unit determinant. Furthermore the matrix product of two elements of the Lie algebra is *not* an element of the Lie algebra (the actual proof by means of (3.3.8) is left to the reader), while the commutator is again an element of the Lie algebra, as we have shown in chapter 2 (the reader should check this too). We find by a short explicit calculation that

$$\left[\overset{\circ}{M}_i, \overset{\circ}{M}_j\right] = -\sum_{k=1}^{3} \epsilon_{ijk}\overset{\circ}{M}_k \qquad (3.3.9)$$

where ϵ_{ijk} is the totally antisymmetric tensor

$$\epsilon_{ijk} = \begin{cases} +1 & \text{if } i, j, k \text{ is an even permutation of } 1, 2, 3 \\ -1 & \text{if } i, j, k \text{ is an odd permutation of } 1, 2, 3 \\ 0 & \text{otherwise.} \end{cases} \quad (3.3.10)$$

Therefore, the structure constants of the rotation group are given by

$$C_{ij}^k = -\epsilon_{ijk}. \quad (3.3.11)$$

The minus sign is in fact irrelevant, because we could as well have taken the active transformations and then all basis elements would have obtained a factor -1 which would have propagated to the C_{ij}^k. As we have discussed in chapter 2, we are free to change the basis of the Lie algebra and with it also the structure constants—but this is a rather trivial change.

We shall indeed right now change the definition of the generators by the following argument: as we know from chapter 2, we generate the local group by exponentiation of the Lie algebra:

$$M_p(\eta) = \exp\left[\sum_{k=1}^3 \eta_k \overset{\circ}{M}_k \right] \qquad \eta \equiv (\eta_1, \eta_2, \eta_3). \quad (3.3.12)$$

We shall soon show explicitly that this indeed leads back to the group we started from. For the time being we observe that multiplying η by i and $\overset{\circ}{M}$ by $-$i in the exponent would change nothing. We give the product $-i\overset{\circ}{M}_R$ the new name M_R (without the superscript $\circ$); then we obtain

$$M_p(\eta) = \exp\left[i\sum_{k=1}^3 \eta_k M_k \right] \qquad M_a(\eta) = \exp\left[-i\sum_{k=1}^3 \eta_k M_k \right]$$

$$M_1 = \begin{pmatrix} 0 & 0 & 0 \\ 0 & 0 & -i \\ 0 & i & 0 \end{pmatrix} \qquad M_2 = \begin{pmatrix} 0 & 0 & i \\ 0 & 0 & 0 \\ -i & 0 & 0 \end{pmatrix} \qquad M_3 = \begin{pmatrix} 0 & -i & 0 \\ i & 0 & 0 \\ 0 & 0 & 0 \end{pmatrix}$$

$$[M_i, M_j] = i\sum_{k=1}^3 \epsilon_{ijk} M_k \qquad C_{ij}^k = i\epsilon_{ijk}.$$

$$(3.3.13)$$

This change of definition may seem superfluous, but it brings us close to the currently used notation of angular momentum. We therefore shall adopt from now on the set M_1, M_2, M_3 of (3.3.13) as 'the generators of the rotation group' although this definition differs slightly (by the factor i) from the one given earlier (chapter 2).

3.3.3   The local group

If we 'exponentiate the Lie algebra', then, according to the general considerations in chapter 2, we must find a group which is at least locally isomorphic to the group we started from, i.e., to the group of matrices $M_p(\eta)$. Let us call the result of exponentiation $\tilde{M}(\eta)$:

$$\tilde{M}(\eta) = \exp[i\eta \cdot M] \qquad \eta \cdot M \equiv \sum_{k=1}^{3} \eta_k M_k \qquad (3.3.14)$$

and see what the relation between the local group (elements $\tilde{M}$) and the rotation group (elements M_p) is.

Let us denote, as usual, $\eta = \eta\, n$ with $n = (n_1, n_2, n_3)$. We obtain

$$i\eta \cdot M = \eta \cdot \begin{pmatrix} 0 & n_3 & -n_2 \\ -n_3 & 0 & n_1 \\ n_2 & -n_1 & 0 \end{pmatrix} \equiv \eta \cdot N \qquad (3.3.15)$$

where N is defined by the last identity. Then

$$\tilde{M}(\eta) = \exp(\eta \cdot N) = \sum_{k=0}^{\infty} \frac{\eta^k}{k!} N^k.$$

One sees from an explicit calculation that

$$N^2 = N' - 1 \qquad N' = \begin{pmatrix} n_1^2 & n_1 n_2 & n_1 n_3 \\ n_1 n_2 & n_2^2 & n_2 n_3 \\ n_1 n_3 & n_2 n_3 & n_3^2 \end{pmatrix} \qquad (3.3.16)$$

and that

$$N^3 = -N, \quad \text{since } NN' = 0.$$

Denoting $N' - 1$ by N'', we have the sequence

$$N$$
$$N^2 = N''$$
$$N^3 = -N$$
$$N^4 = -N''$$
$$N^5 = N, \text{ and so on.}$$

Hence all even terms (except the zeroth) contain N'' while all odd terms contain N. Thus the expansion of $\exp[\eta N]$ splits into a sine series and a cosine series and we obtain

$$\tilde{M}(\eta) = \exp[i\eta \cdot M] = N \cdot \sin\eta + N' \cdot (1 - \cos\eta) + 1 \cdot \cos\eta \qquad (3.3.17)$$

which, by a glance at (3.3.4), is seen to be identical to $M_p(\eta)$. We thus see that for the proper rotation group (reflections excluded) the local group coincides with the whole group: $\tilde{M}(\eta) = M_p(\eta)$; this proves the first line of (3.3.13) to be correct.

3.3.4   Canonical parameters of the first and the second kind

Writing

$$M_p(\eta) = M_p(\eta n) = \exp[i\eta \cdot M] = \exp[i\eta M(n)] \tag{3.3.18}$$

shows at once that the parametrization by $\eta = \eta n$ is canonical. More exactly η_1, η_2, η_3 are canonical parameters of the first kind; indeed for a fixed direction n we obtain an Abelian subgroup and therefore group multiplication is equivalent to addition of rotation angles. The three-dimensional space of the canonical parameters can be identified with the three-dimensional physical space in which the rotations are carried out, and then the straight lines through the origin in the parameter space are identical with the axes of rotation in the physical space.

Apart from this parametrization we use the one with the Euler angles; the latter even has definite advantages, which will show up later. For the moment we copy formula (3.2.12):

$$M_p(\alpha, \beta, \gamma) = M_p(\gamma e_3) M_p(\beta e_2) M_p(\alpha e_3)$$

and write each of these matrices in exponential form. We find at once from (3.3.18)

$$M_p(\alpha, \beta, \gamma) = e^{-i\gamma M_3} e^{-i\beta M_2} e^{-i\alpha M_3} \tag{3.3.19}$$

which shows that α, β and γ are almost canonical parameters of the second kind. The word 'almost' is used here, because the definition of canonical parameters of the second kind is

$$M_p(a, b, c) = e^{-icM_3} e^{-ibM_2} e^{-iaM_1}.$$

Here all three generators of the group are involved; in the Euler parametrization only M_3 and M_2 are used; hence α, β, γ are only almost of the type of a, b, c. But the essential difference between canonical parameters of the first and the second kind is that the first appear in the combination $\sum \eta_k M_k$ in the exponent of one single exponential function, whereas each canonical parameter of the second kind (see subsection 2.3.4) appears separately in an exponential function, these functions then being multiplied.

Finally we work out the relation between (n_1, n_2, n_3) and (α, β, γ). We write a rotation both ways and equate the matrices: from (3.3.4)

$$M_p(\eta) = \sin\eta \begin{pmatrix} 0 & n_3 & -n_2 \\ -n_3 & 0 & n_1 \\ n_2 & -n_1 & 0 \end{pmatrix}$$

$$+ (1 - \cos\eta) \begin{pmatrix} n_1^2 & n_1 n_2 & n_1 n_3 \\ n_2 n_1 & n_2^2 & n_2 n_3 \\ n_3 n_1 & n_3 n_2 & n_3^2 \end{pmatrix} \tag{3.3.20}$$

$$+ \cos\eta \cdot 1$$

whereas (3.3.19) can be written (see (3.3.5))

$M_p(\alpha, \beta, \gamma)$

$$= \begin{pmatrix} \cos\gamma & \sin\gamma & 0 \\ -\sin\gamma & \cos\gamma & 0 \\ 0 & 0 & 1 \end{pmatrix} \begin{pmatrix} \cos\beta & 0 & -\sin\beta \\ 0 & 1 & 0 \\ \sin\beta & 0 & \cos\beta \end{pmatrix} \begin{pmatrix} \cos\alpha & \sin\alpha & 0 \\ -\sin\alpha & \cos\alpha & 0 \\ 0 & 0 & 1 \end{pmatrix}$$

(3.3.21)

or

$M_p(\alpha, \beta, \gamma) =$

$$\begin{pmatrix} \cos\alpha\cos\beta\cos\gamma - \sin\alpha\sin\gamma & \sin\alpha\cos\beta\cos\gamma + \cos\alpha\sin\gamma & -\sin\beta\cos\gamma \\ -\cos\alpha\cos\beta\sin\gamma - \sin\alpha\cos\gamma & -\sin\alpha\cos\beta\sin\gamma + \cos\alpha\cos\gamma & \sin\beta\sin\gamma \\ \cos\alpha\sin\beta & \sin\alpha\sin\beta & \cos\beta \end{pmatrix}$$

(3.3.22)

Comparing (3.3.20) and (3.3.22) yields for M_{33}

$$\cos\beta = \cos\eta + n_3^2(1 - \cos\eta). \tag{3.3.23}$$

This fixes β, since $0 \le \beta \le \pi$. Furthermore for M_{31} and M_{32}

$$\sin\alpha\sin\beta = -n_1\sin\eta + (1 - \cos\eta)n_2n_3$$

$$\cos\alpha\sin\beta = n_2\sin\eta + (1 - \cos\eta)n_3n_1$$

$$\tan\alpha = \frac{-n_1\sin\eta + (1 - \cos\eta)n_2n_3}{n_2\sin\eta + (1 - \cos\eta)n_3n_1}.$$

(3.3.24)

Since $\sin\beta \ge 0$, the signs of $\sin\alpha$ and $\cos\alpha$ are those of the numerator and denominator respectively. Thus α is uniquely determined. The corresponding rule holds for γ in the result of comparing M_{23} and M_{13} respectively:

$$\tan\gamma = \frac{n_1\sin\eta + (1 - \cos\eta)n_2n_3}{-n_1\sin\eta + (1 - \cos\eta)n_3n_1} \tag{3.3.25}$$

by which we have expressed α, β and γ as analytic functions of n_1, n_2, n_3 and η.

3.4 The unitary representation $U(R)$ induced by the three-dimensional rotation R

(Compare the general considerations in subsection 2.3.7.) The Wigner theorem (2.2.11) asserts that there exists a unitary representation (up to a factor of modulus 1) $U(R)$ of the space rotations, such that the rotated state vectors are obtained by the unitary transformation

$$|\psi\rangle \rightarrow |\psi'\rangle = U(R)|\psi\rangle. \tag{3.4.1}$$

If we use in $U(R)$ the same parameters as for $M(R)$, then we obtain $U(\eta)$ and $U(\alpha, \beta, \gamma)$ respectively. Let us consider an active rotation. From (3.3.13) we have

$$M_a(\eta) = \exp[-i\eta \cdot M].$$

We introduce Hermitian operators J_1, J_2 and J_3 by writing the unitary representation of $M_a(\eta)$ in the same form as $M_a(\eta)$ itself, namely

$$U_a(\eta) = \exp[-i\eta \cdot J] \equiv \exp\left[-i\sum_{R=1}^{3} \eta_k J_k\right]. \qquad (3.4.2)$$

If the physical system under observation has rotation symmetry, then these three operators commute with the Hamiltonian and then they are constant observables. We shall show in the next chapter that J_1, J_2 and J_3 are the operators of angular momentum in the x, y and z directions respectively. As (3.4.2) is formally the same as (3.3.13), and as in particular it uses the same canonical parameters, it follows at once that the Lie algebra derived from $U_a(\eta)$ is a representation of the Lie algebra of the rotation group; i.e. any matrix representation of the three operators J_1, J_2 and J_3 is also a representation of the Lie algebra generated by M_1, M_2 and M_3 respectively. In particular, if the dimension of the matrices J_i is k, then for $k^2 \geq 3$, i.e. for $k \geq 2$, these three matrices are even linearly independent and span a *faithful* representation of the Lie algebra of the matrices M_1, M_2 and M_3. We shall show that representations for any $k \geq 1$ indeed exist. As all representations with $k \geq 2$ of J have to follow the same commutation relations as the matrices M, we can immediately copy down the last line of (3.3.13), where we only replace M by J:

$$[J_j, J_k] = i\sum_{l=1}^{3} \epsilon_{jkl} J_l. \qquad (3.4.3)$$

This indeed is the familiar commutation rule of angular momentum. We shall derive the same equation once more later on in section 4.2. Then it will become clear that it holds for the operators J *in abstracto* whereas here we can only say that it holds for all finite-dimensional representations and of course also for the direct sum[2] of such representations. The physicist may be satisfied by this statement but strictly speaking we have not given a proof that (3.4.3) is true for the operators. The commutation relations (3.4.3) will later serve as the main tool to construct all finite-dimensional representations of the rotation group.

We finish this section by writing down the unitary operator in the Euler parametrization.

From (3.4.2) follows immediately

$$U_a(\eta e_k) = e^{-i\eta J_k} \qquad (3.4.4)$$

---

[2] See section 4.3 for the definition of the direct sum.

and since (see (3.2.4))

$$R_a(\alpha, \beta, \gamma) = R_a(\alpha e_3) R_a(\beta e_2) R_a(\gamma e_3)$$

we obtain

$$U_a(\alpha, \beta, \gamma) = e^{-i\alpha J_3} e^{-i\beta J_2} e^{-i\gamma J_3}. \tag{3.4.5}$$

Note that this is *not* equal to $e^{-i(\alpha J_3 + \beta J_2 + \gamma J_3)}$! The fact that we obtain in the η-parametrization one exponential operator with a sum in the exponent arises from building up the rotation $U(\eta)$ as a sequence of infinitesimal rotations, all about the same axis $n = \eta/\eta$ (remember the derivation of (2.3.11)). On the other hand, in building up $U(\alpha, \beta, \gamma)$ from infinitesimal rotations, we have to keep the three rotations distinct since they do not go about one single axis. Therefore each must be built up individually. This is just the essential difference between canonical parameters of the first (η) and the second (α, β, γ) kind respectively.

4

ANGULAR MOMENTUM OPERATORS AND EIGENSTATES

4.1 The operators of angular momentum J_1, J_2 and J_3

4.1.1 The physical significance of J

We know that the three operators J_k are Hermitian and (in a rotationally invariant system) commute with the Hamiltonian; therefore they are *observable constants of the motion*. We shall now find out what their physical significance is. As we know from classical mechanics, the constant of the motion arising from rotational invariance is the angular momentum. We naturally expect the same here.

It is clearly sufficient to establish the physical significance of J_3, since that of J_2 and J_1 follows then simply from the supposed symmetry (and, to be pedantic, a permutation of x, y and z).

We consider a spinless particle in a state $|\psi\rangle$. Its Schrödinger function is

$$\psi(x, y, z) = \langle x|\psi\rangle \text{ and } \psi'(x, y, z) = \langle x|\psi'\rangle \qquad (4.1.1)$$

before and after an active rotation respectively.

If we rotate the physical system by the infinitesimal angle η about the z-axis, then the r.h.s. of (4.1.1) becomes

$$\langle x|\psi'\rangle = \langle x|U_a(\eta e_3)|\psi\rangle = \langle x|(1 - i\eta J_3)|\psi\rangle. \qquad (4.1.2)$$

This is the wave function of the *new* state at the *old* position x. What happens to the l.h.s. of (4.1.1)? We must calculate the value of the *new* wave function at the *old* position (see figure 4.1). The whole wave function is bodily rotated by the angle η. Therefore, if R_a is the rotation, the value of ψ at the point $P = (x, y, z)$ is carried to the point $R_a P$, while its value at the point $R_a^{-1} P$ is carried to the point P where it becomes $\psi'(P)$. Therefore the new wave function at the point P is equal to the old one at the point $R_a^{-1} P$. In symbols

$$\psi'(P) = \psi(R_a^{-1} P) = \psi(R_p P).$$

The last part of this equation says that we have to calculate ψ at a point with those coordinates which would result from rotation by $+\eta$ of the coordinate system. According to (3.3.5) we have then $\psi'(x, y, z) = \psi(x'y'z')$ where for

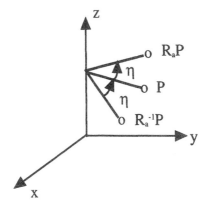

Figure 4.1. Rotation of a wave function.

the considered infinitesimal rotation $x' = M_p x$

$$M_p = \begin{pmatrix} 1 & \eta & 0 \\ -\eta & 1 & 0 \\ 0 & 0 & 1 \end{pmatrix}.$$

Hence with (4.1.2)

$$\psi'(x, y, z) = \psi(x + \eta y, y - \eta x, z) = \left[1 + \eta \left(y \frac{\partial}{\partial x} - x \frac{\partial}{\partial y} \right) \right] \psi(x, y, z)$$

$$= \langle x | (1 - i\eta J_3) | \psi \rangle.$$

It follows that

$$\langle x | J_3 \psi \rangle = \frac{1}{i} \left(x \frac{\partial}{\partial y} - y \frac{\partial}{\partial x} \right) \psi(x, y, z)$$

$$= \langle x | (x p_y - y p_x) | \psi \rangle$$

$$\equiv \langle x | L_z | \psi \rangle.$$

(4.1.3)

We thus have found the physical significance of the vector operator J in the *special case of a spinless particle*:

$$J = L = r \times p = \text{orbital angular momentum}. \qquad (4.1.4)$$

However, since we know that the whole of physics will certainly not be described by scalar wave functions $\psi(x, y, z)$, we may expect that J may be more complicated in more general cases. We see this immediately if we consider a situation where a particle is described by three wave functions which transform as the components of a vector field: we would call such a particle a vector particle. If its state is again written $|\psi\rangle$, then its Schrödinger function is a vector field:

$$\begin{pmatrix} \psi_1(x) \\ \psi_2(x) \\ \psi_3(x) \end{pmatrix} = \psi(x, y, z) = \langle x | \psi \rangle \tag{4.1.5}$$

and after rotating the physical system bodily by η around e_3

$$\psi'(x, y, z) = \langle x | (1 - i\eta J_3) | \psi \rangle.$$

As figure 4.1 shows, the new vector $\psi'(x, y, z)$ is found by rotating the vector $\psi(R_a^{-1}P)$ to the point $P = (x, y, z)$. Hence

$$\psi'(x, y, z) = R_a \psi(R_a^{-1} p) = R_p^{-1} \psi(R_p P).$$

Since ψ is supposed to transform as a vector, i.e. exactly the same as $P \equiv x$, we have with (3.3.5)

$$M_p(\eta e_3) = \begin{pmatrix} 1 & \eta & 0 \\ -\eta & 1 & 0 \\ 0 & 0 & 1 \end{pmatrix} \equiv M \qquad M^{-1} = M_p(-\eta e_3) \tag{4.1.6}$$

$$\psi'(x, y, z) = M^{-1} \psi(Mx).$$

From the previous discussions we already know that

$$\psi(Mx) = \left[1 - i\eta \frac{1}{i} \left(x \frac{\partial}{\partial y} - y \frac{\partial}{\partial x} \right) \right] \psi(x) = [1 - i\eta L_z] \psi(x)$$

where the coefficient of $-i\eta$ was the orbital angular momentum L_z. But here we have still to apply the matrix M^{-1} to the vector $\psi(Mx)$. We *define* S_z by $M_3 = S_z$; hence according to (3.3.13)

$$M^{-1} = 1 - i\eta \begin{pmatrix} 0 & -i & 0 \\ i & 0 & 0 \\ 0 & 0 & 0 \end{pmatrix} \equiv 1 - i\eta S_z.$$

We then have

$$M^{-1} \psi(Mx) = \psi'(x, y, z) = [1 - i\eta S_z][1 - i\eta L_z] \psi(x).$$

Retaining linear terms in η only, we obtain

$$\psi'(x) = [1 - i\eta(S_z + L_z)] \psi(x) = \langle x | (i - i\eta J_3) | \psi \rangle. \tag{4.1.7}$$

We thus have found that in the *special case of a vector particle*:

$$J = L + S$$

$$L = r \times p$$

$$S_v = (S_x, S_y, S_z)_v$$

with (see (3.3.13))

$$S_{x,v} = \begin{pmatrix} 0 & 0 & 0 \\ 0 & 0 & -i \\ 0 & i & 0 \end{pmatrix} \qquad S_{y,v} = \begin{pmatrix} 0 & 0 & i \\ 0 & 0 & 0 \\ -i & 0 & 0 \end{pmatrix}$$

$$S_{z,v} = \begin{pmatrix} 0 & -i & 0 \\ i & 0 & 0 \\ 0 & 0 & 0 \end{pmatrix}.$$

(4.1.8)

In this case J consists of two parts, one of which, L, can still be written as $r \times p$, whereas the other, S_v, cannot. S is the operator of the intrinsic angular momentum or spin.

We now proceed to find explicit matrix forms and eigenstates of these operators J, which we henceforth shall call simply 'angular momentum operators'. We learn from these examples that the operators J_k have a more general meaning than the quantum mechanical equivalent of the orbital angular momentum $L = r \times p$.

4.1.2   The angular momentum component in a direction n

We have just seen that the operator J_3 in the unitary transformation $U_a(\eta e_3) = e^{-i\eta e_3 \cdot J} = e^{-i\eta J_3}$ has the physical significance of being the operator of the z-component of the angular momentum. Now in fact the direction which we choose to call e_3 is arbitrary. Therefore, if $n = (n_1, n_2, n_3)$ is an arbitrary unit vector, then

$$U(\eta n) = e^{-i\eta n \cdot J} = e^{-i\eta J(n)}$$

$$J(n) \equiv n \cdot J = \sum_{i=1}^{3} n_i J_i$$

(4.1.9)

$$= \text{angular momentum component in direction } n.$$

4.2   Commutation relations for angular momenta

The 'abstract rotations' $R_a(\eta)$ form a group; the induced unitary transformations $U(R_a(\eta))$ give rise to the Lie algebra of the

- generators of the rotation group
- operators of infinitesimal rotations
- angular momentum operators.

All these are but different names for the one vector operator $J = (J_1, J_2, J_3)$ in $U_a(\eta) = e^{-i\eta \cdot J}$.

In the Lie algebra the commutator $J_k J_l - J_l J_k$ of the operators J_k and J_l plays the role of multiplication, whereas multiplication of group elements is defined as the simple product of the corresponding operators U_a.

In order to derive once more the commutation relations for the three components of J we ask which group operation corresponds to the commutator $[J_1, J_2]$. As the commutator is again an element of the Lie algebra, such a group element must clearly exist.

We obtain the answer by comparing two descriptions of the same active infinitesimal rotation by an angle α about the z-axis, applied to the operator $n \cdot J$ (4.1.9).

(i)   We rotate the direction of n by α:

$$n' = (n_1 - \alpha n_2, n_2 + \alpha n_1, n_3) \tag{4.2.1}$$

which gives the 'rotated operator' $(n \cdot J)'$:

$$(n \cdot J)' = n \cdot J - \alpha n_2 J_1 + \alpha n_1 J_2. \tag{4.2.2}$$

(ii) On the other hand, the same can be achieved by the corresponding unitary transformation. Namely, let U_a be the unitary representation of an active rotation. Whether or not there is rotational symmetry, everything remains untouched if U transforms both states and operators. Requiring that

$$\langle \psi' | A' | \varphi' \rangle = \langle \psi | A | \varphi \rangle \tag{4.2.3}$$

for any operator A and states $|\psi\rangle$, $|\varphi\rangle$, ... implies

$$|\varphi'\rangle = U_a |\varphi\rangle$$
$$|\psi'\rangle = U_a |\psi\rangle \tag{4.2.4}$$
$$A' = U_a A U_a^\dagger.$$

We may then call A' the 'rotated operator A'. In particular, with $U_a = 1 - i\alpha J_3$ representing the above infinitesimal rotation by α about the z-axis, the rotated operator $n \cdot J$ becomes

$$(n \cdot J)' = (1 - i\alpha J_3) n \cdot J (1 + i\alpha J_3). \tag{4.2.5}$$

(iii) Multiplying out up to linear terms in α and comparing with (4.2.2) yields immediately

$$[J_2, J_3] = iJ_1 \qquad [J_3, J_1] = iJ_2. \tag{4.2.6}$$

The third commutator $[J_1, J_2] = iJ_3$ follows by cyclic permutation (or by re-arranging the coordinate axes).

We have already found these relations from a more formal consideration (see (3.4.3)), but there we tacitly assumed that the J were finite-dimensional matrices: namely we used the results of chapter 2, which were not proven there to hold for other than finite-dimensional spaces. Here, however, we only used the existence of unitary representations (Wigner's theorem) and the fact that a unitary operator can be written in exponential form. In fact we need not even assume U_a and the J to be given as matrices—they are just operators. In particular nothing has been supposed about the structure of J; it may or may not split into an orbital part and spin part; if a spin part is present, then the above-used Schrödinger function has several components.

From now on we shall consider the commutation relations (4.2.6) as the *definition* of angular momentum. We have indeed seen in chapter 2 that these relations are the essence of the Lie algebra and that they uniquely define the group in a neighbourhood of unity; and as far as the rotation group goes, this neighbourhood is the full group. Of course, if by chance another three-dimensional space—different from the 'ordinary' physical space—should turn up in the description of physics, and if by chance rotations in that space had to be considered, then, although the whole mathematics can be literally taken over, the physical significance of the generators of that rotation group would be different and a new name, replacing the word 'angular momentum', would have to be introduced. As it happens, such a three-dimensional space, called isospace, turns up in strong-interaction physics, and rotational invariance in that space is one of the characteristics of strong interactions. Much of what is said in this book can be applied immediately there. The new name is then 'isospin operator' and these isospin operators are the generators of the rotation group in isospace. The word isospin is an unlucky choice because it suggests a physical similarity where there is only a formal one.

We should mention here a very important fact, which in general is considered trivial, but which is worth some thought. We have seen that the angular momentum may split up into two parts: orbital and intrinsic. The orbital part can always be described by the (x, y, z)-dependence of a wave function, whereas the intrinsic part is by definition contained in the transformation property of the wave function; more precisely in that transformation property which persists if one in thought reduces the wave function to simply a constant (independent of x, y, z), that is, if $(r \times p)|\psi\rangle = 0$. Such a wave function may still have components and transform under rotations like a scalar, vector or tensor. In the case of a single component (scalar), no intrinsic angular momentum exists, as we have seen. In all other cases, the wave function has at least two components; thus under rotations it will transform with an at least two-dimensional matrix $D(R)$, which reshuffles its components. In chapter 2 it was proved that for $k^2 \geq r$ (k = dimension of the representation space; r = dimension of the group = number of real parameters), the Lie algebra derived from these representations $D(R)$ is a representation of the original Lie algebra of the group and that in particular the commutation relations are the same. The three basis elements of the Lie algebra derived from the matrices $D(R)$ will be called the 'spin operators' S_1, S_2 and S_3. The matrices $D(R)$ need not be in an irreducible form[1]; but we shall speak of a wave function representing a *particle* only if the matrices $D(R)$ transforming the wave function *are irreducible*. Irreducible or not, the commutation relations are those of the original Lie algebra:

$$[S_j, S_k] = iS_l \qquad (j, k, l = 1, 2, 3 \text{ cyclic}). \qquad (4.2.7)$$

On the other hand, from $L = r \times p$, it follows by means of the commutation relations

$$[x_j, p_k] = i\delta_{jk} \qquad (4.2.8)$$

that

$$[L_j, L_k] = iL_l \qquad (j, k, l = 1, 2, 3 \text{ cyclic}) \qquad (4.2.9)$$

but quite generally $J = L + S$ and $[J_j, J_k] = iJ_l$. Hence

$$[L_j, L_k] + [S_j, S_k] + [L_j, S_k] + [S_j, L_k] = iL_l + iS_l$$

where now (4.2.7) and (4.2.9) imply that

$$[L_j, S_k] + [S_j, L_k] = 0$$

for $j \neq k$. This suggests that the operators L and S commute. Indeed we may always use that basis in the Hilbert space which leads to the description of states by means of Schrödinger functions $\psi(x, y, z)$. Then $L = -ir \times \nabla$ acts only on the arguments x, y and z of the wave function and S only reshuffles its components. Now clearly it is irrelevant which of the two operations is carried out first: they act in different spaces. Hence indeed S and L commute; they do so even if considered as abstract operators, because the description of states by Schrödinger functions is completely general.

We collect the results of this consideration in the following formulae.

If $J = L + S$, then independently

$$\left.\begin{array}{l} [J_1, J_2] = iJ_3 \\ [L_1, L_2] = iL_3 \\ [S_1, S_2] = iS_3 \end{array}\right\} \text{ and cycl. perm.}$$

$$(4.2.10)$$

$$[L_i, S_k] = 0 \qquad \text{for all } i \text{ and } k.$$

Note that the validity of (4.2.10) is independent of whether L and S interact or not; it also does not require that any one of the various operators is conserved. We will come back to this point.

Another relation similar to (4.2.10) proves to be important. Suppose we divide up—if only conceptually—a physical system S into different subsystems

$S_1, \ldots, S_k{}^2$. Let us carry out an active rotation of S. We achieve this by rotating each subsystem individually, and in doing so the order in which this happens is irrelevant. Hence we can write for the state of the rotated system

$$e^{-i\alpha n \cdot J}|\psi\rangle = e^{-i\alpha n \cdot j^{(i_1)}} e^{-i\alpha n \cdot j^{(i_2)}} \ldots e^{-i\alpha n \cdot j^{(i_k)}}|\psi\rangle$$

where the $j^{(i)}$ are the operators which rotate the system S_i (leaving the others untouched) and where $i_1, \ldots, i_k$ is any permutation of the numbers $1, \ldots, k$. This means, however, that the unitary operators $e^{-i\alpha n \cdot j^{(i)}}$ commute with each other because $|\psi\rangle$ is arbitrary; therefore we can write

$$e^{-i\alpha n \cdot J} = e^{-i\alpha n \cdot (j^{(1)} + j^{(2)} + j^{(3)} + \ldots j^{(k)})}$$

and as this is true for arbitrary n, it follows that

$$J = \sum_{i=1}^{k} j^{(i)}. \tag{4.2.11}$$

To each of the individual subsystems we may apply the argument which leads to the commutation rules of J; these rules hold therefore the individual operators $j^{(i)}$.

If $J = \sum_{i=1}^{k} j^{(i)}$ then independently

$$\left.\begin{array}{l}[J_x, J_y] = iJ_z \\ [j_x^{(i)}, j_y^{(i)}] = ij_z^{(i)} \qquad (i = 1, \ldots, k)\end{array}\right\} \text{ and cycl. perm.} \tag{4.2.12}$$

$$[j_l^{(i)}, j_m^{(i')}] = 0 \text{ for } i \neq i'; \; l, m = x, y, z.$$

Proof of the last equation. In the case $J = L + S$ we have found, in addition to L and S being angular momenta, that L and S commute. We should expect that also $j^{(a)}$ and $j^{(b)}$ for $a \neq b$ commute. This is indeed so.

Let us rotate subsystem a about α and subsystem b about β. Whatever the effect of such an operation on the whole system S may be (in general it will become a completely different system S') it is clear that we can first rotate subsystem a by α and afterwards b by β or first b by β and then a by α: the final result will be the same because a and b are different subsystems. Hence the new state is

$$\begin{aligned}|\psi\rangle' &= e^{-i\alpha \cdot j^{(a)}} e^{-i\beta \cdot j^{(b)}}|\psi\rangle \\ &= e^{-i\beta \cdot j^{(b)}} e^{-i\alpha \cdot j^{(a)}}|\psi\rangle\end{aligned} \tag{4.2.13}$$

and since that is true for any $|\psi\rangle$, we find that for any two subsystems $a \neq b$ the two operators commute. Let now $\alpha = \vartheta n$ and $\beta = \vartheta m$; that is, the axes are different but the angles the same. It follows from (4.2.13) that

$$e^{-i\vartheta n \cdot j^{(a)}} e^{-i\vartheta m \cdot j^{(b)}} e^{i\vartheta n \cdot j^{(a)}} e^{i\vartheta m \cdot j^{(b)}} = 1.$$

---

[2] The exact meaning of this subdivision will be elaborated in the next two sections.

Using (2.3.12) up to terms in ϑ^2 leads to

$$1 - \vartheta^2[\boldsymbol{n} \cdot \boldsymbol{j}^{(a)}, \boldsymbol{n} \cdot \boldsymbol{j}^{(b)}] = 1.$$

Hence

$$[\boldsymbol{n} \cdot \boldsymbol{j}^{(a)}, \boldsymbol{n} \cdot \boldsymbol{j}^{(b)}] = 0. \tag{4.2.14}$$

In other words: *all components of $\boldsymbol{j}^{(a)}$ commute with all components of $\boldsymbol{j}^{(b)}$*.

This proves the last line of (4.2.12) to be correct. Nevertheless, the reader should feel somewhat unsafe here and in need of more explanation—unless he knows already that what we have just done is mathematically nothing else than considering the system S as the 'direct product of its subsystems'. In the next paragraph we shall explain this a little further.

4.3   Direct sum and direct product

In the preceding paragraph we have 'divided a system S into subsystems $S^{(1)}$, ..., $S^{(n)}$'. What does this mean mathematically?

There are—apart from the vector addition and the scalar product—two other ways of combining two vectors: the direct sum and the direct product. Both occur in the theory of angular momentum, in fact whenever symmetry is involved.

We consider two Hilbert spaces, $\mathcal{H}^{(a)}$ and $\mathcal{H}^{(b)}$ respectively:

$$\mathcal{H}^{(a)} \text{ contains states } |\psi^{(a)}\rangle = \sum_i a_i |\psi_i^{(a)}\rangle$$

$$\mathcal{H}^{(b)} \text{ contains states } |\varphi^{(b)}\rangle = \sum_i b_i |\varphi_i^{(b)}\rangle. \tag{4.3.1}$$

$\mathcal{H}^{(a)}$ and $\mathcal{H}^{(b)}$ may have different dimensions; scalar product of and addition of pairs of vectors belonging to different spaces are not defined.

We may, however, combine these two spaces into one new space—and that by means of two essentially different operations, called the direct sum and the direct product.

(i)   *The direct sum $\mathcal{H}_\oplus = \mathcal{H}^{(a)} \oplus \mathcal{H}^{(b)}$.* This space is spanned by the ensemble of the basis vectors $\{|\varphi_i^{(a)}\rangle, |\varphi_i^{(b)}\rangle\}$, hence its states are

$$|\psi_\oplus\rangle = |\psi^{(a)} \oplus \varphi^{(b)}\rangle = \sum_i \left\{ a_i |\psi_i^{(a)}\rangle + b_i |\varphi_i^{(b)}\rangle \right\} \tag{4.3.2}$$

that is the row vector $\langle\psi_\oplus|$ would be written (finite-dimensional example) $\langle\psi_\oplus| = \{a_1 a_2 \cdots a_n b_1 b_2 \cdots b_m\}$ with $n + m$ components; and the dimension of $\mathcal{H}^{(a)} \oplus \mathcal{H}^{(b)}$ is the *sum* of the dimensions $\mathcal{H}^{(a)}$ and $\mathcal{H}^{(b)}$.

Linear operators $M^{(a)}$ and $N^{(b)}$ are combined into $M_\oplus = M^{(a)} \oplus N^{(b)}$ such that

$$M_\oplus |\psi_\oplus\rangle = |M^{(a)}\psi^{(a)} \oplus N^{(b)}\varphi^{(b)}\rangle$$

$$M_\oplus = \left(\begin{array}{c|c} M^{(a)} & 0 \\ \hline 0 & N^{(b)} \end{array} \right). \tag{4.3.3}$$

The general matrix element becomes

$$\langle\psi'_\oplus|M_\oplus|\psi_\oplus\rangle = \langle\psi'^{(a)} \oplus \varphi'^{(b)}|M^{(a)}\psi^{(a)} \oplus N^{(b)}\varphi^{(b)}\rangle$$

$$= \langle\psi'^{(a)}|M^{(a)}|\psi^{(a)}\rangle + \langle\varphi'^{(b)}|N^{(b)}|\varphi^{(b)}\rangle. \tag{4.3.4}$$

From (4.3.3) we read off the law for matrix multiplication and addition

$$(M^{(a)} \oplus N^{(b)}) \cdot (P^{(a)} \oplus Q^{(b)}) = M^{(a)} P^{(a)} \oplus N^{(b)} Q^{(b)}$$

$$M^{(a)} \oplus N^{(b)}) + (P^{(a)} \oplus Q^{(b)}) = (M^{(a)} + P^{(a)}) \oplus (N^{(b)} + Q^{(b)}). \tag{4.3.5}$$

(ii) *The direct product* $\mathcal{H}_\otimes = \mathcal{H}^{(a)} \otimes \mathcal{H}^{(b)}$. This space is spanned by the ensemble of the basis vectors $\{|\varphi_i^{(a)}, \varphi_k^{(b)}\rangle\}$ where i and k go separately through their range in $\mathcal{H}^{(a)}$ and $\mathcal{H}^{(b)}$. Consequently the dimension of $\mathcal{H}^{(a)} \otimes \mathcal{H}^{(b)}$ is the *product* of the dimensions of $\mathcal{H}^{(a)}$ and $\mathcal{H}^{(b)}$. Its states are (we suppress the $\otimes$ sign in the states and write a comma instead)

$$|\psi_\otimes\rangle = |\psi^{(a)}\rangle \otimes |\varphi^{(b)}\rangle \equiv |\psi^{(a)}\rangle|\varphi^{(b)}\rangle = \sum_{i,k} a_i b_k |\psi_i^{(a)}, \varphi_k^{(b)}\rangle. \tag{4.3.6}$$

We have to define linear transformations. Let us represent states by their components, i.e.

$$|\psi^{(a)}\rangle \Longleftrightarrow \{a_i\}$$
$$|\varphi^{(b)}\rangle \Longleftrightarrow \{b_i\}$$
$$|\psi^{(a)}\rangle|\varphi^{(b)}\rangle \Longleftrightarrow \{a_i b_k\}. \tag{4.3.7}$$

Then we have

$$M^{(a)}|\psi^{(a)}\rangle = |\psi'^{(a)}\rangle \Longleftrightarrow \left\{ a'_i = \sum_j M_{ij}^{(a)} a_j \right\}$$

$$N^{(b)}|\varphi^{(b)}\rangle = |\varphi'^{(b)}\rangle \Longleftrightarrow \left\{ b'_k = \sum_l N_{kl}^{(b)} b_l \right\}. \tag{4.3.8}$$

We *define* now

$$\left(M^{(a)} \otimes N^{(b)} \right) |\psi^{(a)}, \varphi^{(b)}\rangle \equiv |\psi'^{(a)}, \varphi'^{(b)}\rangle. \tag{4.3.9}$$

Then the matrix elements of $(M^{(a)} \otimes N^{(b)})$ turn out to be given by

$$a'_i b'_k = \sum_{jl} M_{ij}^{(a)} N_{kl}^{(a)} a_j b_l \equiv \sum \left(M^{(a)} \otimes N^{(b)} \right)_{ik,jl} a_j b_l$$

$$\left(M^{(a)} \otimes N^{(b)} \right)_{ik,jl} = M_{ij}^{(a)} N_{kl}^{(b)} \tag{4.3.10}$$

where the *pair* (ik) labels the rows and the *pair* (jl) the columns of the direct product matrix.

We define the scalar product as usual: take two direct products of states:

$$|\psi'^{(a)}, \varphi'^{(b)}\rangle \Longleftrightarrow \{\alpha_i \beta_k\}$$
$$|\psi^{(a)}, \varphi^{(b)}\rangle \Longleftrightarrow \{a_i b_k\}$$

then

$$\langle \psi'^{(a)}, \varphi'^{(b)} | \psi^{(a)}, \varphi^{(b)} \rangle = \sum_{i,k} \alpha_i^* a_i \beta_k^* b_k = \langle \psi'^{(a)} | \psi^{(a)} \rangle \langle \varphi'^{(b)} | \varphi^{(b)} \rangle.$$

With this the general matrix element becomes

$$\langle \psi'_\otimes | M_\otimes | \psi_\otimes \rangle = \langle \psi'^{(a)}, \varphi'^{(b)} | M^{(a)} \otimes N^{(b)} | \psi'^{(a)}, \varphi'^{(b)} \rangle$$
$$= \langle \psi'^{(a)} | M^{(a)} | \psi^{(a)} \rangle \cdot \langle \varphi'^{(b)} | N^{(b)} | \varphi^{(b)} \rangle.$$

$$(4.3.11)$$

Consider the direct product of a vector $|\psi^{(a)}\rangle$ with the sum of two others $|\varphi_1^{(b)}\rangle + |\varphi_2^{(b)}\rangle$. As one sees from the representation (4.3.7), the new vector has components such that one must define

$$|\psi^{(a)}\rangle \otimes \left\{ |\varphi_1^{(b)}\rangle + |\varphi_2^{(b)}\rangle \right\} = |\psi^{(a)}\rangle \otimes |\varphi_1^{(b)}\rangle + |\psi^{(a)}\rangle \otimes |\varphi_2^{(b)}\rangle$$
$$= |\psi^{(a)}, \varphi_1^{(b)}\rangle + |\psi^{(a)}, \varphi_2^{(b)}\rangle.$$

$$(4.3.12)$$

From this we derive immediately the rule for addition in the direct product:

$$M^{(a)} \otimes (N^{(b)} + Q^{(b)}) = M^{(a)} \otimes N^{(b)} + M^{(a)} \otimes Q^{(b)}. \qquad (4.3.13)$$

For matrix multiplication (4.3.9) gives also immediately

$$(M^{(a)} \otimes N^{(b)}) \cdot (P^{(a)} \otimes Q^{(b)}) = M^{(a)} P^{(a)} \otimes N^{(b)} Q^{(b)}. \qquad (4.3.14)$$

In particular

$$\left(M^{(a)} \otimes 1^{(b)} \right) \cdot \left(1^{(a)} \otimes N^{(b)} \right) = M^{(a)} \otimes N^{(b)}. \qquad (4.3.15)$$

Having now described two ways of combining states into new states of an enlarged Hilbert space, we must ask which procedure we have to adopt in the case of two physical systems, $S^{(a)}$ and $S^{(b)}$, described by states in the Hilbert spaces $\mathcal{H}^{(a)}$ and $\mathcal{H}^{(b)}$ respectively, when we wish to describe the system $S = \{S^{(a)}, S^{(b)}\}$ as *one* entity in a Hilbert space $\mathcal{H}$. Are we to take the direct product space or the direct sum space? The answer is dictated by the probability interpretation of quatum mechanics, for if the two systems $S^{(a)}$ and $S^{(b)}$ do *not* interact, the probability of finding $S^{(a)}$ in a state $|\psi^{(a)}\rangle$ and $S^{(b)}$ in a state $|\varphi^{(b)}\rangle$ must *multiply*; under these circumstances only the direct product can be accepted. If the systems $S^{(a)}$ and $S^{(b)}$ *do* interact, then the state must be built up from a superposition of direct product states—but we certainly can never use the direct sum for describing a composite system. That the direct sum nevertheless fulfils useful functions will become obvious when we study the reduction of direct products to direct sums in chapters 5 and 6.

Returning to the question of what we have done in the last paragraph when splitting up S into $S^{(1)}, \ldots, S^{(n)}$, the answer is we have considered the Hilbert space of S as being the direct product space of the Hilbert spaces of $S^{(1)}, \ldots, S^{(n)}$. For instance, in the spin case the wave function

$$\psi(x) = \begin{pmatrix} \psi_1(x) \\ \psi_2(x) \\ \psi_3(x) \end{pmatrix}$$

was nothing else than the Schrödinger function of a direct product state. Namely, in Hilbert space 'a' the basis states are labelled by x and the set of components $\{a_i\}$ characterizing the state is the *set of all values* of the function $\psi(x)$, namely $\{\psi(x)\}$. Hilbert space 'b' is three-dimensional and the basis states may be labelled by $|1\rangle$, $|2\rangle$ and $|3\rangle$. Then the most general direct state is a superposition of the basis states $|1, x\rangle$, $|2, x\rangle$ and $|3, x\rangle$ of the form

$$|\psi\rangle = \sum_{i,x} \psi_i(x)|i, x\rangle$$

which contains three functions.

In the consideration leading to (4.2.11) and (4.2.12), we have used the following argument: the system S, consisting of subsystems $S^{(1)}, \ldots, S^{(k)}$ is rotated by rotating each subsystem individually. The state $|\psi\rangle$ describing the whole system S is a linear combination of direct product states (we repeat the consideration for $n = 2$)

$$|\psi\rangle = \sum_{i,k} c_{ik} |\varphi_i^{(a)}, \varphi_k^{(b)}\rangle.$$

Rotating each subsystem individually amounts to writing

$$e^{-i\alpha n \cdot J} = e^{-i\alpha n \cdot J^{(a)}} e^{-i\alpha n \cdot J^{(b)}}$$

where $J^{(a)}$ should act only on the states $\varphi^{(a)}$ and $J^{(b)}$ only on the states $\varphi^{(b)}$; with respect to $\varphi^{(a)}$ the operator $J^{(a)}$ should behave as the unit operator, as should $J^{(b)}$ with respect to the states $\varphi^{(a)}$. We thus see that in the above formula we should have written more correctly

$$J^{(a)} \otimes 1^{(b)} \text{ instead of } J^{(a)}$$

and

$$1^{(a)} \otimes J^{(b)} \text{ instead of } J^{(b)}.$$

Then the operator $\exp(-i\alpha n \cdot J)$ becomes

$$e^{-i\alpha n \cdot J} = e^{-i\alpha n \cdot (J^{(a)} \otimes 1^{(b)})} e^{-i\alpha n \cdot (1^{(a)} \otimes J^{(b)})}.$$

Now we can also formally prove that the two exponentials commute: we expand each of them and use the rules (4.3.13) and (4.3.14); we abbreviate $(-i\alpha n \cdot J)$ by J and obtain

$$e^{(J^{(a)} \otimes 1^{(b)})} e^{(1^{(a)} \otimes J^{(b)})} = \left[\sum_n \frac{(J^{(a)} \otimes 1^{(b)})^n}{n!} \right] \left[\sum_k \frac{(1^{(a)} \otimes J^{(b)})^k}{k!} \right]$$

$$= \left[\left(\sum_n \frac{J^{(a)n}}{n!} \right) \otimes 1^{(b)} \right] \left[1^{(a)} \otimes \left(\sum_k \frac{J^{(b)k}}{k!} \right) \right]$$

$$= \left[e^{J^{(a)}} \otimes 1^{(b)} \right] \left[1^{(a)} \otimes e^{J^{(b)}} \right]$$

$$= e^{J^{(a)}} \otimes e^{J^{(b)}}.$$

We thus can simply write

$$e^{J^{(a)} \otimes 1^{(b)}} e^{1^{(a)} \otimes J^{(b)}} = e^{J^{(a)} \otimes 1^{(b)} + 1^{(a)} \otimes J^{(b)}}.$$

Indeed, expanding the right-hand side gives

$$\sum_n \frac{1}{n!} [J^{(a)} \otimes 1^{(b)} + 1^{(a)} \otimes J^{(b)}]^n$$

$$= \sum_n \frac{1}{n!} \left[\sum_{k=1}^{n} \binom{n}{k} (J^{(a)} \otimes 1^{(b)})^k (1^{(a)} \otimes J^{(b)})^{n-k} \right]$$

$$= \sum_n \frac{1}{n!} \left[\sum_{k=1}^{n} \binom{n}{k} (J^{(a)k} \otimes 1^{(b)}) (1^{(a)} \otimes J^{(b)n-k}) \right]$$

$$= \sum_n \frac{1}{n!} \left[\sum_{k=1}^{n} \binom{n}{k} J^{(a)k} \otimes J^{(b)n-k} \right].$$

Putting $\binom{n}{k} = \dfrac{n!}{k!(n-k)!}$ and $(n-k) = j$ we obtain

$$= \sum_{k,j} \frac{1}{k!} \cdot \frac{1}{j!} J^{(a)k} \otimes J^{(a)j} = e^{J^{(a)}} \otimes e^{J^{(b)}}.$$

We combine these results in the formulae

$$e^{-i\alpha n \cdot J} = e^{-i\alpha n \cdot J^{(a)}} \otimes e^{-i\alpha n \cdot J^{(b)}} = e^{-i\alpha n \cdot (J^{(a)} \otimes 1^{(b)} + 1^{(a)} \otimes J^{(b)})}$$

$$J = J^{(a)} \otimes 1^{(b)} + 1^{(a)} \otimes J^{(b)}$$

(4.3.16)

from which we see that (4.2.11)

$$J = \sum_{i=1}^{k} j^{(i)} \tag{4.3.17}$$

should be written more correctly as follows:

$$J = \sum_{i=1}^{k} 1^{(1)} \otimes \cdots \otimes 1^{(i-1)} \otimes j^{(i)} \otimes 1^{(i+1)} \otimes \cdots \otimes 1^{(k)}. \tag{4.3.18}$$

The unit operators may be combined into one unit operator which is the direct product of all unit operators belonging to the spaces $\mathcal{H}^{(1)}$ to $\mathcal{H}^{(k)}$, with $1^{(i)}$ being omitted. Let us denote this product by $1'^{(k.i)}$, i.e.

$$1'^{(k.i)} \equiv 1^{(1)} \otimes \cdots \otimes 1^{(i-1)} \otimes 1^{(i+1)} \otimes \cdots \otimes 1^{(k)}. \tag{4.3.19}$$

Then

$$J = \sum_{i=1}^{k} j^{(i)} \otimes 1'^{(k.i)}. \tag{4.3.20}$$

This is the exact meaning of (4.2.11). The reader should keep this in mind, because it is in current use in the physical literature to employ the simpler, but not strictly correct, notation

$$J = \sum j^{(i)} \tag{4.3.21}$$

where $j^{(i)}$ means in fact $j^{(i)} \times 1'^{(k.i)}$.

Having warned the reader, we shall from now on follow the common use and write $j^{(i)}$ instead of $j^{(i)} \otimes 1'^{(k.i)}$, sometimes recalling the present discussion by a remark.

4.4   Angular momenta of interacting systems

Having laid down the physical significance and the commutation relations of the angular momentum operators, we can start to build up the whole theory, but, before we do so, it might be good to stop for a moment and contemplate what we have done and try to understand it correctly in terms of physics.

We started by considering the consequences of invariance of a physical system under rotations and found that the rotations led to Hermitian operators, constant in time, which were the generators of the induced unitary transformations and which we called angular momenta. But what if the system is not invariant under rotations (think of a particle in an external field)? Does that mean that angular momentum cannot be defined? It can. The point is this: a general postulate, which cannot be proved or disproved except by experiment and which we have so far no reason to disbelieve, is that space is isotropic. If therefore our system is not yet invariant under rotations, then that means that it is not yet isolated in empty space but instead is part of a large (isolated) system, which as a whole is invariant. Thus by a suitable extension of our system we can always achieve full invariance. Let then $S^{(a)}$ be our system. If it is not invariant under rotations, we may take a sufficiently large part of its environment—call it $S^{(b)}$—and add it to $S^{(a)}$ such that a new system S results, which is (sufficiently) isolated in empty space and therefore invariant, while $S^{(a)}$ and $S^{(b)}$ separately are not.

On the other hand, we might have—equally well—*removed* the environment of $S^{(a)}$ such that $S^{(a)}$ would have been left in 'empty space' and then $S^{(a)}$ would have been invariant. We see that it is not truly $S^{(a)}$ itself which is lacking invariance: it is $S^{(a)}$ plus something else relating it to $S^{(b)}$; this whole thing, namely $S^{(a)}$ including its relation to $S^{(b)}$, we have—wrongly—taken to be $S^{(a)}$. Giving from now on $S^{(a)}$ and $S^{(b)}$ a more restricted sense and calling $S^{(ab)}$ the 'interaction part' between $S^{(a)}$ and its environment $S^{(b)}$, we state that under rotations (and in fact, as we believe, under the Poincaré group)

$$S^{(a)} \text{ and } S^{(b)} \qquad\qquad \text{are invariant}$$

$$S = S^{(a)} + S^{(b)} + S^{(ab)} \qquad \text{is invariant} \qquad\qquad (4.4.1)$$

$$S^{(a)} + S^{(ab)} \text{ or } S^{(b)} + S^{(ab)} \quad \text{are not invariant.}$$

This means that whenever we find that a physical system is not invariant under rotations, we have a system of the type $S^{(a)} + S^{(ab)}$ and we are compelled to find out what $S^{(a)}$ is and what $S^{(ab)}$ is. Then, having done this, there will be nothing inherent in $S^{(a)}$ which would make it non-invariant. Hence the operator of angular momentum, $J^{(a)}$, can be defined as before for $S^{(a)}$, where $S^{(a)}$ is considered as an isolated system and described in a Hilbert space $\mathcal{H}^{(a)}$.

Let us now consider $S^{(a)}$ in interaction with its symmetry-breaking environment, that is $S^{(a)} + S^{(ab)}$. It is still described in the same Hilbert space, because the system $S^{(b)}$ is not yet included; $S^{(ab)}$ is considered so far as fixed external perturbation. It leads, however, to additional terms in the equations of motion (Hamiltonian) of $S^{(a)}$. In this old Hilbert space of $S^{(a)}$ our operator $j^{(a)}$ is still a decent Hermitian operator; however, under the influence of $S^{(ab)}$ this operator will no longer commute with the Hamiltonian $H^{(a)} + H^{(ab)}$. In the same way we define the operator $j^{(b)}$ which describes the rotations of the system $S^{(b)}$ in its Hilbert space $\mathcal{H}^{(b)}$. If the presence of $S^{(b)}$ becomes felt, $j^{(b)}$ no longer commutes with $H^{(b)} + H^{(ab)}$.

We now proceed to the description of $S = S^{(a)} + S^{(b)} + S^{(ab)}$ as a whole. The states are states of the direct product space $\mathcal{H} = \mathcal{H}^{(a)} \otimes \mathcal{H}^{(b)}$; the total Hamiltonian is $H = H^{(a)} + H^{(b)} + H^{(ab)}$; the total angular momentum is $J = j^{(a)} + j^{(b)}$ (more correctly $j^{(a)} \otimes 1^{(b)} + 1^{(a)} \otimes j^{(b)}$).

We have then

$$\text{states of } S \qquad\qquad |\psi_{ik}\rangle = |\psi_i^{(a)} \psi_k^{(b)}\rangle$$

$$(4.4.2)$$

$$\text{angular momentum of } S \quad J = j^{(a)} + j^{(b)}.$$

Although by these definitions the range of the operators $j^{(a)}$ (more correctly $1^{(a)} \otimes j^{(b)}$) is the whole product space $\mathcal{H} = \mathcal{H}^{(a)} \otimes \mathcal{H}^{(b)}$; although therefore the operators $j^{(a)}$ and $j^{(b)}$ have become operators having a physical significance with respect to the whole system S—although this is so, it follows from the meaning of the direct product that this physical significance is essentially the same as before: $j^{(a)}$ (better $j^{(a)} \otimes 1^{(b)}$) and $j^{(a)}$ (better $1^{(a)} \otimes j^{(b)}$) are the angular momentum operators of the subsystem $S^{(a)}$ and $S^{(b)}$ respectively, when these two are understood to be integrated into the one system $S = S^{(a)} + S^{(b)} + S^{(ab)}$.

Since, by writing $S^{(ab)}$ explicitly, the components $S^{(a)}$ and $S^{(b)}$ have become separately rotation invariant, it is clear that under a partial rotation of S, say of $S^{(a)}$ alone, only $S^{(b)}$ is changed and that by carrying out the same rotation on $S^{(b)}$ the old $S^{(b)}$ will be restored since S, $S^{(a)}$ and $S^{(b)}$ are invariant. Carrying out different rotations on $S^{(a)}$ and $S^{(b)}$ will change S into something new, however.

These considerations show that, although the angular momenta were defined as the generators of the unitary transformations related to rotational invariance, they retain their physical significance even if this invariance is not present, because $j^{(a)}$ (for instance) may be regarded as belonging to a system $S^{(a)}$ which would have invariance, were it not for the presence of something else ($S^{(b)}$) in relation to which the invariance is broken. We have seen that the angular momentum, once defined for the unperturbed (invariant) subsystem, can afterwards be carried over to the total Hilbert space (describing the system and the perturbation as a new entity) without losing its original physical meaning. The same is true if we renounce enlarging our system and simply remain in the original Hilbert space, considering the symmetry-breaking environment as 'external perturbation'. The same holds for the full symmetry group of space-time. We may say that, by starting from the invariance corresponding to the full symmetry group, we have been able to introduce the most natural coordinate system in Hilbert space; namely that one which is labelled by the quantum numbers of the conserved operators. As long as we describe systems which are indeed invariant under the whole symmetry group, these coordinates (i.e. the quantum numbers) are invariant in time. If we describe systems which are not invariant ($S^{(a)} + S^{(ab)}$), then the coordinate system introduced is still useful and legitimate; the states of such non-invariant systems have, however, varying coordinates.

4.5 Irreducible representations; Schur's lemma

Before we build up the eigenstates of angular momentum, in the next section, we shall try to give a motivation for why we just do it this way and no other. In fact, we shall do it exactly as everybody else does it; but, in the light of what we have learned on the relation between a group and its Lie algebra and between their corresponding representations, this usual construction gains a new aspect in which it will lose some of its apparent arbitrariness (when one first learns about angular momentum in quantum mechanics and sees how the eigenstates and matrices are constructed, a natural first reaction would be that if physics required one to be able to invent such things out of the air, then it might be better to give up physics at once). What in fact, is done?

One constructs finite-dimensional, *irreducible representations* of the Lie algebra of the rotation group; from these, as we know, the representations of the local group follow by exponentiation. To each finite-dimensional irreducible representation there belongs a finite-dimensional *irreducible invariant subspace*; the basic states spanning this irreducible subspace are the angular momentum eigenstates: under rotations they are transformed among themselves (i.e. within that subspace) and the corresponding transformation matrices make up just the irreducible representation which leaves this subspace invariant.

What are irreducible invariant subspaces and why do we construct irreducible representations?

Because, in a certain sense, they are the simplest ones: they cannot be reduced to anything simpler. Let us briefly explain this.

Consider a matrix M transforming the vector x of an n-dimensional space R_n. If there exists a subspace $R^{(1)}$ of dimension $n_1 < n$ such that the image Mx of any vector $x \in R^{(1)}$ lies again in $R^{(1)}$ then M is called *reducible* and that subspace is called an *invariant subspace*. (The invariant subspace is not invariant on its own account; to be invariant is not its inherent property; it is invariant *with respect to M*; therefore, the word 'invariant subspace', which suggests the invariance to be an attribute of that subspace, is misleading. As it is, however, commonly used, we shall use it too.) We may arrange the basis in our space such that the invariant subspace $R^{(1)}$ is spanned by the first n_1 basis vectors. In this basis M must take the form

$$M = \left(\begin{array}{c|c} m_1 & m \\ \hline 0 & m_2 \end{array} \right) \tag{4.5.1}$$

because then and only then will the image of any vector x of the invariant subspace lie again in it, although this image may contain contributions from the images of vectors outside the invariant subspaces. It may be that the whole space R decomposes into a direct sum $R^{(1)} \oplus R^{(2)}$ where not only $R^{(1)}$ but also $R^{(2)}$ is invariant. In that case the matrix m in the upper right corner of (4.5.1) must be zero and M decomposes into a direct sum $M = m_1 \oplus m_2$. We then call M *fully reducible*.

$$M = \left(\begin{array}{c|c} m_1 & 0 \\ \hline 0 & m_2 \end{array} \right) = m_1 \oplus m_2. \tag{4.5.2}$$

Now m_1 and m_2 may be again reducible or fully reducible; but eventually this procedure ends, namely when the blocks m_i are *irreducible* (i.e. not reducible). When we start with one single matrix M, then most frequently (for instance if M is Hermitian or unitary or has all different eigenvalues) the process of reduction ends up with a full decomposition into a direct sum of one-dimensional matrices—and these complicated words mean nothing more or less than that we have brought the matrix to diagonal form:

$$
M = \begin{pmatrix} \mu_1 & & & 0 \\ & \mu_2 & & \\ & & \ddots & \\ 0 & & & \mu_n \end{pmatrix}.
\tag{4.5.3}
$$

For a single matrix M, therefore, reduction is in most cases the familiar diagonalization. The situation greatly changes when a system of matrices $M(\alpha)$ is considered. Let α label the set; α may be discrete or continuous and may also denote a set of parameters. A group representation would be one example of such a set $M(\alpha)$; a representation of a Lie algebra another one. As is well known, a set of matrices can be brought simultaneously to diagonal form only if all matrices of the set commute with each other. A generalization (see (4.5.9)) of this statement is that a set of matrices is reducible if a matrix A exists which commutes with all matrices of the set and which is not a multiple of the unit matrix (because $A = \lambda \cdot 1$ of course commutes with every matrix).

Reducibility of a whole matrix system $M(\alpha)$ means that there exists a matrix B such that the similarity transformation $BM(\alpha)B^{-1}$ transforms *all* $M(\alpha)$ *simultaneously into the same form*

$$
M'(\alpha) = BM(\alpha)B^{-1} = \left(\begin{array}{c|c} m_1(\alpha) & m(\alpha) \\ \hline 0 & m_2(\alpha) \end{array} \right) \quad \text{for all } \alpha.
\tag{4.5.4}
$$

The matrix system $M(\alpha)$ is called reducible if (4.5.4) holds; it is called fully reducible if $m(\alpha)$ is zero:

$$
M'(\alpha) = BM(\alpha)B^{-1} = \left(\begin{array}{c|c} m_1(\alpha) & 0 \\ \hline 0 & m_2(\alpha) \end{array} \right) = m_1(\alpha) \oplus m_2(\alpha) \quad \text{for all } \alpha.
\tag{4.5.5}
$$

Also here $m_1(\alpha)$ and/or $m_2(\alpha)$ may be further reducible, but the process comes to an end and then the blocks $m_i(\alpha)$ are called irreducible. This stage is reached much earlier for matrix systems than for a single matrix—it leads to full diagonalization only when all $M(\alpha)$ commute with each other.

One now sees immediately why reducible representations lead to simplification: suppose $M(g)$ is a group representation by $n \times n$ matrices and let the representation be reducible:

$$
M(g) = \left(\begin{array}{c|c} m_1(g) & m(g) \\ \hline 0 & m_2(g) \end{array} \right).
$$

Group multiplication becomes, with $M(g \cdot h) = M(g) \cdot M(h)$

$$
M(g \cdot h) = \left(\begin{array}{c|c} m_1(g \cdot h) & m(g \cdot h) \\ \hline 0 & m_2(g \cdot h) \end{array} \right)
$$

$$
= \left(\begin{array}{c|c} m_1(g) \cdot m_1(h) & m_1(g) \cdot m(h) + m(g) \cdot m_2(h) \\ \hline 0 & m_2(g) \cdot m_2(h) \end{array} \right)
$$

so that

$$m_1(g \cdot h) = m_1(g) \cdot m_1(h)$$
$$m_2(g \cdot h) = m_2(g) \cdot m_2(h).$$

(4.5.6)

Thus two new representations are furnished by the matrix systems $m_1(g)$ and $m_2(g)$—both are of a smaller dimension as M was. As to the matrix $m(g)$ in the upper right corner, it is obvious that it is not a representation; but what is important is that it in no way prevents $m_1(g)$ and $m_2(g)$ being representations. This means that, in order to give rise to at least two new lower-dimensional representations, the representation $M(g)$ need not be fully reducible; simple reducibility suffices. If, however, $M(g)$ is fully reducible, then we can write $M(g) = m_1(g) \oplus m_2(g)$.

Exactly the same statements hold for representations of a Lie algebra, because if in the above considerations we replace ordinary matrix products by commutators, we find, instead of (4.5.6)

$$m_1([\overset{\circ}{g}, \overset{\circ}{h}]) = [m_1(\overset{\circ}{g}), m_1(\overset{\circ}{h})]$$

$$m_2([\overset{\circ}{g}, \overset{\circ}{h}]) = [m_2(\overset{\circ}{g}), m_2(\overset{\circ}{h})]$$

(4.5.7)

so that reducibility (not necessarily full reducibility) of a representation $M(\overset{\circ}{g})$ of a Lie algebra yields two new, lower-dimensional representations $m_1(\overset{\circ}{g})$ and $m_2(\overset{\circ}{g})$.

Clearly then, if we can find *all irreducible representations* of a group (or of its Lie algebra) then we have the building blocks from which all representations are made. This explains why the construction of all irreducible representations is a central problem. In many cases—and in fact in rotational symmetry—the representations which come up are fully reducible. That means that we shall encounter only situations in which

$$M(\alpha) = \begin{pmatrix} m_1(\alpha) & & & 0 \\ & m_2(\alpha) & & \\ & & m_3(\alpha) & \\ 0 & & & \ddots \end{pmatrix} = m_1(\alpha) \oplus m_1(\alpha) \oplus \ldots$$

$$\mathcal{H} = \mathcal{H}^{(1)} \oplus \mathcal{H}^{(2)} \oplus \ldots$$

$$|\psi\rangle = |\psi^{(1)}\rangle \oplus |\psi^{(2)}\rangle \oplus \ldots$$

$$M(\alpha)|\psi\rangle = m_1(\alpha)|\psi^{(1)}\rangle \oplus m_2(\alpha)|\psi^{(2)}\rangle \oplus \ldots.$$

(4.5.8)

Finally we prove a theorem which is a very powerful tool in the search for irreducible representations; it is the famous lemma of Schur, of which we state and prove only its simplest form.

If a matrix A commutes with all matrices of an irreducible system $M(\alpha)$, then $A = \lambda \cdot 1$.

(4.5.9)

Proof. Every matrix, whether we can diagonalize it or not, possesses eigenvalues λ and eigenvectors x. Let then $Ax = \lambda x$. The supposition that A commutes with all matrices of the set $M(a)$ means

$$A[M(\alpha)x] = M(\alpha)Ax = \lambda[M(\alpha)x].$$

In other words, $M(\alpha)x$ is again an eigenvector of A with the same eigenvalue λ. This holds for all vectors belonging to the subspace spanned by the eigenvectors of A with this same eigenvalue λ. Hence this subspace is not only an invariant subspace with respect to A, but also with respect to the whole set $M(\alpha)$: every x belonging to that subspace is transformed by $M(\alpha)$ into another x' belonging to that same subspace. This contradicts the assumption that $M(\alpha)$ is irreducible— except if this invariant subspace is the whole space; then $A = \lambda \cdot 1$, as the theorem asserts.

This proof also shows as a by-product

> if a matrix A commutes with all matrices of a system $M(\alpha)$ and A can be diagonalized, then the system is fully reducible into a direct sum; to each different eigenvalue λ_i of A (4.5.10) belongs one irreducible invariant subspace and one irreducible representation.

Note that it is tacitly assumed that no other matrix A' with a set of different eigenvalues inside the invariant subspaces of A exists and commutes with all $M(\alpha)$; if this is the case, then A splits $M(\alpha)$ into a direct sum of representations which are still further reducible by A'.

We shall use this fact in the construction of eigenstates and representations of angular momentum; the operator $J^2 = J_x^2 + J_y^2 + J_z^2$—called the *Casimir operator*—will play the role of the matrix A. In what follows, we shall not always stress these general aspects of what we do; the reader is urged to take a glance over and over again back to chapter 2 (symmetry)—just as he would from time to time reassure himself by a look at the map when he hikes in an unknown region where his view is barred by many nearby hills, trees, buildings or even by fog and clouds.

4.6 Eigenstates of angular momentum

We consider a system S with total angular momentum J. What are the eigenstates of these operators?

As follows from the commutation relations

$$[J_1, J_2] = iJ_3 \text{ and cycl. perm.} \tag{4.6.1}$$

each component J_i commutes with the operator ('Casimir operator')

$$J^2 \equiv J_1^2 + J_2^2 + J_3^2. \tag{4.6.2}$$

Consequently any component—but only one—and J^2 can be taken to possess a common system of eigenstates.

We define the 'ladder operators'

$$
\begin{array}{ll}
\text{'raising operator'} & J_+ = J_1 + iJ_2 \\
\text{'lowering operator'} & J_- = J_1 - iJ_2 \\
& J_- = J_+^\dagger
\end{array}
\tag{4.6.3}
$$

and find by explicit calculation the rules

$$
[J^2, J_\pm] = [J^2, J_{1,2,3}] = 0
$$
$$
[J_1, J_\pm] = \mp J_3
$$
$$
[J_2, J_\pm] = -iJ_3
$$
$$
[J_3, J_\pm] = \pm J_\pm
\tag{4.6.4}
$$
$$
[J_+, J_-] = 2J_3
$$
$$
J^2 = J_3^2 + \tfrac{1}{2}(J_+ J_- + J_- J_+) = J_3(J_3 - 1) + J_+ J_-
$$
$$
= J_3(J_3 + 1) + J_- J_+.
$$

Let us use J^2 and J_3 as commuting observables. If all other observables commuting with these two and among each other are taken together in one symbol Γ then the basis in the Hilbert space is made up by the orthonormal states $|\gamma, \lambda, \mu\rangle$ where

$$
\Gamma|\gamma\lambda\mu\rangle = \gamma|\gamma\lambda\mu\rangle
$$
$$
J^2|\gamma\lambda\mu\rangle = \lambda|\gamma\lambda\mu\rangle
$$
$$
J_3|\gamma\lambda\mu\rangle = \mu|\gamma\lambda\mu\rangle
\tag{4.6.5}
$$
$$
\langle\gamma'\lambda'\mu'|\gamma\lambda\mu\rangle = \delta(\gamma'\gamma; \lambda'\lambda; \mu'\mu)
$$
$$
\sum_{\gamma\lambda\mu} |\gamma\lambda\mu\rangle\langle\gamma\lambda\mu| = 1.
$$

The $\delta(\gamma'\gamma; \lambda'\lambda; \mu'\mu)$ is Kronecker's or Dirac's according to circumstances.

Whereas (by definition of Γ) it is clear from (4.6.5) that operating with Γ, J^2, J_3 on $|\gamma\lambda\mu\rangle$ will not change any eigenvalues, it is not clear that application of J_1, J_2, J_+ or J_- on $|\gamma\lambda\mu\rangle$ will leave γ unchanged. It will leave λ unchanged, since all these J commute with J^2; but they need not commute with γ. This fact is generally neglected in textbooks. Therefore we shall keep it in mind. It will turn out that it causes no difficulty. We shall build up the eigenstates of angular momentum by considering matrix elements of the operators and commutators appearing in (4.6.4). Since all operators commute with J^2, we consider λ to be fixed once and for all.

The situation—expressed in the general terms used in the preceding section 4.5—is then this.

The operator J^2 is Hermitian and can thus be diagonalized; it furthermore commutes with J_1, J_2, J_3, that is, with the whole Lie algebra of the rotation group. Therefore theorem (4.5.10) tells us that the representations of the Lie algebra are fully reducible to a direct sum of (as we shall see, finite-dimensional) irreducible representations belonging to irreducible subspaces; in each of these subspaces J^2 is a multiple of the unit matrix. Considering λ to be fixed once for ever means we have chosen the irreducible subspace $\mathcal{H}_\lambda$, labelled by the eigenvalue λ, and there we build up a basis; the basis vectors are written $|\gamma\lambda\mu\rangle$.

All we have to do now is to play around with equations (4.6.4) and (4.6.5). We shall obtain several intermediate results, labelled by (a), (b), ... which then will be combined in (4.6.12), (4.6.13) and (4.6.14). Let us now play the game (4.6.4), (4.6.5).

The relation

$$J^2 - J_3^2 = J_1^2 + J_2^2$$

gives

$$\langle\gamma\lambda\mu|J^2 - J_3^2|\gamma\lambda\mu\rangle = \lambda - \mu^2$$

$$= \langle\gamma\lambda\mu|J_1^2|\gamma\lambda\mu\rangle + \langle\gamma\lambda\mu|J_2^2|\gamma\lambda\mu\rangle \geq 0$$

since each term is ≥ 0. Hence

(a)
$$\mu^2 \leq \lambda. \tag{4.6.6}$$

Similarly, the commutation relation

$$[J_3, J_\pm] = \pm J_\pm$$

gives

$$\langle\gamma'\lambda\mu'|[J_3, J_\pm]|\gamma\lambda\mu\rangle$$
$$= \pm\langle\gamma'\lambda\mu'|J_\pm|\gamma\lambda\mu\rangle$$
$$= (\mu' - \mu)\langle\gamma'\lambda\mu'|J_\pm|\gamma\lambda\mu\rangle$$

or equivalently

$$(\mu' - \mu\mp1)\langle\gamma'\lambda\mu'|J_\pm|\gamma, \lambda, \mu\rangle = 0. \tag{4.6.7}$$

The matrix element is zero unless $\mu' = \mu\pm1$.

Now we shall first consider the situation $[J_\pm, \Gamma] = 0$. Then the action of $J_\pm$ does not change γ, and the previous equation gives

(b)
$$J_\pm|\gamma, \lambda, \mu\rangle = C^\pm(\lambda\mu)|\gamma, \lambda, \mu\pm1\rangle. \tag{4.6.8}$$

The constant

(c)

$$C^{(\pm)}(\lambda\mu) = \langle \gamma, \lambda, \mu\pm1|J_\pm|\gamma, \lambda, \mu\rangle \qquad (4.6.9)$$

can be determined from the relation

$$J_\mp J_\pm = J^2 - J_3(J_3\pm1).$$

Since $J_- = J_+^\dagger$ we have on the l.h.s. a positive operator and therefore

(d)

$$\langle\gamma\lambda\mu|J_\mp J_\pm|\gamma\lambda\mu\rangle = \langle\gamma\lambda\mu|J^2 - J_3(J_3\pm1)|\gamma\lambda\mu\rangle = \lambda - \mu(\mu\pm1) \geq 0.$$

Obviously, by complex conjugation of (4.6.9) we obtain

(e)

$$C^{(+)*}(\lambda, \mu) = C^{(-)}(\lambda, \mu + 1)$$

and inserting (d) into (4.6.8) we obtain the important relation

(f)

$$|C^{(\pm)}(\lambda\mu)|^2 = \lambda - \mu(\mu\pm1). \qquad (4.6.10)$$

It is useful to fix the phases of the basis in such a way that $C^{(\pm)}(\lambda\mu)$ are real. Then we shall have

(g)

$$J_\pm|\gamma, \lambda, \mu\rangle = \sqrt{\lambda - \mu(\mu\pm1)}|\gamma, \lambda, \mu\pm1\rangle. \qquad (4.6.11)$$

We call J_+ the raising operator, and the J_- the lowering operator, according to their effect on $|\gamma, \lambda, \mu\rangle$.

We finally determine the possible values of λ, μ. From (a), we see that

$$\lambda \geq 0 ; \quad -\sqrt{\lambda} \leq \mu \leq \sqrt{\lambda}.$$

This condition may be violated by a repeated application of J_+ and J_- (respectively) to a state $|\gamma\lambda\mu\rangle$ (see (b)), unless the series of states thus created breaks off when some limit values $\max(\mu) = \bar{\mu}$ and $\min(\mu) = \underline{\mu}$ (respectively) are reached. Hence we must have

$$J_+|\lambda, \bar{\mu}\rangle = 0$$
$$J_-|\lambda, \underline{\mu}\rangle = 0.$$

Applying J_- to the upper and J_+ to the lower equation gives (see (d))

$$J_-J_+|\lambda, \bar{\mu}\rangle = [\lambda - \bar{\mu}(\bar{\mu} + 1)]|\lambda, \bar{\mu}\rangle = 0$$
$$J_+J_-|\lambda, \underline{\mu}\rangle = [\lambda - \underline{\mu}(\underline{\mu} - 1)]|\lambda, \underline{\mu}\rangle = 0.$$

Since $|\lambda, \bar{\mu}\rangle$ and $|\lambda, \underline{\mu}\rangle$ are not zero, we conclude

(h)

$$\lambda = \overline{\mu}(\overline{\mu} + 1) = \underline{\mu}(\underline{\mu} - 1).$$

This equation has two solutions

$$\underline{\mu} = -\overline{\mu} \quad \text{and} \quad \underline{\mu} = \overline{\mu} + 1$$

where the second drops out since by definition $\underline{\mu} \le \overline{\mu}$.

The subspace $\mathcal{H}_\lambda$ is thus equally well characterized by the largest eigenvalue $\overline{\mu} = -\underline{\mu}$ of J_3 in that space; this number is called the 'weight' of this irreducible subspace (or of the corresponding irreducible representation); in the mathematical literature on Lie groups the weight is designated by the letter l; physicists prefer (for angular momentum) j. Thus we put

$$j = \overline{\mu} = -\underline{\mu} \quad \text{and} \quad \mu = m;$$

hence $-j \le m \le j$ where the limits $\pm j$ are indeed assumed. Since by application of $J_\pm$ one goes in integer steps from $\mp j$ to $\pm j$, we have $2j =$ number of steps $=$ integer. Our earlier eigenvalue of J^2 becomes now $\lambda = j(j + 1)$.

We have constructed for given j (integer or half-integer) a set of $2j + 1$ eigenstates J^2 and J_3. This was done by our raising and lowering operators. But what if we had by mistake overlooked the fact that there were other states between these, which could not be reached by our $J_\pm$ since their spacing in the label m is not 1? (For instance, could it not be that there exist raising and lowering operators, Λ_+ and Λ_-, say, which we had not yet discovered and which would change the eigenvalue μ not by 1 but by $1/n$ or any other number < 1?) This is impossible. Assume we had such a state, then by repeated application of J_+ or J_- we would run into the contradiction with $-\sqrt{\lambda} \le \mu \le \sqrt{\lambda}$ if the sequence did not break off. If, however, it breaks off, we are back to our $2j + 1$ states. Therefore this set of $2j + 1$ states is complete in the subspace $\mathcal{H}_{j\gamma}$ for fixed j and γ.

The same argument excludes the existence of any continuous eigenvalues of J and J^2 and thereby ensures that we now possess all the irreducible unitary representations of the Lie algebra of the rotation group, namely when we allow all integer and half-integer values of j; all these representations are finite-dimensional, as their corresponding irreducible subspaces have dimension $2j+1$.

We thus obtain (in the subspace of fixed γ and j) a complete system of $2j + 1$ eigenstates $|\gamma jm\rangle$ with $m = -j, -j + 1, \ldots, j$ and $2j =$ integer; the phases, relative to each other, of these $2j + 1$ states are uniquely defined by the factors $\sqrt{j(j + 1) - m(m \pm 1)}$ in the raising and lowering operation (see (4.6.11)).

$$\Gamma|\gamma jm\rangle = \gamma|\gamma jm\rangle$$

$$J^2|\gamma jm\rangle = j(j+1)|\gamma jm\rangle$$

$$J_3|\gamma jm\rangle = m|\gamma jm\rangle$$

$$J_\pm|\gamma jm\rangle = \sqrt{j(j+1)-m(m\pm1)}|\gamma jm\pm1\rangle$$

$$= \sqrt{(j\mp m)(j\pm m+1)}|\gamma jm\pm1\rangle$$

$$j = 0, 1/2, 1, 3/2, \ldots; \quad -j \le m \le j.$$

(The second form is currently used; the first is easier to memorize.)

$$\sum_{jm}|\gamma jm\rangle\langle\gamma jm| = P(\gamma) = \text{ projection onto the subspace with } \gamma \text{ fixed}$$

$$\sum_{m}|\gamma jm\rangle\langle\gamma jm| = P(\gamma, j) = \text{projection onto the subspace}$$

$$\text{with } \gamma, j \text{ fixed}$$

$$\sum_{\gamma,m}|\gamma jm\rangle\langle\gamma jm| = P(j) = \text{ projection onto the subspace with}$$

$$\text{total angular momentum } j.$$

$$(4.6.12)$$

The matrix representations of J^2, J_3, Γ in this basis are diagonal with the eigenvalues showing up in (4.6.12); J_1 and J_2, however, are not diagonal. We find from $J_1 \pm iJ_2 = J_\pm$ that

$$\langle\gamma' j'm'|J_1|\gamma jm\rangle = \tfrac{1}{2}\delta(\gamma'\gamma; j'j)\,[\delta_{m',m+1}\sqrt{j(j+1)-m(m+1)}$$

$$+ \delta_{m',m-1}\sqrt{j(j+1)-m(m-1)}]$$

$$(4.6.13)$$

$$\langle\gamma' j'm'|J_2|\gamma jm\rangle = \tfrac{1}{2}\delta(\gamma'\gamma; j'j)\,[\delta_{m',m+1}\sqrt{j(j+1)-m(m+1)}$$

$$- \delta_{m',m-1}\sqrt{j(j+1)-m(m-1)}].$$

Note that these formulae with the factor $\delta(\gamma'\gamma)$ hold only as long as $[\Gamma, J_{1,2}] = 0$. Otherwise, the situation must be dealt with using some care along the lines of pages 97–98.

In an explicit matrix representation of an operator A it is customary to label as follows:

$$
\left(A^{(j)}_{m'm}\right) =
\begin{array}{c|ccccc}
 & m\;j & j-1 & \cdots\,m & \cdots -j \\
m' & & & & \\
\hline
j & A_{j,j} & A_{j,j-1} & \cdots A_{j,m} & \cdots A_{j,-j} \\
\vdots & \vdots & & \vdots & \vdots \\
m' & A_{m',j} & A_{m',j-1} & \cdots A_{m',m} & \cdots A_{m',-j} \\
\vdots & \vdots & & \vdots & \vdots \\
-j & A_{-j,j} & A_{-j,j-1} & \cdots A_{-j,m} & \cdots A_{-j,-j}
\end{array}
\tag{4.6.14}
$$

With (4.6.12) and (4.6.13), we have obtained all irreducible representations of the Lie algebra of the rotation group. The (canonical) representations of the group follow then by exponentiation. This problem will be taken up in section 4.8 of this chapter (spin $\frac{1}{2}$) and in chapter 6 (general case).

Let us finally discuss the case $[J_\pm, \Gamma] \neq 0$. According to (4.6.7) the action of $J_\pm$ on $|\gamma, \lambda, \mu\rangle$ will change μ to $\mu\pm1$, and, in general, γ will change too. Instead of the property (b) we shall now have

$$
J_\pm|\gamma\lambda\mu\rangle = \sqrt{\lambda - \mu(\mu\pm1)} \sum_{\gamma'} F^{(\pm)}_{\gamma'\gamma}(\lambda\mu)|\gamma', \lambda, \mu\pm1\rangle
\tag{4.6.15}
$$

where for convenience we wrote the square root appearing in (4.6.11) explicitly. On the r.h.s. now there appears a matrix in γ-indices:

$$
\mathcal{F}^{(\pm)}(\lambda\mu) = (F^{(\pm)}_{\gamma'\gamma}(\lambda\mu))
\tag{4.6.16}
$$

where the matrix elements are given by

$$
\langle\gamma', \lambda, \mu\pm1|J_\pm|\gamma\lambda\mu\rangle = \sqrt{\lambda - \mu(\mu\pm1)}\,F^{(\pm)}_{\gamma'\gamma}(\lambda\mu).
\tag{4.6.17}
$$

Since $J_- = J_+^\dagger$, it follows from (4.6.17) that

$$
\mathcal{F}^{(+)\dagger}(\lambda, \mu) = \mathcal{F}^{(-)}(\lambda, \mu+1).
\tag{4.6.18}
$$

Moreover, repeating the argument leading to (4.6.10) we obtain

$$
\mathcal{F}^{(\pm)\dagger}(\lambda, \mu)\mathcal{F}^{(\pm)}(\lambda, \mu) = 1
\tag{4.6.19}
$$

where 1 denotes the unit matrix in γ-indices. (4.6.19) means that $\mathcal{F}^{(\pm)}(\lambda, \mu)$ are unitary matrices. For the matrix elements this can be expressed as

$$
\sum_{\gamma'} F^{(\pm)*}_{\gamma''\gamma'}(\lambda\mu) F^{(\pm)}_{\gamma'\gamma}(\lambda\mu) = \delta_{\gamma''\gamma}.
\tag{4.6.20}
$$

The raising and lowering operators we *define* (in accordance with (4.6.11)) by

$$
I_\pm|\gamma\lambda\mu\rangle = \sqrt{\lambda - \mu(\mu\pm1)}|\gamma, \lambda, \mu\pm1\rangle.
\tag{4.6.21}
$$

Below we show that they are given as

$$I_\pm = F^{(\pm)\dagger} J_\pm \tag{4.6.22}$$

where $F^{(\pm)\dagger}$ are operators defined by

$$F^{(\pm)\dagger} |\gamma, \lambda, \mu\pm1\rangle = \sum_{\gamma'} F_{\gamma'\gamma}^{(\pm)*}(\lambda\mu) |\gamma', \lambda, \mu\pm1\rangle. \tag{4.6.23}$$

To prove (4.6.21) for the operators $I_\pm$ given in (4.6.22), we shall act on an arbitrary state $|\gamma, \lambda, \mu\rangle$, and we shall successively use (4.6.15), (4.6.23) and (4.6.20):

$$\begin{aligned}
I_\pm |\gamma, \lambda, \mu\rangle &= F^{(\pm)\dagger} J_\pm |\gamma, \lambda, \mu\rangle \\
&= \sqrt{\lambda - \mu(\mu\pm1)} \sum_{\gamma'} F_{\gamma'\gamma}^{(\pm)}(\lambda\mu) F^{(\pm)\dagger} |\gamma', \lambda, \mu\pm1\rangle \\
&= \sqrt{\lambda - \mu(\mu\pm1)} \sum_{\gamma'\gamma''} F_{\gamma'\gamma}^{(\pm)}(\lambda\mu) F_{\gamma''\gamma'}^{(\pm)*}(\lambda\mu) |\gamma'', \lambda, \mu\pm1\rangle \\
&= \sqrt{\lambda - \mu(\mu\pm1)} \sum_{\gamma''} \delta_{\gamma''\gamma} |\gamma'', \lambda, \mu\pm1\rangle \\
&= \sqrt{\lambda - \mu(\mu\pm1)} |\gamma, \lambda, \mu\pm1\rangle.
\end{aligned}$$

In the first step we used (4.6.15), in the second (4.6.23), then (4.6.20) and the last line follows from the definition of the $\delta_{\gamma''\gamma}$. This completes the proof of (4.6.21).

Remark 1. The operators $F^\pm$ have the following properties:

$$\begin{aligned}
[F^{(\pm)}, J^2] &= [F^{(\pm)}, J_3] = 0 \\
F^{(\pm)\dagger} F^{(\pm)} &= F^{(\pm)} F^{(\pm)\dagger} = 1
\end{aligned} \tag{4.6.24}$$

where on the r.h.s. 1 stands for the unit operator in the full Hilbert space.

The first property (4.6.24) is a direct consequence of the fact that the action of $F^{(\pm)}$ on $|\gamma, \lambda, \mu\rangle$ changes neither λ, nor μ (see (4.6.23)). The second property is just a straightforward consequence of (4.6.19): the subspaces $\mathcal{H}_{\lambda\mu}$ spanned by $|\gamma\lambda\mu\rangle$ with λ, μ fixed, are mutually orthogonal, and $F^{(\pm)}$ are in each $\mathcal{H}_{\lambda\mu}$ represented by the unitary matrix $\mathcal{F}^{(\pm)}(\lambda, \mu)$.

Remark 2. Defining

$$I_1 = \frac{1}{2}(I_+ + I_-) \qquad I_2 = \frac{1}{2i}(I_+ - I_-)$$

it is shown by straightforward calculation that

$$\begin{aligned}
[I_1, I_2] &= iJ_3 \\
[I_2, J_3] &= iI_1 \\
[J_3, I_1] &= iI_2 \\
[I_1, J^2] &= [I_2, J^2] = 0
\end{aligned} \tag{4.6.25}$$

for which the commutativity of $F^{(\pm)}$ with J^2 and J_3 is essential.

These redefined angular momentum operators have thus the same commutation relations as the old ones, but in addition all three of them commute with Γ.

We shall not follow this line any further, because these new operators cannot serve as generators of the rotation group, having been forced to commute with operators Γ which we supposed to be not invariant under rotations. Operators which generate the rotation group, must not, however, commute with operators which are non-invariant under rotations; thus I_1 and I_2 cannot be used when the rotation group is considered. We shall from now on forget this complication. The discussion aimed only to show that raising and lowering operators without any undesired side-effects can be constructed, whether J_1 and/or J_2 commute with Γ or not. Thus, in any case a complete set of eigenfunctions of Γ, J^2, J_3 can be built up (along the lines of what follows) even if the system in question is not invariant under rotations. In this case, however, these eigenfunctions do not represent stationary states. But even if the system is rotationally invariant, it need not be that all operators of the set Γ commute with $J_1 \pm i J_2$, i.e. that all observables γ used for labelling the states $|\gamma\lambda\mu\rangle$ are invariant under rotations. If they are not invariant, then the above unitary operators $F^{(\pm)}$ are different from 1. If they are invariant, we are left with $F = e^{i\alpha(\lambda,\mu)}$ and may then put $\alpha = 0$ to obtain $F \equiv 1$ (this is the usual choice of the arbitrary phase $\alpha(\lambda, \mu)$).

As we shall see, the angular momentum eigenstates exhibit most clearly what will happen under a rotation of the system. It is therefore convenient to use for further characterization of the states only such operators Γ which *are* invariant under rotation.

A basis $|\gamma jm\rangle$ belonging to a set Γ of observables which commute with J_1, J_2 and J_3 is called a 'standard basis' for the description of angular momentum. It is a basis of this kind which has to be taken when general rotations are considered. The unitary transformation relating a non-standard basis to a standard one can be found.

We therefore shall assume in the whole book that we have adopted a standard basis: having discussed what happens if this is not so, we put $F^{(\pm)} = 1$ in the following and omit γ altogether.

4.7 Orbital angular momentum

An important particular realization of the abstract algebra of angular momenta as laid down in (4.6.12) is furnished by the x-representation in the case of a scalar particle. The Schrödinger function of such a particle in the state $|\psi\rangle$ is

$$\psi(x) = \langle x|\psi\rangle \qquad (4.7.1)$$

and the angular momentum operators become (see (4.1.4))

$$J = L = x \times P = -i x \times \nabla = \text{orbital angular momenta}. \qquad (4.7.2)$$

We shall see that the eigenfunctions are

$$\psi_{\gamma lm}(x) = \langle x|\gamma lm\rangle = R_\gamma(r)Y_{lm}(\vartheta, \varphi) \tag{4.7.3}$$

where $x = r(\sin\vartheta\cos\varphi, \sin\vartheta\sin\varphi, \cos\vartheta)$, $Y_{lm}(\vartheta, \varphi)$ are the usual spherical harmonics and γ is the ensemble of all other quantum numbers.

4.7.1 Angular momentum operators in polar coordinates

We shall work in polar coordinates. We first calculate L in these coordinates. Figure 4.2 shows three unit vectors

(i)

$$e_1' = e_\vartheta$$
$$e_2' = e_\varphi$$
$$e_3' = e_r.$$

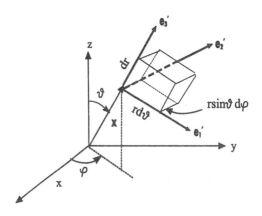

Figure 4.2. Polar coordinates.

For any function $F(r, \vartheta, \varphi)$ we have

(ii)

$$\mathrm{d}F = \frac{\partial F}{\partial\vartheta}\,\mathrm{d}\vartheta + \frac{\partial F}{\partial\varphi}\,\mathrm{d}\varphi + \frac{\partial F}{\partial r}\,\mathrm{d}r \equiv \nabla F \cdot \mathrm{d}x.$$

We choose as coordinate system the three axes e_1', e_2' and e_3' at x and find

(iii)

$$\mathrm{d}x' = (r\,\mathrm{d}\vartheta, r\sin\vartheta\,\mathrm{d}\varphi, \mathrm{d}r).$$

Hence, with (ii), x, ∇ and L become in this system

$$\nabla' = \left(\frac{1}{r}\frac{\partial}{\partial\vartheta}, \; \frac{1}{r\sin\vartheta}\frac{\partial}{\partial\varphi}, \; \frac{\partial}{\partial r} \right)$$

$$x' = (0, 0, r) \tag{4.7.4}$$

$$L' = -ix' \times \nabla' = \left(\frac{i}{\sin\vartheta}\frac{\partial}{\partial\varphi}, \; -i\frac{\partial}{\partial\vartheta}, \; 0 \right).$$

These components are valid in the coordinate system defined by e'_1, e'_2 and e'_3 in figure 4.2.

Next we transform to the x, y, z-coordinates. (4.7.4) gives the components in the e'_1, e'_2, e'_3-system. We obtain the components in the x, y, z-system by carrying out a passive rotation: we rotate the e'_1, e'_2, e'_3-frame first by $-\vartheta$ about e'_2 and then by $-\varphi$ about the new axis e''_3, which is the z-axis. We can copy down the matrix for this rotation from (3.3.5): it is

$$M = \begin{pmatrix} \cos\varphi & -\sin\varphi & 0 \\ \sin\varphi & \cos\varphi & 0 \\ 0 & 0 & 1 \end{pmatrix} \begin{pmatrix} \cos\vartheta & 0 & \sin\vartheta \\ 0 & 1 & 0 \\ -\sin\vartheta & 0 & \cos\vartheta \end{pmatrix}$$

$$= \begin{pmatrix} \cos\vartheta\cos\varphi & -\sin\varphi & \sin\vartheta\cos\varphi \\ \cos\vartheta\sin\varphi & \cos\varphi & \sin\vartheta\sin\varphi \\ -\sin\vartheta & 0 & \cos\vartheta \end{pmatrix}.$$

The x, y, z-components of x, L and ∇ expressed in r, ϑ and φ now follow by multiplying (4.7.4) by M; for L we obtain

$$L_x = i\left(\cot\vartheta\cos\varphi\frac{\partial}{\partial\varphi} + \sin\varphi\frac{\partial}{\partial\vartheta} \right)$$

$$L_y = i\left(\cot\vartheta\sin\varphi\frac{\partial}{\partial\varphi} - \cos\varphi\frac{\partial}{\partial\vartheta} \right) \tag{4.7.5}$$

$$L_z = -i\frac{\partial}{\partial\varphi}$$

and from this

$$L_\pm = L_x \pm iL_y = e^{\pm i\varphi}\left(i\cot\vartheta\frac{\partial}{\partial\varphi} \pm \frac{\partial}{\partial\vartheta} \right) \tag{4.7.6}$$

and, with the help of (4.6.4), after a short calculation

$$L^2 = L_z^2 - L_z + L_+L_- = -\left(\frac{1}{\sin^2\vartheta}\frac{\partial^2}{\partial\varphi^2} + \frac{1}{\sin\vartheta}\frac{\partial}{\partial\vartheta}\sin\vartheta\frac{\partial}{\partial\vartheta} \right). \tag{4.7.7}$$

4.7.2   Construction of the eigenfunctions

Since the operators L do not contain r, we can disregard the part $R_y(r)$ of the Schrödinger function: it simply drops out of all equations. We are then left with the eigenvalue equations (we call the eigenfunctions Y_{lm} but of course assume for the moment that we know nothing about them)

$$L^2 Y_{lm} = l(l+1)Y_{lm}$$
$$L_z Y_{lm} = m Y_{lm}$$ (4.7.8)

and with the raising and lowering equations

$$L_\pm Y_{lm} = \sqrt{l(l+1) - m(m \pm 1)} Y_{l,m\pm1}.$$ (4.7.9)

The Y_{lm} will be normalized:

$$\int Y_{lm}^*(\vartheta, \varphi) Y_{l'm'}(\vartheta, \varphi) \sin \vartheta \, d\vartheta \, d\varphi = \delta_{ll'} \delta_{mm'}.$$ (4.7.10)

That they are orthogonal for $l \neq l'$ or $m \neq m'$ follows from the general property that $\langle l'm'|lm \rangle = \delta_{ll'} \delta_{mm'}$. We have also seen that for l fixed m can take the values $-l \leq m \leq l$ and that the $2l+1$ functions are complete in the subspace of given l. Therefore (see 4.6.12)

$$\sum_{l=0}^{\infty} \sum_{m=-l}^{l} Y_{lm}^*(\vartheta'\varphi') Y_{lm}(\vartheta, \varphi) = \delta(\varphi' - \varphi) \delta(\cos \vartheta' - \cos \vartheta).$$ (4.7.11)

There are now two easy ways to obtain the Y_{lm} explicitly. One is to observe that the φ-dependence can be given at once:

$$L_z Y_{lm} = -i \frac{\partial}{\partial \varphi} Y_{lm} = m Y_{lm}$$ (4.7.12)

$$Y_{lm} = e^{im\varphi} f_{lm}(\cos \vartheta).$$ (4.7.13)

At this place most textbooks require that m (and therefore l) should be integer in order to make Y_{lm} unique. This argument is insufficient, because $|Y_{lm}|^2$ would still be unique even if Y_{lm} were double valued, and it is unnecessary as we shall see below. Hence for the time being we allow l to be half-integer as well as an integer number.

One now uses the fact that

$$L_+ Y_{ll} = 0$$ (4.7.14)

which gives with $Y_{ll} = e^{il\varphi} f_{ll}(\cos \vartheta)$ and (4.7.6)

$$L_+ Y_{ll} = e^{i(l+1)\varphi} \left(\frac{d}{d\vartheta} - l \cot \vartheta \right) f_{ll}(\cos \vartheta) = 0.$$ (4.7.15)

This differential equation has the simple solution

$$f_{ll}(\cos \vartheta) = c_l(\sin \vartheta)^l. \tag{4.7.16}$$

From here one obtains by repeated application of L_- all other f_{lm} for $-l \leq m \leq l$.

In order to carry out this programme, we must find $L_\pm^k$. Working out directly the kth power of $L_\pm$ from (4.7.6) leads to very ugly expressions. Fortunately, someone[3] found the way to a concise formula for $L_\pm^k$ which has become standard in the theory of spherical harmonics (and Legendre polynomials) closely related to the eigenfunctions of orbital angular momenta. We shall follow this way here although at first it may seem artificial.

As our eigenfunctions are of the form $Y_{l\mu} = e^{i\mu\varphi} f_{l\mu}(\cos \vartheta)$ (4.7.13) we must look to see what effect has $L_\pm$ on such a function. With

$$\frac{\partial}{\partial \vartheta} = -\sin \vartheta \frac{\partial}{\partial \cos \vartheta} \qquad -\cot \vartheta = \frac{d \sin \vartheta}{d \cos \vartheta}$$

and (4.7.6) we obtain, e.g.

$$L_+ e^{i\mu\varphi} f_{l\mu}(\cos \vartheta) = e^{i\varphi} \left[\mu \frac{d \sin \vartheta}{d \cos \vartheta} - \sin \vartheta \frac{d}{d \cos \vartheta} \right] e^{i\mu\varphi} f_{l\mu}(\cos \vartheta)$$

$$= -e^{i\varphi} \sin^{1+\mu} \vartheta \left[-\mu(\sin^{-\mu-1} \vartheta) \frac{d \sin \vartheta}{d \cos \vartheta} + \sin^{-\mu} \vartheta \frac{d}{d \cos \vartheta} \right] e^{i\mu\varphi} f_{l\mu}(\cos \vartheta)$$

$$= -e^{i\varphi} \sin^{1+\mu} \vartheta \frac{d}{d \cos \vartheta} \sin^{-\mu} \vartheta [e^{i\mu\varphi} f_{l\mu}(\cos \vartheta)].$$

Similarly for L_-. The net result is written as an operator acting on the eigenfunction $e^{i\mu\varphi} f_{l\mu}(\cos \vartheta)$; this operator has become, by the above manipulations, explicitly μ-dependent (the original $L_\pm$ is not)[4]

$$L_\pm(\mu) = \mp e^{\pm i\varphi} \sin^{1\pm\mu} \vartheta \frac{d}{d \cos \vartheta} \sin^{\mp\mu} \vartheta. \tag{4.7.17}$$

This explicit μ-dependence implies that this operator is a raising (lowering) operator *only if applied to an eigenfunction* $e^{i\mu\varphi} f_{l\mu}(\cos \vartheta)$ *with the same* μ. Consider L_+^2 applied to $e^{i\mu\varphi} f_{l\mu}(\cos \vartheta)$: if we use for L_+ the above $L_+(\mu)$, then the first application leaves us with an eigenfunction with the new eigenvalue $\mu + 1$ and the second application (if it is to be again a raising operation) must then be $L_+(\mu + 1)$. Generally $L_+^k e^{i\mu\varphi} f_{l\mu}(\cos \vartheta) = L_+(\mu + k - 1) \cdots L_+(\mu + 1) L_+(\mu) e^{i\mu\varphi} f_{l\mu}(\cos \vartheta)$. Exactly this circumstance has the lucky effect of producing a concise closed formula; one easily checks by working out

[3] Who is never quoted.
[4] We write one with, the other without, argument μ.

$L_\pm^2 = L_\pm(\mu+1)L_\pm(\mu)$ and proceeding up to $L_\pm^k$ that

$$
L_+^k\left[e^{i\mu\varphi}f_{l\mu}(\cos\vartheta)\right] = (-1)^k e^{ik\varphi}\sin^{k+\mu}\vartheta\,\frac{d^k}{d^k\cos\vartheta}\sin^{-\mu}\vartheta
$$
$$
\times\left[e^{i\mu\varphi}f_{l\mu}(\cos\vartheta)\right] \tag{4.7.18}
$$
$$
L_-^k\left[e^{i\mu\varphi}f_{l\mu}(\cos\vartheta)\right] = e^{-ik\varphi}\sin^{k-\mu}\vartheta\,\frac{d^k}{d^k\cos\vartheta}\sin^{\mu}\vartheta\left[e^{i\mu\varphi}f_{l\mu}(\cos\vartheta)\right].
$$

Note the $(-1)^k$ which appears only in the first equation. Note further that (4.7.18) makes sense only with the argument $e^{i\mu\varphi}f_{l\mu}(\cos\vartheta)$ (and not with any other argument-function $g(\vartheta,\varphi)$), because of the explicit μ-dependence of the differential operator on the r.h.s.

Having obtained the powerful tool (4.7.18) one can now start with $Y_{ll} = c_l\,e^{il\varphi}\sin^l\vartheta$ and work down (using (4.7.9) to obtain the correct normalization) the whole ladder until one arrives at $Y_{l,-l}$. Finally one fixes c_l such that the normalizing condition (4.7.10) is fulfilled and $Y_{l0}(0,0)$ is real and positive, which is the most commonly used phase convention (here we have anticipated that l is integer and that therefore $\mu = 0$ appears in the spectrum).

We leave it as an exercise to the reader to carry this through in detail, because we shall not follow this line but use below another, slightly different one, which emphasizes the close relation of the orbital momentum eigenfunction to the Legendre polynomials.

4.7.3   Orbital angular momenta have only integer eigenvalues

Before we proceed to calculate the Y_{lm}, we show that l must be an integer.

From the general formalism of the angular momentum it follows that

$$
L_-^{2l}Y_{ll} = \text{constant } Y_{l,-l}
$$

and then that

$$
L_-^{2l+n}Y_{ll} \equiv 0 \text{ for all } n \geq 1. \tag{4.7.19}
$$

We know, on the other hand, from (4.7.16), that

$$
Y_{ll} = \text{constant } e^{il\varphi}\sin^l\vartheta. \tag{4.7.20}
$$

Inserting (4.7.20) into (4.7.18), where we put $k = 2l+n$, we obtain on account of (4.7.19)

$$
L_-^{2l+n}Y_{ll} = \text{constant } e^{-i(l+n)\varphi}(\sin^{l+n}\vartheta)\frac{d^{2l+n}}{d^{2l+n}\cos\vartheta}(\sin^{2l}\vartheta) \equiv 0
$$

which requires that for all $n \geq 1$

$$
\frac{d^{2l+n}}{d^{2l+n}\cos\vartheta}(\sin^{2l}\vartheta) \equiv 0.
$$

Putting $\cos \vartheta = x$ we find

$$\frac{d^{2l+n}}{dx^{2l+n}}[(1-x^2)^l] \equiv 0 \text{ for all } n \geq 1.$$

This clearly requires l to be an integer. We thus have shown

> orbital angular momenta have only integer eigenvalues; therefore the Y_{lm} are single valued. (4.7.21)

Note that in this proof no assumptions were made which were not already contained in the general formalism of angular momentum (namely (4.7.8) and (4.7.9)) and in the definition of orbital angular momentum (namely (4.1.4)). These two have, roughly speaking, the following consequences.

(i) The general formalism requires $L_-^{2l+n} Y_{ll} \equiv 0$, for all $n \geq 1$.
(ii) The definition of orbital momentum requires L_- to be a differential operator; L_-^{2l+n} contains an integer number of differentiations, even if l is half-integer.
(iii) If an integer number of differentiations is to make a function identically zero, then this function must contain only a finite number of integer powers of its variable: it is a polynomial.

4.7.4 Spherical harmonics

We shall determine the Y_{lm} now in a way slightly different from that indicated in section 4.7.2; it is not simpler but exhibits from the beginning the relation to the Legendre polynomials $P_l(\cos \vartheta)$, which we shall suppose the reader to be familiar with. We observe that for $m = 0$ (4.7.13) reads

$$Y_{l0} = f_{l0}(\cos \vartheta). \tag{4.7.22}$$

Therefore, the operator L^2 (4.7.7) gives

$$L^2 f_{l0}(\cos \vartheta) = -\frac{1}{\sin \vartheta} \frac{d}{d\vartheta} \sin \vartheta \frac{d}{d\vartheta} f_{l0}(\cos \vartheta)$$

$$= l(l+1) f_{l0}(\cos \vartheta).$$

By means of $d/d\vartheta = -\sin \vartheta \, d/d\cos \vartheta$ this can be written

$$\left[\frac{d}{d\cos \vartheta} \sin^2 \vartheta \frac{d}{d\cos \vartheta} + l(l+1)\right] f_{l0}(\cos \vartheta) = 0. \tag{4.7.23}$$

This is the differential equation for the Legendre polynomials $P_l(\cos \vartheta)$; consequently we put

$$f_{l0}(\cos \vartheta) = C_l P_l(\cos \vartheta) \equiv C_l \frac{(-1)^l}{2^l \, l!} \left(\frac{d}{d\cos \vartheta}\right)^l (\sin \vartheta)^{2l}. \tag{4.7.24}$$

Normalization requires (4.7.10)

$$\int Y_{l0}^* Y_{l0} \sin \vartheta \, d\vartheta \, d\varphi = 2\pi |C_l|^2 \int_{-1}^{+1} P_l^2(\cos \vartheta) \, d\cos \vartheta = 1.$$

Putting $\cos \vartheta = \xi$, we obtain with (4.7.24)

$$\int_{-1}^{+1} P_l(\xi) P_{l'}(\xi) \, d\xi = \frac{(-1)^{l+l'}}{2^{l+l'} l! l'!} \int_{-1}^{+1} \frac{d^l}{d\xi^l}(1-\xi^2)^l \frac{d^{l'}}{d\xi^{l'}}(1-\xi^2)^{l'} \, d\xi.$$

Assume $l \geq l'$ and integrate l times by parts:

$$\int_{-1}^{+1} P_l(\xi) P_{l'}(\xi) \, d\xi = \frac{(-1)^{l+l'}(-1)^l}{2^{l+l'} l! l'!} \int_{-1}^{+1} (1-\xi^2)^l \frac{d^{l+l'}}{d\xi^{l+l'}}(1-\xi^2)^{l'} \, d\xi.$$

This is zero for $l > l'$, since $(1-\xi^2)^{l'}$ is a polynomial in ξ of order $2l'$ which is differentiated $l + l' > 2l'$ times. This proves orthogonality for $l \neq l'$ (although this proof is not necessary, P_l being an eigenfunction of a Hermitian operator). For $l = l'$, we obtain for the l.h.s. above the expression

$$\frac{(-1)^l}{2^{2l}(l!)^2} \int_{-1}^{+1} (1-\xi^2)^l \underbrace{\frac{d^{2l}}{d\xi^{2l}}(1-\xi^2)^l}_{(-1)^l \cdot (2l)!} \, d\xi = \frac{(2l)!}{2^{2l}(l!)^2} \int_{-1}^{+1} (1-\xi^2)^l \, d\xi.$$

Integrating l times by parts gives

$$\int_{-1}^{+1} (1-\xi^2)^l \, d\xi = \frac{2^l l!}{1 \cdot 3 \cdot 5 \cdots (2l-1)} \int_{-1}^{+1} \xi^{2l} \, d\xi = \frac{2}{(2l+1)} \frac{2^l l!}{1 \cdot 3 \cdot 5 \cdots (2l-1)}.$$

Hence finally

$$\int_{-1}^{+1} P_l^2(\xi) \, d\xi = \frac{2}{(2l+1)} \frac{2^l l!}{1 \cdot 3 \cdot 5 \cdots (2l-1)} \frac{(2l)!}{2^{2l}(l!)^2} = \frac{2}{2l+1}.$$

This gives

$$|C_l|^2 = \frac{2l+1}{4\pi}$$

and since $P_l(1) = 1$, we obtain $Y_{l0}(0,0) = +1$ by putting $|C_l| = \sqrt{(2l+1)/4\pi}$, hence

$$Y_{l0} = \sqrt{\frac{2l+1}{4\pi}} P_l(\cos \vartheta) = \sqrt{\frac{2l+1}{4\pi}} \frac{(-1)^l}{2^l l!} \left(\frac{d}{d\cos \vartheta}\right)^l \sin^{2l} \vartheta. \qquad (4.7.25)$$

Next we obtain from this by application of $L_\pm$ all the other eigenfunctions Y_{lm}. Since we have the factors $\sqrt{l(l+1) - m(m \pm 1)}$ in (4.7.9), the other eigenfunctions are automatically normalized.

If one writes

$$l(l+1) - \mu(\mu \pm 1) = (l \mp \mu)(l \pm \mu + 1)$$

one sees by going a few steps from $\mu = 0$ to $\mu = \pm m$, that

$$Y_{lm} = \sqrt{\frac{(l-m)!}{(l+m)!}} L_+^m Y_{l0}$$

$$Y_{l-m} = \sqrt{\frac{(l-m)!}{(l+m)!}} L_-^m Y_{l0}.$$

(4.7.26)

We now can apply (4.7.18), noting that (see (4.7.22)) $Y_{l0}(\vartheta, \varphi)$ is of the type $e^{i\mu\varphi} f(\cos \vartheta)$ with $\mu = 0$. Hence, combining with (4.7.25), (4.7.26) and (4.7.18) we obtain[5]

$$Y_{lm}(\vartheta, \varphi) = \sqrt{\frac{(l-m)!}{(l+m)!} \frac{(2l+1)}{4\pi}} e^{im\varphi} \left[\frac{(-1)^{l+m}}{2^l l!} \sin^m \vartheta \frac{d^{m+l}}{d^{m+l} \cos \theta} \sin^{2l} \vartheta \right]$$

$$Y_{l,-m}(\vartheta, \varphi) = \sqrt{\frac{(l-m)!}{(l+m)!} \frac{(2l+1)}{4\pi}} e^{-im\varphi} \frac{(-1)^l}{2^l l!} \sin^m \vartheta \frac{d^{m+l}}{d^{m+l} \cos \theta} \sin^{2l} \vartheta.$$

(4.7.27)

The function in square brackets is called the 'associated Legendre polynomial':

$$P_l^m(\cos \vartheta) = (-1)^m \sin^m \vartheta \frac{d^m}{d^m \cos \vartheta} P_l(\cos \vartheta)$$

$$= \frac{(-1)^{l+m}}{2^l l!} \sin^m \vartheta \frac{d^{m+l}}{d^{m+l} \cos \vartheta} \sin^{2l} \vartheta.$$

(4.7.28)

One sees immediately that

$$Y_{l,-m}(\vartheta, \varphi) = (-1)^m Y_{lm}^*(\vartheta\, \varphi)$$

$$P_l^{-m}(\cos \vartheta) = (-1)^m \frac{(l-m)!}{(l+m)!} P_l^m(\cos \vartheta) .$$

(4.7.29)

This result is a by-product of our procedure working from $m = 0$ in both directions to $m = -l$ and $m = +l$ with $L_\pm$ respectively. One can prove (4.7.29) directly by showing that

$$\sin^{-m} \vartheta \frac{d^{l-m}}{d^{l-m} \cos \vartheta} \sin^{2l} \vartheta = (-1)^m \frac{(l-m)!}{(l+m)!} \sin^m \vartheta \frac{d^{l+m}}{d^{l+m} \cos \vartheta} \sin^{2l} \vartheta.$$

---

[5] We assume m to be a positive integer and write explicit expressions for $-m$. Thereby we avoid the use of $|m|$, which appears in the formulae of some authors.

4.7.5   The phase convention

As we have stated earlier in section 4.6, the definition of the factor $\sqrt{j(j+1) - m(m \pm 1)}$ with the raising and lowering operators is not unique—an arbitrary phase factor $e^{i\alpha(j,m)}$ is left open. Choosing the positive square root without another $e^{i\alpha}$ makes a definite choice as far as the relation among the eigenfunctions $Y_{lm}(\vartheta, \varphi)$ for l fixed is concerned.

The last choice is then the common phase of the entire family Y_{lm} for l fixed. This choice is made by requiring $Y_{l0}(0, 0)$ to be real and positive, in fact according to (4.7.25)

$$Y_{l0}(0, 0) = \sqrt{\frac{2l+1}{4\pi}}. \qquad (4.7.30)$$

This particular choice of the phase of the $Y_{lm}(\vartheta, \varphi)$ is the most common one. It is, however, not the only one in use. Sometimes it is advantageous to use other phase conventions. As should be clear from our discussion this is of no influence on the physical content of the theory and will never show up in any result of calculation when this result refers to observable quantities.

We compare the choice of phase used here to that in a few frequently used texts, which the reader is likely to look up: our Y_{lm} have to be multiplied by $e^{i\alpha(l,m)}$ to obtain those of the listed references.

It should be mentioned that in some of the references the notation $Y_{lm}(\vartheta, \varphi)$ is not used. The difference in phase lies in that case in the definition of $P_l^m(\cos \vartheta)$—see (4.7.28). Also the normalization is not always the same, even if the same symbols are used. The best warranty against mistakes is to compare the definitions (4.7.29) and (4.7.28) to the corresponding ones in the other text one wants to use; these formulae are invariably given wherever extensive use of spherical harmonics is made.

4.7.6   Parity

Changing x to $-x$ is changing φ into $\varphi + \pi$ and ϑ into $\pi - \vartheta$. Hence from (4.7.27)

$$Y_{lm}(\pi - \vartheta, \varphi + \pi) = e^{im\pi}(-1)^{m+1}Y_{lm}(\vartheta, \varphi) = (-1)^l Y_{lm}(\vartheta, \varphi).$$

Eigenstates $|lm\rangle$ of orbital angular momentum have definite parity: $\langle -x|lm\rangle = (-1)^l \langle x|lm\rangle$.

The $P_l^m(\cos \vartheta)$ do not contain the factor $e^{im\varphi}$, hence their 'parity' is $(-1)^{l+m}$:

$$P_l^m(-\cos \vartheta) = (-1)^{l+m} P_l^m(\cos \vartheta). \qquad (4.7.31)$$

4.7.7   Particular cases

We list in table 4.2 the spherical harmonics for $l = 0, 1, 2, 3, 4$ and $m = 0, \ldots, l$. For $m < 0$ one uses (4.7.29).

Table 4.1. Phase factors $e^{i\alpha(l,m)}$ relating our definition of spherical harmonics Y_{lm} to that appearing in the listed references. In references denoted by * only P_{lm} are defined and we give the phase relating their definition to ours. In Handbuch der Physik (1957) the definition of P_{lm} is the same as ours but the definition of Y_{lm} differs.

$e^{i\alpha(l,m)}$	Reference
$(-1)^m$	Bethe and Salpeter (1957)
	Jahnke-Emde (1948)*
	Handbuch der Physik (1957)
	Madelung (1950)
	Morse and Feshbach (1953)
i^l	Landau and Lifschitz (1981)
$(-1)^{\frac{(m+|m|)}{2}}$	Bohm (1989)
$+1$	Blatt and Weisskopf (1959)
	Brink and Satchler (1968)
	Condon and Shortley (1953)
	Edmonds (1957)
	Galindo and Pascual (1990)
	Gasiorowicz (1996)
	Hamilton (1959)
	Källen (1964)
	Merzbacher (1961)
	Messiah (1970)
	Rose (1957)
	Schiff (1955)
	de-Shalit and Talmi (1963)
	Varshalovich *et al* (1988)
	Review of Particle Properties (1996)
	Abramowitz and Stegun (1970)*
	Gradshteyn and Ryzhik (1984)*
	Magnus and F. Oberhettinger (1949)

The spherical harmonics of order one are particularly interesting. This is seen by rewriting them in terms of x, y, z.

$$Y_{11} = -\sqrt{\frac{3}{8\pi}}(\sin\vartheta\cos\varphi + i\sin\vartheta\sin\vartheta) = \frac{1}{r}\sqrt{\frac{3}{4\pi}}\left[-\frac{1}{\sqrt{2}}(x+iy)\right]$$

$$Y_{10} = \sqrt{\frac{3}{4\pi}}\cos\vartheta = \frac{1}{r}\sqrt{\frac{3}{4\pi}}z$$

$$Y_{1-1} = -Y_{11}^* = \frac{1}{r}\sqrt{\frac{3}{4\pi}}\left[\frac{1}{\sqrt{2}}(x-iy)\right].$$

Table 4.2. Spherical harmonics for $l = 0, 1, 2, 3, 4$.

$l = 0$ $Y_{00} = \dfrac{1}{\sqrt{4\pi}}$

$l = 1$ $Y_{10} = \sqrt{\dfrac{3}{4\pi}} \cos \vartheta$

$\qquad\quad Y_{11} = -\sqrt{\dfrac{3}{8\pi}}\, e^{i\varphi} \sin \vartheta$

$l = 2$ $Y_{20} = \sqrt{\dfrac{5}{4\pi}} (\tfrac{3}{2} \cos^2 \vartheta - \tfrac{1}{2})$

$\qquad\quad Y_{21} = -\sqrt{\dfrac{15}{8\pi}}\, e^{i\varphi} \sin \vartheta \cos \vartheta$

$\qquad\quad Y_{22} = \dfrac{1}{4}\sqrt{\dfrac{15}{2\pi}}\, e^{2i\varphi} \sin^2 \vartheta$

$l = 3$ $Y_{30} = \sqrt{\dfrac{7}{4\pi}} (\tfrac{5}{2} \cos^2 \vartheta - \tfrac{3}{2} \cos \vartheta)$

$\qquad\quad Y_{31} = -\dfrac{1}{4}\sqrt{\dfrac{21}{4\pi}}\, e^{i\varphi} (5 \cos^2 \vartheta - 1) \sin \vartheta$

$\qquad\quad Y_{32} = \dfrac{1}{4}\sqrt{\dfrac{105}{2\pi}}\, e^{2i\varphi} \sin^2 \vartheta \cos^2 \vartheta$

$\qquad\quad Y_{33} = -\dfrac{1}{4}\sqrt{\dfrac{35}{4\pi}}\, e^{3i\varphi} \sin^3 \vartheta$

$l = 4$ $Y_{40} = \sqrt{\dfrac{9}{4\pi}} (\tfrac{35}{8} \cos^4 \vartheta - \tfrac{15}{4} \cos^2 \vartheta + \tfrac{3}{8})$

$\qquad\quad Y_{41} = -\dfrac{3}{4}\sqrt{\dfrac{5}{4\pi}}\, e^{i\varphi} (7 \cos^3 \vartheta - 3 \cos \vartheta) \sin \vartheta$

$\qquad\quad Y_{42} = \dfrac{3}{4}\sqrt{\dfrac{5}{8\pi}}\, e^{2i\varphi} (7 \cos^2 \vartheta - 1) \sin^2 \vartheta$

$\qquad\quad Y_{43} = -\dfrac{3}{4}\sqrt{\dfrac{35}{4\pi}}\, e^{3i\varphi} \cos \vartheta \sin^3 \vartheta$

$\qquad\quad Y_{44} = \dfrac{3}{8}\sqrt{\dfrac{35}{4\pi}}\, e^{4i\varphi} \sin^4 \vartheta$

That is, the components of the radius vector in the 'spherical basis'

$$r \equiv \begin{pmatrix} -\dfrac{1}{\sqrt{2}}(x + iy) \\ z \\ \dfrac{1}{\sqrt{2}}(x - iy) \end{pmatrix} \quad \text{transform as} \quad \begin{pmatrix} Y_{11} \\ Y_{10} \\ Y_{1,-1} \end{pmatrix} \qquad (4.7.32)$$

under rotations and reflections.

4.7.8 Further formulae

In chapter 6 the representation matrices $D^{(j)}$ of the rotation group will be shown to be related to spherical harmonics. Formulae given there yield many more formulae for spherical harmonics which we do not list here. An excellent collection of formulae and theorems is found in Magnus and F. Oberhettinger (1949) and in Varshalovich *et al* (1988); numerical values are given in Jahnke-Emde (1948) (see table 4.1).

4.8 Spin-$\frac{1}{2}$ eigenstates and operators

If, in (4.6.13), we put $j = \frac{1}{2}$, then $m = \pm\frac{1}{2}$ and the three matrices are immediately found to be

$$J_1 = \frac{1}{2}\begin{pmatrix} 0 & 1 \\ 1 & 0 \end{pmatrix} \quad J_2 = \frac{1}{2}\begin{pmatrix} 0 & -i \\ i & 0 \end{pmatrix} \quad J_3 = \frac{1}{2}\begin{pmatrix} 1 & 0 \\ 0 & -1 \end{pmatrix} \tag{4.8.1}$$

or

$$J = \tfrac{1}{2}\sigma \tag{4.8.2}$$

where

$$\sigma = (\sigma_1, \sigma_2, \sigma_3) = \left(\begin{pmatrix} 0 & 1 \\ 1 & 0 \end{pmatrix}, \begin{pmatrix} 0 & -i \\ i & 0 \end{pmatrix}, \begin{pmatrix} 1 & 0 \\ 0 & -1 \end{pmatrix} \right) \tag{4.8.3}$$

is Pauli's spin operator. We shall now write down the spin-$\frac{1}{2}$ operator for measuring the spin component in direction n and the corresponding eigenstates for spin $\frac{1}{2}$ in direction $\pm n$. We put

$$n = (n_1, n_2, n_3) = (\sin\vartheta\cos\varphi, \sin\vartheta\sin\varphi, \cos\vartheta). \tag{4.8.4}$$

According to (4.1.9) the component in direction n is 'measured' by the operator

$$J_n = n \cdot J = \tfrac{1}{2}n \cdot \sigma = \frac{1}{2}\begin{pmatrix} n_3 & n_1 - in_2 \\ n_1 + in_2 & -n_3 \end{pmatrix}$$
$$= \frac{1}{2}\begin{pmatrix} \cos\vartheta & e^{-i\varphi}\sin\vartheta \\ e^{i\varphi}\sin\vartheta & -\cos\vartheta \end{pmatrix}. \tag{4.8.5}$$

Similarly to what we did in section 3.3.3, we shall use the explicit form (4.8.5) to calculate $U_a(\eta) = \exp[-i\eta\, n \cdot J]$ for the spin-$\frac{1}{2}$ case.

We write $i\eta\, n \cdot J = \tfrac{1}{2}(i\eta\, 2n \cdot J)$ and observe from (4.8.5) that

$$(2n \cdot J)^2_{1/2} = 1.$$

Thus all even powers of $(2n \cdot J)_{1/2}$ are equal to unity and all odd powers are equal to $(2n \cdot J)_{1/2}$. The power series for the exponential then becomes

$$\exp\left[-\frac{i}{2}\eta\, 2n \cdot J\right]_{1/2} = \sum_{k=0}^{\infty} \frac{(-i\eta/2)^k}{k!} (2n \cdot J)_{1/2}^k$$

$$= 1 \cos\frac{\eta}{2} - i\begin{pmatrix} n_3 & n_1 - in_2 \\ n_1 + in_2 & -n_3 \end{pmatrix} \sin\frac{\eta}{2}. \qquad (4.8.6)$$

In particular

$$[U_a(\varphi e_3)]_{1/2} = [e^{(-i\varphi J_3)}]_{1/2} = \begin{pmatrix} e^{-i\varphi/2} & 0 \\ 0 & e^{i\varphi/2} \end{pmatrix}$$

$$[U_a(\beta e_2)]_{1/2} = [e^{(-i\beta J_2)}]_{1/2} = \begin{pmatrix} \cos\dfrac{\beta}{2} & -\sin\dfrac{\beta}{2} \\ \sin\dfrac{\beta}{2} & \cos\dfrac{\beta}{2} \end{pmatrix} \qquad (4.8.7)$$

which gives by combination the matrix for rotations by the three Euler angles α, β and γ

$$[U_a(\alpha, \beta, \gamma)]_{1/2} = [e^{-i\alpha J_3}e^{-i\beta J_2}e^{-i\gamma J_3}]_{1/2}$$

$$= \begin{pmatrix} e^{-i\alpha/2}\cos\dfrac{\beta}{2}e^{-i\gamma/2} & -e^{-i\alpha/2}\sin\dfrac{\beta}{2}e^{i\gamma/2} \\ e^{i\alpha/2}\sin\dfrac{\beta}{2}e^{-i\gamma/2} & e^{i\alpha/2}\cos\dfrac{\beta}{2}e^{i\gamma/2} \end{pmatrix}. \qquad (4.8.8)$$

The eigenstates of the spin-operators (called 'spinors') for the direction n are easily found from (4.8.1); we first choose an eigenstate of J_3, namely the state

$$|\tfrac{1}{2}, \tfrac{1}{2}\rangle = \begin{pmatrix} 1 \\ 0 \end{pmatrix}. \qquad (4.8.9)$$

This already fixes everything, because the state $|\tfrac{1}{2}, -\tfrac{1}{2}\rangle$ is now completely determined:

$$|\tfrac{1}{2}, -\tfrac{1}{2}\rangle = J_-|\tfrac{1}{2}, \tfrac{1}{2}\rangle = \begin{pmatrix} 0 \\ 1 \end{pmatrix} \qquad (4.8.10)$$

and the eigenstates of the spin operator in director n follow from the states $|\tfrac{1}{2}, \tfrac{1}{2}\rangle$ and $|\tfrac{1}{2}, -\tfrac{1}{2}\rangle$ respectively by rotating them by means of $U_a(\alpha, \beta, \gamma)$ (4.8.8). In doing that we find, however, that the eigenstates with spin up or down in a given direction n depend on three angles—namely the three Euler angles—whereas the direction n contains only two, namely ϑ and φ: $n = (\sin\vartheta \cos\varphi, \sin\vartheta \sin\varphi, \cos\vartheta)$. Indeed the 'states with spin up and down respectively in the $\varphi, \vartheta, \gamma$-orientation' are the following ones (see (4.8.8)):

$$|\tfrac{1}{2}, \tfrac{1}{2}\rangle_{\varphi,\vartheta,\gamma} = [e^{-i\varphi J_3} e^{-i\vartheta J_2} e^{-i\gamma J_3}]_{1/2} \begin{pmatrix} 1 \\ 0 \end{pmatrix}$$

$$= e^{-i\gamma/2} \begin{pmatrix} e^{-i\varphi/2} \cos(\vartheta/2) \\ e^{i\varphi/2} \sin(\vartheta/2) \end{pmatrix}$$

$$|\tfrac{1}{2}, -\tfrac{1}{2}\rangle_{\varphi,\vartheta,\gamma} = [e^{-i\varphi J_3} e^{-i\vartheta J_2} e^{-i\gamma J_3}]_{1/2} \begin{pmatrix} 0 \\ 1 \end{pmatrix}$$

$$= e^{i\gamma/2} \begin{pmatrix} -e^{-i\varphi/2} \sin(\vartheta/2) \\ e^{i\varphi/2} \cos(\vartheta/2) \end{pmatrix}.$$

(4.8.11)

    Usually one puts $\gamma = 0$, because the first rotation by γ only rotates the z-axis in itself and does not affect the statement 'spin up or down in the z-direction'. The second and third rotations (ϑ and φ) bring the space into the position such that a pointer in the z-direction, moving with the space, would now point in the n-direction, but we are not forced to put $\gamma = 0$ and we have to accept (4.8.11) as the general form. In fact, the two factors $e^{-i\gamma/2}$ and $e^{i\gamma/2}$ are nothing other than phase factors, which have to be tolerated, as only rays—not state vectors—correspond to physical states. The *relative* phase between the two states (4.8.11) cannot, however, be fixed arbitrarily, because the state $|\tfrac{1}{2}, \tfrac{1}{2}\rangle_{\varphi,\vartheta,\gamma}$ must be transformed into $|\tfrac{1}{2}, -\tfrac{1}{2}\rangle_{\varphi,\vartheta,\gamma}$ by the rotated lowering operator $J_-(\varphi, \vartheta, \gamma) = U_a(\varphi, \vartheta, \gamma) J_- U_a^{-1}(\varphi, \vartheta, \gamma)$ and that fixes the relative phase of the two states:

$$J_-(\varphi, \vartheta, \gamma)|\tfrac{1}{2}, \tfrac{1}{2}\rangle_{\varphi,\vartheta,\gamma} = U_a(\varphi, \vartheta, \gamma) J_- U_a^{-1}(\varphi, \vartheta, \gamma)$$

$$\times \left[U_a(\varphi, \vartheta, \gamma) \begin{pmatrix} 0 \\ 1 \end{pmatrix} \right]$$

$$= \left[U_a(\varphi, \vartheta, \gamma) \begin{pmatrix} 0 \\ 1 \end{pmatrix} \right] = |\tfrac{1}{2}, -\tfrac{1}{2}\rangle_{\varphi,\vartheta,\gamma}.$$

(4.8.12)

We add one more remark.

    One could think of defining only one state, say $|\tfrac{1}{2}, \tfrac{1}{2}\rangle_{\varphi,\vartheta,\gamma}$, by a rotation and then obtaining the corresponding spin-down state by reversing the direction of the axis n. Reversing $n \to -n$ is equivalent to the substitutions

$$\varphi \to \varphi + \pi$$
$$\vartheta \to \pi - \vartheta$$
$$\gamma \to \gamma', \text{ where } \gamma' \text{ remains currently undetermined.}$$

The substitutions imply the following changes of functions:

$$e^{\pm i(\varphi/2)} \rightarrow \pm e^{i(\pi/2)} e^{\pm i(\varphi/2)}$$
$$\cos(\vartheta/2) \rightarrow \sin(\vartheta/2) \text{ and } \sin(\vartheta/2) \rightarrow \cos(\vartheta/2)$$
$$e^{\pm i(\gamma/2)} \rightarrow e^{\pm i(\gamma'/2)}.$$

If we carry out these substitutions on (4.8.11) we obtain

$$|\tfrac{1}{2}, \tfrac{1}{2}\rangle_{\varphi+\pi,\pi-\vartheta,\gamma'} = e^{-i((\gamma+\gamma'-\pi)/2)} |\tfrac{1}{2}, -\tfrac{1}{2}\rangle_{\varphi,\vartheta,\gamma}.$$

We see that we indeed obtain the correct spin-down state, if we choose $\gamma + \gamma' = \pi \pmod{4\pi}$. This then determines $\gamma' = \pi - \gamma$; hence

$$|\tfrac{1}{2}, \tfrac{1}{2}\rangle_{\varphi+\pi,\pi-\vartheta,\pi-\gamma} = |\tfrac{1}{2}, -\tfrac{1}{2}\rangle_{\varphi,\vartheta,\gamma}.$$

And now comes the surprise: one might think that by exactly the same operation carried out on the spin-down state, one would obtain the spin-up state. Wrong! We obtain

$$|\tfrac{1}{2}, -\tfrac{1}{2}\rangle_{\varphi+\pi,\pi-\vartheta,\pi-\gamma} = |\tfrac{1}{2}, \tfrac{1}{2}\rangle_{\varphi+2\pi,\vartheta,\gamma} = -|\tfrac{1}{2}, \tfrac{1}{2}\rangle_{\varphi,\vartheta,\gamma}. \qquad (4.8.13)$$

There appears an extra minus sign (which is related to the double valuedness of the representation—see the next section). We therefore see that we cannot replace the use of the ladder operators $J_\pm$ by a prescription which only manipulates the angles, in other words, by rotations. Indeed, although the ladder operators act on *some* states *as if* they were rotations, they are basically different from rotations, as for instance $J_-|\tfrac{1}{2}, -\tfrac{1}{2}\rangle = 0$ shows: no rotation acting on any state can give zero. Explicitly one sees it clearly with the following example: the two matrices

$$[U_a(\pi e_2)]_{\frac{1}{2}} = \begin{pmatrix} 0 & -1 \\ 1 & 0 \end{pmatrix} \text{ and } [J_-]_{\frac{1}{2}} = \begin{pmatrix} 0 & 0 \\ 1 & 0 \end{pmatrix}$$

have the same effect on the state $\begin{pmatrix} 1 \\ 0 \end{pmatrix}$: they transform it into the state $\begin{pmatrix} 0 \\ 1 \end{pmatrix}$.

But on $\begin{pmatrix} 0 \\ 1 \end{pmatrix}$ they act differently: $U_a(\pi e_2)$ transforms it into $-\begin{pmatrix} 1 \\ 0 \end{pmatrix}$ whereas J_- annihilates it.

4.9 Double-valued representations; the covering group $SU(2)$

The rotation operators (3.3.17) and (4.8.6) or (4.8.8) are nothing other than representations of the rotation group, generally called $D^{(1)}$ and $D^{(\frac{1}{2})}$. There is one important difference between them, which is immediately seen by considering a rotation by 2π about an arbitrary direction: with $\eta = 2\pi$ (3.3.17) gives

$$\tilde{M}(2\pi) = 1$$

whereas (4.8.6) yields

$$\exp\left[-2\pi i n \cdot J\right]_{\frac{1}{2}} = -1. \tag{4.9.1}$$

This property is shared by all half-integer j representations. In fact these representations are not, in a global sense, representations of the three-dimensional rotation group (as the operators $\tilde{M}(\eta)$ of (3.3.17) indeed are), but they are representations of the group called $SU(2)$, namely the two-dimensional unitary group with determinant $+1$, of which (4.8.6) and (4.8.8) are the original representation defining the group. The fact that by using the Lie algebra of the rotation group we have obtained two types of representation, one which is single valued and one which is not, need not disturb us: we know that by exponentiating the Lie algebra of the group we obtain a matrix group, which is *locally*, but not necessarily *globally*, isomorphic to the group one starts from. As we concluded from (4.8.6) and (4.9.1), there exist in $SU(2)$ two elements, $+1$ and -1, which correspond to the unit element of the three-dimensional rotation group $0^+(3)$ ($0^+(3)$ means orthogonal three-dimensional with determinant $+1$); therefore these two elements ± 1 constitute a normal subgroup N of $SU(2)$ and the factor group $SU(2)/N$ is isomorphic to $0^+(3)$. That double-valued representations of the rotation group arise comes from the twofold connectedness of the parameter space: one sees easily when one characterizes a rotation by αn, namely as points inside a sphere of radius π, that there exist two kinds of closed curve, those lying entirely in the interior and those touching the surface; the first type can be continuously contracted to a point, the second only under certain conditions: when the curve reaches the surface, it corresponds to a rotation πn; however, $-\pi n$ denotes the same rotation and therefore we can draw a continuous curve, which starts somewhere in the interior, reaches the surface, jumps to the antipodal point and returns to the starting point. It is 'continuous' and closed, but cannot be contracted into one point. It is easy to see that each closed curve can be deformed into either one or the other type; indeed all curves with an even number (including zero) of jumps to the antipodal points can be contracted into one point; those with an odd number cannot. Therefore the rotation group is twofold connected. Each group element g of the rotation group can thus be reached in two essentially different ways by continuous paths from 1 to g: by a path on which an odd number of jumps occurs or by a path with an even number of jumps. If one considers a new group with elements which consist of a group element g of $0^+(3)$ *together with* an indication of how g was reached, then to each element g of $0^+(3)$ there correspond two elements of this new group because of the two classes of paths:

$$g : \begin{cases} g_0 & \text{for paths with an even number of jumps} \\ g_1 & \text{for paths with an odd number of jumps.} \end{cases} \tag{4.9.2}$$

The group with these elements is called the universal covering group of the rotation group. It is isomorphic to the group $SU(2)$ and the half-integer representations of $0^+(3)$ are single-valued representations of its covering group $SU(2)$. We may now alternatively consider the half-integer representations as

- single-valued representations of the covering group $SU(2)$
- double-valued representations of $0^+(3)$.

We shall choose the second alternative. It is clear then that we cannot throw away half of the representation matrices—we have to accept two, which simply differ in sign for each rotation. The possible occurrence of such *representations up to a factor* was already anticipated in our general discussion in chapter 2; see in particular the statement (2.2.7) and the following text, up to Wigner's theorem (2.2.11).

4.10   Construction of the general j, m-state from spin-$\frac{1}{2}$ states

In section 4.8 we found explicit formulae for the spin-$\frac{1}{2}$ states and their transformation properties under rotations. As $j = \frac{1}{2}$ is the smallest non-vanishing j-value, we expect that all $|jm\rangle$ states can be built up from spin-$\frac{1}{2}$ states or, in a perhaps better formulation, that for any given j and m it should be possible to form such a direct product of spin-$\frac{1}{2}$ states that the resultant state belongs to the subspace $\mathcal{H}_{jm}$. Such a $|jm\rangle$ state will then be a particular realization of the abstractly defined $|jm\rangle$ state, just as the spherical harmonics $Y_{lm}(\vartheta, \varphi)$ are a particular realization of the abstract $|lm\rangle$ states.

It is evident that such a realization, which (in contradistinction to the Y_{lm}) covers integer and half-integer j, will prove to be very useful because we know so much about the spin-$\frac{1}{2}$ states. It is indeed possible to build up the whole theory of angular momentum from the two states $|\frac{1}{2}, \frac{1}{2}\rangle$ and $|\frac{1}{2}, -\frac{1}{2}\rangle$, which in turn are either abstractly defined (Schwinger (1952)) or realized by analytic functions (Bargmann (1962))[6].

We shall consider now the construction of $|jm\rangle$ as a direct product of spin-$\frac{1}{2}$ states, because it seems to be up to now the only known way to construct the general $|jm\rangle$ state in a form in which its transformation properties are explicit.

We shall assume the orthonormal states $|\frac{1}{2}, \frac{1}{2}\rangle$ to be abstractly given; we know what happens when J_+, J_- or J_3 act on them; that is sufficient. We introduce the following notation:

$$u_+ \equiv |\tfrac{1}{2}, \tfrac{1}{2}\rangle$$

$$u_- \equiv |\tfrac{1}{2}, -\tfrac{1}{2}\rangle$$

(4.10.1)

---

[6] Both reprinted in Biedenharn and van Dam (1956), where also the Jordan–Schwinger construction (presented in our chapter 7) can be found.

and have with (4.8.2) and (4.8.3)

$$J_+ u_+ = 0$$

$$J_+ u_- = u_+$$

$$J_- u_+ = u_-$$

$$J_- u_- = 0 \tag{4.10.2}$$

$$J_3 u_+ = \tfrac{1}{2} u_+$$

$$J_3 u_- = -\tfrac{1}{2} u_-.$$

The behaviour under rotations is laid down in (4.8.8) but is currently not of interest.

Let us consider $2p$ systems of $j = \frac{1}{2}$, labelled by a superscript $(i) = (1), (2), \ldots, (2p)$ and combining into a single supersystem. A definite state of the combined system is given if each of the $2p$ subsystems is in a specified state, either $u_+^{(i)}$ or $u_-^{(i)}$. Hence

$$|(1, 2, \ldots, q)_+ (q + 1, \ldots, 2p)_-\rangle \equiv u_+^{(1)} \otimes \cdots \otimes u_+^{(q)} \otimes u_-^{(q+1)} \otimes \cdots \otimes u_-^{(2p)}$$
$$\tag{4.10.3}$$

is a definite normalized (because the u are normalized) state of the combined system, namely the one in which

- the systems with label $1, 2, \ldots, q$ are in a state u_+
- the systems with label $q + 1, \ldots, 2p$ are in a state u_-.

This state does not change if the labels of the u_+-systems are permuted among each other and/or if the labels of the u_--systems are permuted among each other; it does change, however, if any label is exchanged between a u_+-system and a u_--system: if, for example, the labels 1 and $q + 1$ are interchanged, then that means that in the new state system 1 has changed from spin up to spin down and system $q + 1$ has changed from spin down to spin up; this new state is orthogonal to the old one, because $\langle u_-^{(1)} | u_+^{(1)} \rangle = \langle u_+^{(q+1)} | u_-^{(q+1)} \rangle = 0$ and both these appear as a factor in

$$\langle (q + 1, 2, 3, \ldots, q)_+ (1, q + 2, \ldots, 2p)_- | (1, \ldots, q)_+ (q + 1, \ldots, 2p)_- \rangle.$$

Considering then all $(2p)!$ permutations of the $2p$ labels in (4.10.3) and keeping in mind that permutations inside the sets of $+$ labels and $-$ labels do not give a new state, we see that there exist

$$\frac{(2p)!}{q!(2p - q)!} = \binom{2p}{q}$$

different classes of normalized states, each containing $q!(2p - q)!$ states. States belonging to different classes are orthogonal to each other. A typical state is then

$|(i_1, \ldots, i_q)_+(i_{q+1}, \ldots, i_{2p})_-\rangle$, where the set $\{|(i_1, \ldots, i_q)_+(i_{q+1}, \ldots, i_{2p})_-\rangle\}$ is a permutation of the numbers $1, 2, \ldots, 2p$.

We now show that each of these states is an eigenstate of $J_3 = J_3^{(1)} + J_3^{(2)} + \cdots + J_3^{(2p)}$ (where this sum has to be understood in the sense of our discussion in section 4.3, in particular (4.3.16) and (4.3.18): namely, $J_3^{(n)}$ acts only on subsystem n and is the unit operator for all other subsystems). Hence

$$J_3|(i_1, \ldots, i_q)_+(i_{q+1}, \ldots, i_{2p})_-\rangle$$

$$= \left\{ \sum_n J_3^{(n)} \right\} |(i_1, \ldots, i_q)_+(i_{q+1}, \ldots, i_{2p})_-\rangle$$

$$= \tfrac{1}{2}[q - (2p - q)]|(i_1, \ldots, i_q)_+(i_{q+1}, \ldots, i_{2p})_-\rangle$$

$$= (q - p)|(i_1, \ldots, i_q)_+(i_{q+1}, \ldots, i_{2p})_-\rangle.$$

Putting $q - p = m$, we see that $|(i_1, \ldots, i_{p+m})_+(i_{p+m+1}, \ldots, i_{2p})_-\rangle$ is an eigenstate of J_3 with eigenvalue m; the same holds for all the $(2p)!$ states of this type.

Next we consider the effect of $J_+^{(n)}$ on such a state. As $J_+^{(n)}$ acts only on the subsystem n, it annihilates the state unless the label n occurs in the set $|(i_{p+m+1}, \ldots, i_{2p})_-\rangle$; if n occurs there, then this subsystem undergoes the change

$$J_+^{(n)} u_-^{(n)} = u_+^{(n)}.$$

In other words if $n \in (i_{p+m+1}, \ldots, i_{2p})_-$ then $J_+^{(n)}$ transfers it to the set $(i_1, \ldots, i_{p+m})_+$:

$$J_+^{(n)}|(i_1, \ldots, n, \ldots, i_{p+m})_+(i_{p+m+1}, \ldots, i_{2p})_-\rangle = 0$$

$$J_+^{(n)}|(i_1, \ldots, i_{p+m})_+(i_{p+m+1}, \ldots, n, \ldots, i_{2p})_-\rangle \qquad (4.10.4)$$

$$= |(i_1, \ldots, i_{p+m}n)_+(i_{p+m+1}, \ldots, \ldots, i_{2p})_-\rangle.$$

This shows that the resultant state is either zero or it is one of the states with eigenvalue $m + 1$ of J_3. As this is so for each $J_+^{(n)}, n = 1, \ldots, 2p$, it follows that the operator $J_+ = J_+^{(1)} + J_+^{(2)} + \ldots + J_+^{(2p)}$ acting on $|(i_1, \ldots, i_{p+m})_+(i_{p+m+1}, \ldots, i_{2p})_-\rangle$ produces a sum of $p - m$ different states with eigenvalue $m + 1$ of J_3. This suggests considering not a single such state, but rather the sum over all $(2p)!$ states generated from an arbitrary one by all possible permutations. Consequently we define the (not yet normalized) state

$$|(p, m)\rangle_S \equiv \frac{1}{2p!} \sum_P |(1, 2, \ldots, p+m)_+(p+m+1, \ldots, 2p)_-\rangle$$

$$\equiv (u_+^{p+m} u_-^{p-m})_\otimes. \qquad (4.10.5)$$

Here S stands for 'symmetrized', P for 'all permutations of the numbers $1, \ldots, 2p$' and the second line represents 'completely symmetrized direct product of $p + m$ states u_+ and $p - m$ states u_-'.

The normalization of that state is easily found by remembering that the terms of the sum group themselves into $(2p)!/[(p+m)!(p-m)!]$ mutually orthogonal classes, each containing $(p+m)!(p-m)!$ equal states. Thus, taking a representative of each class, we can write

$$|(p, m)\rangle_S = \frac{(p+m)!(p-m)!}{(2p)!} \sum_{\text{classes}} |\text{representative}\rangle.$$

All states in the sum over classes are now mutually orthogonal. Hence

$$\langle(p, m)|(p, m)\rangle_S = \left[\frac{(p+m)!(p-m)!}{(2p)!}\right]^2 \frac{(2p)!}{(p+m)!(p-m)!}$$

so that the normalized states are (the subscript N stands for normalization)

$$|(p, m)\rangle_{S.N} = \sqrt{\frac{(2p)!}{(p+m)!(p-m)!}}|(p, m)\rangle_S. \tag{4.10.6}$$

We now show that these normalized states are indeed the usual states $|jm\rangle$ with $j = p$. That is seen by considering the effect of J_+ on the state $(p, m)\rangle_S$. If $m = p$, then in (4.10.5) no u_- occurs and thus $J_+|(p, p)\rangle_S = 0$. Assume $m < p$. Then

$$J_+|(p, m)\rangle_S = \frac{1}{(2p)!} \sum_P J_+|(1, 2, \ldots, p+m)_+(p+m+1, \ldots, 2p)_-\rangle$$

$$= \frac{1}{(2p)!} \sum_P \sum_n J_+^{(n)}|(1, 2, \ldots, p+m)_+(p+m+1, \ldots, 2p)_-\rangle$$

$$= \frac{1}{(2p)!} \sum_P \sum_{n=p+m+1}^{2p} |(1, 2, \ldots, p+m, n)_+(p+m+1, \ldots, \ldots, 2p)_-\rangle$$

$$= \frac{1}{(2p)!} \sum_{n=p+m+1}^{2p} \sum_P |(1, 2, \ldots, p+m, n)_+(p+m+1, \ldots, \ldots, 2p)_-\rangle.$$

Here $\sum_P$ already contains all permutations, no matter what n is; hence

$$J_+|(p, m)\rangle_S = (p - m) \frac{1}{(2p)!} \sum_P |(1, 2, \ldots, p+m)_+(p+m+1, \ldots, 2p)_-\rangle.$$

Thus

$$J_+|(p, m)\rangle_S = (p - m)|(p, m+1)\rangle_S \tag{4.10.7}$$

and with the help of (4.10.6) we find for the properly normalized states

$$J_+|(p, m)\rangle_{S.N} = \sqrt{\frac{(2p)!}{(p + m)!(p - m)!}}$$

$$\times (p - m) \sqrt{\frac{(p + m + 1)!(p - m - 1)!}{(2p)!}}|(p, m + 1)\rangle_{S,N}$$

$$= \sqrt{(p + m + 1)(p - m)}|(p, m + 1)\rangle_{S,N}$$

$$= \sqrt{p(p + 1) - m(m + 1)}|(p, m + 1)\rangle_{S,N}.$$

Comparison with (4.6.12) shows that $|(p, m)\rangle_{S,N}$ is indeed a realization of the abstract state $|jm\rangle$ with $j = p$. It is also properly normalized. We summarize:

A properly normalized realization of the abstract $|jm\rangle$ state is given by the totally symmetrized direct product of spin-$\frac{1}{2}$ states u_+ and u_-:

$$|jm\rangle = \sqrt{\frac{(2j)!}{(j + m)!(j - m)!}}(u_+^{j+m}u_-^{j-m})_\otimes$$

$$(u_+^{j+m}u_-^{j-m})_\otimes \equiv \frac{1}{(2j)!}\sum_P |(1, 2, \ldots, j + m)_+(j + m + 1, \ldots, 2p)_-\rangle \quad (4.10.8)$$

$$= \frac{1}{(2j)!}\sum_P u_+^{(1)} \otimes \ldots u_+^{(j+m)} \otimes u_-^{(j+m+1)} \ldots \otimes u_-^{(2j)}.$$

It should perhaps be remarked that the present mathematical construction is independent of whether or not such state vectors occur as state vectors in a theory describing physical objects existing in nature; in fact the Pauli principle requires that a system of spin-$\frac{1}{2}$ particles is always in a completely antisymmetric state[7]—but this in no way prevents us from realizing the *mathematical object* $|jm\rangle$ by the above construction and using it to study explicitly its transformation properties under rotations. Note that in most cases the symmetrization is an unnecessary luxury and may be omitted (cancel $1/(2j)!$ and $\sum_P$). In chapter 7 on the Jordan–Schwinger construction, further relevant information can be found.

---

[7] With respect to all its variables, not only spin.

5

ADDITION OF ANGULAR MOMENTA

In this chapter we shall discuss the addition of angular momenta in the physicist's way; the relation of the present discussion to the group representations will become clear in the next chapters.

5.1 The general problem

We have seen in sections 4.2–4.4 that a system S may be thought of as being composed of several subsystems $S^{(i)}$. Each of these subsystems has its own angular momentum $J^{(i)}$; the total angular momentum is $J = \sum_{i=1}^{n} J^{(i)}$. It is important now that the $J^{(i)}$ of different subsystems commute and that for each $J^{(i)}$ as well as for J the commutation relations for the three components are the familiar ones. In this respect also the orbital and spin angular momenta L and S may be considered to belong to two 'subsystems'; namely we have shown in sections 4.1 and 4.2 that $J = L + S$, that L and S commute and that the components of L and of S fulfil the usual commutation rules. Thus for the following the nature of the $J^{(i)}$, which are summed up to yield J, is irrelevant: there may be orbital and spin parts in any combination.

In dividing up a system S into subsystems we arrive finally at an end when each subsystem is a single 'elementary object'. The meaning of 'elementary' is uniquely defined by convention; for instance in the Hamiltonian: to each pair of conjugate variables q_i, p_i one has to attach one 'elementary object'. As far as the mathematical structure is concerned, a macroscopic symmetric top may then be in this sense an 'elementary object'.

Let us now consider the expression (remember (4.3.16), (4.3.18) and (4.3.20) in section 4.3)

$$J = \sum_{i=1}^{n} J^{(i)} \tag{5.1.1}$$

for the total angular momentum (n is the number of 'elementary objects' plus the number of spins). It is at the moment irrelevant which—if any—of all the operators occurring on both sides of this equation commute with the Hamiltonian; this question enters only later.

From the commutation relations of the components of each $J^{(i)}$ and from the fact that different $J^{(i)}$ commute, it follows that the set

$$\{J^{(1)^2}, J_z^{(1)}; J^{(2)^2}, J_z^{(2)}; \ldots; J^{(n)^2}, J_z^{(n)}\} \tag{5.1.2}$$

is—as far as the angular momentum is concerned—a complete set of commuting observables. This is so because it leads to the most detailed description we can hope for: the angular momentum of each 'elementary object' of our system is described as fully as possible.

However, as (5.1.1) indicates, this is by no means the only possible description. J^2 and J_z are two commuting operators which belong to another such set. We cannot add these two to our old set (5.1.2). Namely, although $J_z = \sum_{i=1}^{n} J_z^{(i)}$ commutes with all operators of (5.1.2), it is not worthwhile to add it, being simply a linear combination of operators already there. J^2, on the other hand, does commute with each $J^{(l)2}$, but with no $J_z^{(l)}, \ldots$. Namely

$$J^2 = \sum_{l,l'} (J_x^{(l)} J_x^{(l')} + J_y^{(l)} J_y^{(l')} + J_z^{(l)} J_z^{(l')}) \tag{5.1.3}$$

gives after a short calculation

$$[J^2, J_z^{(k)}] = 2\mathrm{i}\left(J_x^{(k)} \sum_{j\neq k} J_y^{(j)} - J_y^{(k)} \sum_{j\neq k} J_x^{(j)} \right) \neq 0.$$

Consider now the other mentioned set which begins with J^2 and J_z. Which other operations can we add to these two? $J_z = \sum J_z^{(l)}$ commutes with every $J^{(k)2}$, since each $J_z^{(i)}$ does already. Likewise J^2 also commutes with every $J^{(k)2}$, since each single term of (5.1.3) commutes already. Hence

$$[J_z, J^{(k)^2}] = [J^2, J^{(k)^2}] = 0. \tag{5.1.4}$$

This shows that we can add to the second set of commuting observables, which begins with J^2 and $J_z = \sum J_z^{(i)}$, the squares of each single operator $J^{(i)^2}$. That gives a set of (so far) $n+2$ operators

$$\{J^2, J_z; J^{(1)^2}, J^{(2)^2}, \ldots, J^{(n)^2}\} \tag{5.1.5}$$

whereas (5.1.2) contained $2n$ operators. Therefore we expect that we might be able to complete this set to $2n$ commuting (independent) observables. We shall do this now in the most general way, leading to the construction of all such sets containing J^2 and J_z. We shall not, however, start with the set (5.1.5), but only with J^2 and J_z. The $J^{(1)^2}$ will be found automatically then.

5.2 Complete sets of mutually commuting (angular momentum) observables

The construction is easily performed by a graphical representation. Let us represent our system $S = \sum_{i=1}^{n} S^{(i)}$ by n little boxes in one row:

Undivided parts, like a or b, will be called 'connected ensembles'. We may subdivide our system at any place into

$$S = S^{(a)} + S^{(b)}.$$

Then

$$J = J_a = J_b.$$

We now know from (5.1.4) that J_a^2 and J_b^2 commute with J^2, J_z and among each other (since they belong to different parts of the system). This first division gives us (besides J^2 and J_z which we have from the beginning) two new operators for our set: J_a^2 and J_b^2.

Next we divide $S^{(a)}$ in an arbitrary way:

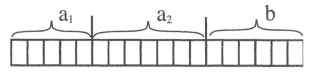

$$J = J_{a_1} + J_{a_2} + J_b.$$

Again we know from (5.1.4) that $J_{a_1}^2$, $J_{a_2}^2$ and J_a^2 commute with each other, with J_b^2 and with J^2 and J_z. Hence the two new operators $J_{a_1}^2$ and $J_{a_2}^2$ may be added to our set.

Proceeding in this way one sees that every new subdivision yields two new operators, which commute among themselves and with all those constructed so far. Each such subdivision will be marked in our graph by a new division line. We will invariably end up with the same final pattern, whatever sequence of subdivisions we choose, namely with this one:

In this final pattern are then all the possible $n-1$ division lines. A sequence of subdivisions ending up with this pattern cannot be continued and will be called a 'complete sequence of subdivisions'. It leads in fact to a complete set of commuting (angular momentum) observables. Namely, the $n-1$ subdivisions have yielded altogether $2(n-1)$ operators of the type J_α^2 (α indicating any connected ensemble) which all commute among each other and with J^2 and J_z. Including J^2 and J_z, we have just $2n$ commuting observables; this, as we know from the set (5.1.2), is the maximum number we should expect. By whichever sequence of subdivisions we proceed, we will necessarily find $J^{(1)^2}$, $J^{(2)^2}, \ldots, J^{(n)^2}$ (no one missing!) among our $2n$ commuting observables.

We have seen that by any sequence of subdivisions we obtain a set of $2n$ commuting observables and we have said that this number $2n$ is the one we expected for a 'complete set', but do we have a guarantee that we cannot add any more (angular momentum) observables which commute with the $2n$ already there and which at the same time are not simply functions of them? That is, can we show that there are at most $2n$ observables in each set? And furthermore, can we say when two sequences of subdivisions will lead to different sets of $2n$ observables? We shall deal with the last question first.

We call two sequences of subdivisions different, if they lead to two sets of commuting observables which are not identical.

It is clear that we may subdivide in different orders and yet obtain the same set of operators. Let a certain sequence of subdivisions be given. We show the consecutive divisions and give each new division line a number. This is on the left-hand column of the following graph; the right-hand column shows another sequence which leads to the same operators:

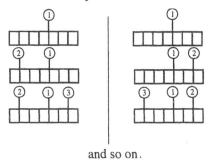

and so on.

As one sees, it does not matter *which* connected ensemble of blocks is next split into two; it does, however, matter *how* each such ensemble is split, i.e. *where* in a connected ensemble the next dividing line is drawn. Namely, let the following graph correspond to any connected ensemble A occurring in a sequence and let us subdivide it twice differently by one further dividing line (left- and right-hand side):

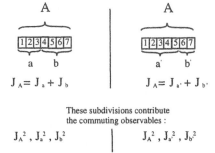

We see that they contribute different operators to our set of $2n$ commuting observables; thus these two sets are different. Furthermore, the two sets as a whole will no longer commute since, e.g., J_b^2 and $J_{a'}^2$ do not commute. This we shall show now. We draw the two different subdivisions in one graph, where we distinguish the two division lines (one pointing up, the other down):

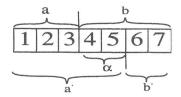

We see that a' and b have some boxes α in common, whereas a and b' have nothing in common. Therefore J_a^2 and $J_{b'}^2$ will commute, wherereas the boxes α will spoil the commutativity of $J_{a'}^2$ and J_b^2.

We write

$$J_{a'} = J_a + J_\alpha \qquad J_b = J_{b'} + J_\alpha$$
$$J_{a'}^2 = J_a^2 + J_\alpha^2 + 2J_a \cdot J_\alpha$$
$$J_b^2 = J_{b'}^2 + J_\alpha^2 + 2J_{b'} \cdot J_\alpha.$$

Evaluating the commutator $[J_{a'}^2, J_b^2]$ we can, step by step, eliminate operators which commute with what is on the other side of the comma. Operators sharing no common boxes commute. Thus

$$[J_{a'}^2, J_b^2] = [J_a^2 + J_\alpha^2 + 2J_a \cdot J_\alpha, J_b^2]$$
$$= [J_\alpha^2 + 2J_a \cdot J_\alpha, J_{b'}^2 + J_\alpha^2 + 2J_{b'} \cdot J_\alpha]$$
$$\left.\begin{array}{l} = [J_\alpha^2 + 2J_a \cdot J_\alpha, J_\alpha^2 + 2J_{b'} \cdot J_\alpha] \\ = 2[J_a \cdot J_\alpha, J_\alpha^2 + 2J_{b'} \cdot J_\alpha] \end{array}\right\} \text{(since } [J_\alpha^2, J_{\alpha,i}] = 0, i = x, y, z)$$
$$= 4 \sum_{j,k} J_{a,j} \underbrace{[J_{\alpha,j}, J_{\alpha,k}]}_{i J_{\alpha,l} \ (j,k,l \text{ cycl.})} J_{b',k}$$

(since J_α commutes with J_a and J_b). Hence if a and b' are different systems

$$\begin{aligned}[(J_a + J_\alpha)^2, (J_{b'} + J_\alpha)^2] = {}&4i\{J_{\alpha x}(J_{ay}J_{b'z} - J_{az}J_{b'y}) \\ &+ J_{\alpha y}(J_{az}J_{b'x} - J_{ax}J_{b'z}) \\ &+ J_{\alpha z}(J_{ax}J_{b'y} - J_{ay}J_{b'x})\} \\ &\neq 0.\end{aligned} \qquad (5.2.1)$$

We have seen that two complete sequences of subdivisions which lead to different sets of commuting observables lead in fact to two sets which do not commute with each other; i.e. it is always possible to find a pair of non-commuting observables such that one member of this pair is taken from the first and the other one from the second set.

This then also settles the question of completeness, the question being whether it is possible to enlarge (in a non-trivial way) such a set of $2n$ commuting observables. The answer is no; these sets are complete. The argument goes as follows: the equation below (5.1.3) says $[J^2, J_z^{(k)}] \neq 0$. The same will be true for $J_x^{(k)}$ and $J_y^{(k)}$. Hence no operator of this component type has a chance to enter into our system of $2n$ commuting operators (which contains J^2). The only type of operator which may possibly enter is of the square type J_β^2, where β denotes any subsystem. Now, *either* in a given set of $2n$ commuting operators this J_β^2 is already contained, so by adding it once more to the set nothing is gained; *or* this J_β^2 is not contained in the set; in that case the connected ensemble β of boxes does not occur in the complete sequence of subdivisions leading to the given set. Then there must be in this sequence at least one connected ensemble β' which spoils the occurrence of β by splitting β up and sharing some boxes with it (without being contained in or containing β); graphically

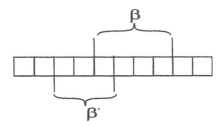

Then $[J_{\beta'}^2, J_\beta^2] \neq 0$ according to (5.2.1).

We have thus found the following result:

If a system S can be subdivided into n subsystems (n = number of 'elementary objects' + number of spins), then the total angular momentum operator is

$$J = \sum_{l=1}^{n} J^{(l)} \quad \text{(recall section 4.3)}.$$

There are different complete sets of commuting angular momentum operators available, of which one important one is the 'uncoupled set'

$$\{J^{(1)^2}, J_z^{(1)}; \ J^{(2)^2}, J_z^{(2)}; \ldots; \ J^{(n)^2}, J_z^{(n)}\}.$$

The other ones which are of importance are those which contain the square J^2 and the z-component J_z of the total angular momentum, and besides these two also all squares $J^{(i)^2} (i = 1, \ldots, n)$. They will contain a further $n-2$ squares $J_\alpha^2, J_\beta^2, \ldots$ of angular momenta of subsystems $S_\alpha, S_\beta, \ldots$ of S; the complete set has the form

$$\{J^2, J_z, J^{(1)^2}, J^{(2)^2}, \ldots, J^{(n)^2}; \ J_\alpha^2, J_\beta^2, \ldots\}.$$

Each of these sets is generated by a certain complete sequence of subdivisions of S and each of these sets will contain the same first $n+2$ operators $J^2, J_z, J^{(1)^2}, \ldots, J^{(n)^2}$; they differ from each other by the remaining ones J_α^2, $J_\beta^2, \ldots$. Each of these sets contains $2n$ observables and is—as far as angular momentum is concerned—complete.

All possible complete sequences of of subdivisions[1] of S will generate all possible complete sets of commuting (angular momentum) observables containing J^2 and J_z.

There will of course be other complete sets of commuting (angular momentum) operators besides those constructed above, but they cannot contain J^2 and J_z; therefore they are not of any importance. (One could for instance make any arbitrary (non-complete) subdivision $S = S_{a_1} + S_{a_2} + \ldots + S_{a_p}$ and then treat each S_{a_i} separately: each would be described by its own complete set of observables (either of the uncoupled type or containing $S_{a_i}^2, J_{a_i z}$) and all these sets may be unified into one set of again exactly $2n$ commuting operators which form certainly a complete set, although in general it is of no use.)

From our considerations it follows that there are several ways to construct complete systems of commuting angular momentum observables. Each of these will lead to a complete description of the states of the system by $2n$ quantum numbers; such a state will have the form (as far as angular momentum is concerned)

$$|\psi\rangle = |q_1, q_2, \ldots, q_{2n}\rangle$$

and the ensemble of these states $\{|q_1, \ldots, q_{2n}\rangle\}$ will form a complete basis in the angular momentum part of the Hilbert space.

Choosing another possible complete set (another 'coupling') of commuting angular momentum observables will lead to other states of the form

$$|\varphi\rangle = |r_1, r_2, \ldots, r_{2n}\rangle$$

where the r_i will in general have a different physical significance from the q_i above. The ensemble of these states $\{|r_1, r_2, \ldots, r_{2n}\rangle\}$ will form another complete basis in the angular momentum part of the Hilbert space.

Between such bases, belonging to two complete sets (two different 'couplings') of commuting angular momentum observables, there must exist a unitary transformation. We shall study these transformations in the next sections.

As we have seen, the complete sets of commuting observables are generated by complete sequences of subdivisions. As each single subdivision splits just one subsystem into two new ones, it is clear that the combination of two angular momenta is the basic operation; combination of more than two involves then only repeated applications of this basic operation.

For two angular momenta there are only the two complete sets

$$\{J^{(1)^2}, J_z^{(1)}; J^{(2)^2}, J_z^2\} \qquad \text{the 'uncoupled'}$$
$$\{J^{(1)^2}, J^{(2)^2}; J^2, J_z\} \qquad \text{the 'coupled'}.$$

[1] 'All possible complete sequences of subdivisions of S' means, of course, that we not only consider all $(n-1)!$ possible sequences of division lines, but also all $n!$ ways of associating n systems with n boxes. The resulting $n!(n-1)!$ complete sequences generate certainly all possible sets of commuting operators (containing J^2 and J_z), but these sets will not all be different from each other.

Studying the unitary transformation C between the corresponding states leads to the Clebsch–Gordan coefficients as their matrix elements. For more than two angular momenta there will be more sets and more unitary transformations between the different couplings. The matrix elements of these transformations are called 'recoupling coefficients'. They can be expressed by sums over products of Clebsch–Gordan coefficients; this is to be expected from the fact that repeated coupling of two angular momenta generates any desired coupling scheme. It also is to be expected that the transformations between different coupling schemes become rapidly more complicated with growing numbers of subsystems.

The recoupling coefficients for three angular momenta are the so-called Racah coefficients and/or $6-j$ symbols. A certain type of recoupling coefficient for four angular momenta is called the $9-j$ symbol, etc. The importance of all these transformations and different sets of observables arises from the fact that in general not all these sets commute with the total Hamiltonian. In a reaction (scattering, production) the initial and final states will consist of free particles (asymptotically). These are most frequently described and experimentally analysed by the set $\{J^{(1)^2}, J_z^{(1)}; J^{(2)^2}, J_z^2; \ldots; J^{(n)^2}, J_z^{(n)}\}$. This set will almost certainly not commute with the total Hamiltonian. The Hamiltonian will, however, commute with J^2 and J_z. Therefore a complete set of the type $\{J^2, J_z; J^{(1)^2}, \ldots\}$ might be adequate (but even this need not be, since sometimes H does not commute with $J^{(1)^2}, \ldots, J^{(n)^2}$). Then the S-matrix elements will be labelled by the eigenvalues of J^2 and others, the asymptotic states by those of $J^{(i)^2}$ and J_z^i—hence we must know the unitary transformation between these and perhaps other types of coupling. The $j-j$ and $L-S$ couplings are examples of importance in nuclear physics.

5.3   Combining two angular momenta; Clebsch–Gordan (Wigner) coefficients

5.3.1   Notation

There are many different notations in use and we are going to introduce a further one; it is, however, almost the same as the notation used by many authors.

The point is this: whichever set of observables

$$\{J^{(1)^2}, J_z^{(1)}; \ J^{(2)^2}, J_z^2\} \text{ or } \{J^{(1)^2}, J^{(2)^2}; \ J^2, J_z\} \qquad (5.3.1)$$

we might use, there are always just four quantum numbers. As long as the general state is written down, everything is clear. We might for instance write

$$|j_1, m_1, j_2, m_2\rangle \text{ and } |j_1, j_2, j, m\rangle \qquad (5.3.2)$$

for the states corresponding to the two sets, and everybody would know which states belong to which operators. Also the Clebsch–Gordan (or Wigner) coefficients, which are the scalar products between these states

$$\langle j_1, m_1, j_2, m_2 | j_1, j_2, j, m \rangle$$

would be quite obvious. The trouble is that one frequently uses these expressions with numbers or definite values replacing the symbols, and then what does

$$|1, 0, 1, 1\rangle \text{ or } |j_1 j_1 j_2 j_2\rangle$$

mean? These states can occur in both representations, but designate quite different things there.

We shall adopt the following rules:

(i)   of a pair jm belonging together, j is always written first;
(ii)   in the coupled representation $j_1 j_2$ come before jm and the coupling of j_1 and j_2 will be symbolized by parentheses around j_1 and j_2;
(iii)   in the uncoupled representation no parentheses occur and the sequence is $j_1 m_1 j_2 m_2$;
(iv)   we do not separate the quantum numbers by commas, except if the comma helps to clarify the meaning;
(v)   the Clebsch–Gordan coefficients (hence forward called CGCs) have the property that (see (5.3.8))

$$\langle j_1 m_1 j_2 m_2 | (j_1' j_2') jm \rangle = \delta_{j_1 j_1'} \delta_{j_2 j_2'} \delta_{m_1 + m_2, m}$$
$$\times \langle j_1 m_1 j_2 m_2 | (j_1 j_2) j, m_1 + m_2 \rangle. \tag{5.3.3}$$

Therefore several symbols are redundant. We shall omit $(j_1 j_2)$ from the coupled state appearing in the CGC (but only there). It will be understood that $(j_1 j_2)$ could always be added and, if so, then in exactly the same sequence as in the uncoupled state.

We shall write, according to these rules:

$$|(j_1 j_2) jm\rangle \qquad \text{in the coupled representation}$$
$$|j_1 m_1 j_2 m_2\rangle \qquad \text{in the uncoupled representation} \tag{5.3.4}$$
$$\langle j_1 m_1 j_2 m_2 | jm\rangle \quad \text{for the CGCs.}$$

As the reader might frequently encounter other notations, we collect some of them in table 5.1.

Let us, with a few words, indicate the link between this formal question of notation and the considerations in section 4.3 (direct product spaces) and section 4.5 (irreducible representations; Schur's lemma). It will be seen, then, that our notation expresses the mathematical content of what we mean by 'addition of two angular momenta'.

The state $|j_1 m_1 j_2 m_2\rangle$ is nothing other than the direct product state $|j_1 m_1\rangle \otimes |j_2 m_2\rangle$ and differs from it in notation only by omission of $\rangle \otimes |$; in

Table 5.1. The counterpart of our CGC $\langle j_1 m_1 j_2 m_2 | jm \rangle$ in the quoted references.

Blatt and Weisskopf (1959)	$C_{j_1 j_2}(jm; m_1 m_2)$
Condon and Shortley (1953)	$\langle j_1 j_2 m_1 m_2 | j_1 j_2 jm \rangle$ and $\langle j_1 j_2 m_1 m_2 | jm \rangle$
Edmonds (1957)	$\langle j_1 m_1 j_2 m_2 | j_1 j_2 jm \rangle$
Fano and Racah (1959)	$\langle jm | m_1 m_2 \rangle$ and $\langle j_1 j_2 jm | j_1 m_1 j_2 m_2 \rangle$
Galindo and Pascual (1990)	$\langle j_1 m_1 j_2 m_2 | (j_1 j_2) jm \rangle$ $= C(j_1 j_2 j; m_1 m_2 m)$
Hamilton (1959)	$\langle j_1 j_2 m_1 m_2 | jm \rangle$
Källen (1964)	$\langle j_1, m_1; j_2, m_2 | j, m \rangle$
Messiah (1970)	$\langle j_1 j_2 m_1 m_2 | jm \rangle$
Rose (1957)	$C(j_1 j_2 j; m_1 m_2 m)$ $C(j_1 j_2 j; m_1, m - m_1)$ $C(j_1 j_2 j_3; m_1 m_2 m_3)$
de-Shalit and Talmi (1963)	$a^{jm}_{m_1 m_2}$ and $\langle j_1 m_1 j_2 m_2 | j_1 j_2 jm \rangle$
Varshalovich et al (1988)	$C^{jm}_{j_1 m_1 j_2 m_2}$
Wigner (1959)	$s^{j_1 j_2}_{jm_1 m_2}$

particular the quantum numbers appear in the same order. The states $| j_1 m_1 j_2 m_2 \rangle$ thus span a certain invariant subspace, namely the product space $\mathcal{H}_{j_1 j_2} = \mathcal{H}_{j_1} \otimes \mathcal{H}_{j_2} = \mathcal{H}_{j_2 j_1}$. This space gives rise to a representation of the Lie algebra with operators (4.3.16):

$$J = J^{(j_1)} \otimes 1^{(j_2)} + 1^{(j_1)} \otimes J^{(j_2)} \tag{5.3.5}$$

and as J^2 commutes with each component of J, the representation is fully reducible (see (4.5.10)) and the invariant subspace $\mathcal{H}_{j_1 j_2}$ splits up into a direct sum of irreducible invariant subspaces:

$$\mathcal{H}_{j_1 j_2} = \mathcal{H}_{j_1} \otimes \mathcal{H}_{j_2} = \sum_j \oplus \mathcal{H}_j. \tag{5.3.6}$$

All this will be carried out in detail. The states spanning the irreducible subspaces $\mathcal{H}_j$ cannot simply be written $| jm \rangle$, because one has to indicate that they are subspaces of the particular subspace $\mathcal{H}_{j_1 j_2}$ (it cannot be assumed that the same quantum numbers j and m could not appear also in another product space

$\mathcal{H}_{j'_1 j'_2}$—in fact they generally do appear); that is why we should write these states as $|(j_1 j_2) jm\rangle$ which means a basis state of the invariant subspace $\mathcal{H}_j \in \mathcal{H}_{j_1 j_2}$.

Here we remark that the states $|(j_1 j_2) jm\rangle$ and $|(j_2 j_1) jm\rangle$ are not the same; the order in which the two angular momenta are coupled shows up in a phase factor which is fixed *by convention*. (See (5.3.25) and subsection 5.3.5, equation (5.3.33)).

Finally, the CGCs are written as the scalar products between these states; only the $(j_1 j_2)$ on the right has been omitted as j_1 and j_2 appear already on the left.

5.3.2 Definition and some properties of the Clebsch–Gordan coefficients

The two sets of commuting angular momentum operators

$$\{J^{(1)^2}, J_z^{(1)}, J^{(2)^2}, J_z^{(2)}\} \text{ and } \{J^{(1)^2}, J^{(2)^2}, J^2, J_z\} \qquad (5.3.7)$$

have the three operators $J^{(1)^2}$, $J^{(2)^2}$ and J_z in common (although in the uncoupled set $J_z^{(1)}$ and $J_z^{(2)}$ are even separately diagonal).

Consequently, the two sets of eigenstates

$$\{|j_1 m_1 j_2 m_2\rangle\} \quad \text{with } j_1, j_2 \text{ and } m_1 + m_2 = m \text{ fixed,}$$
$$m_1 \text{ or } m_2 \text{ free to vary}$$

and

$$\{|(j_1 j_2) jm\rangle\} \quad \text{with } j_1, j_2, m \text{ fixed, } j \text{ free to vary}$$

span the same subspace $\mathcal{H}_{j_1 j_2 m}$ of the Hilbert space $\mathcal{H}$. The corresponding projection operators are

$$P_{j_1 j_2 m} = \sum_{m_1 + m_2 = m} |j_1 m_1 j_2 m_2\rangle \langle j_1 m_1 j_2 m_2| = \sum_j |(j_1 j_2) jm\rangle \langle (j_1 j_2) jm|.$$

Leaving m free to vary, these sets span the subspace $\mathcal{H}_{j_1 j_2}$ of $\mathcal{H}$. It is inside $\mathcal{H}_{j_1 j_2 m}$ only that we have to consider the unitary transformations between the two bases: because $j_1 j_2 m$ are common quantum numbers of both sets of operators, it is not allowed that in the transformation

$$|(j_1 j_2) jm\rangle = \sum |j'_1 m_1 j'_2 m_2\rangle \langle j'_1 m_1 j'_2 m_2|(j_1 j_2) jm\rangle$$

any state $|j'_1 m_1 j'_2 m_2\rangle$ with $j'_1 \neq j_1$, $j'_2 \neq j_2$ and $m_1 + m_2 \neq m$ appears on the right-hand side. We thus conclude the first property of the expansion coefficients $\langle j_1 m_1 j_2 m_2|(j_1 j_2) jm\rangle$:

$$\langle j'_1 m_1 j'_2 m_2|(j_1 j_2) jm\rangle = 0 \qquad \text{unless } \begin{cases} j'_1 = j_1 \\ j'_2 = j_2 \\ m_1 + m_2 = m. \end{cases} \qquad (5.3.8)$$

Therefore, in the non-vanishing coefficients, j_1 and j_2 would appear twice; two of them and furthermore either m_1 or m_2 or m are redundant. As laid down in (5.3.4) we shall omit $(j_1 j_2)$ and simply write $\langle j_1 m_1 j_2 m_2 | jm \rangle$.

It follows from (5.3.8) that the transformation

$$|(j_1 j_2) jm \rangle = \sum_{m_1 + m_2 = m} |j_1 m_1 j_2 m_2 \rangle \langle j_1 m_1 j_2 m_2 | jm \rangle \qquad (5.3.9)$$

and its inverse

$$|j_1 m_1 j_2 m_2 \rangle = \sum_j |(j_1 j_2) jm \rangle \langle jm | j_1 m_1 j_2 m_2 \rangle \qquad (5.3.10)$$

involve only summations over m_1 and m_2 such that $m_1 + m_2 = m$ and over j (with limits to be specified later) respectively. The coefficients of these expansions are called Clebsch–Gordan or Wigner coefficients:

$$\langle j_1 m_1 j_2 m_2 | jm \rangle. \qquad (5.3.11)$$

We now prove that (j_1 and j_2 being given) j can vary only between $|j_1 - j_2|$ and $j_1 + j_2$.

(i) In any family of states $|jm\rangle$, m varies between, and indeed assumes, all values from $-j$ to $+j$ (integer steps). Hence, if we know the maximum possible m, then j cannot be larger; otherwise there would be also a larger m value. Consider now (5.3.9) and (5.3.10). The maximal values of m_1 and m_2 are j_1 and j_2 respectively; $m = m_1 + m_2$ implies $\max(m) = j_1 + j_2$. This then is also the maximum value of j.

(ii) As the possible m-values differ by integers, the same holds for the allowed j-values:

$$j_1 + j_2 \geq (j = j_1 + j_2 - k) \geq j_1 + j_2 - n; \quad k = 0, 1, \ldots, n$$

where n is to be determined. To each of these j belong $2j + 1$ states with $j \geq m \geq -j$, so that the total number of states $|(j_1 j_2) jm \rangle$ is

$$\sum_{k=0}^{n} [2(j_1 + j_2 - k) + 1] = (n + 1)(2 j_1 + 2 j_2 + 1 - n) = (2 j_1 + 1)(2 j_2 + 1)$$

where the last equation holds because the total number of states is the same in both the coupled and the uncoupled set. One sees easily that $n = 2 j_1$ and $n = 2 j_2$ are solutions. Hence

$$j_{\min} = j_1 + j_2 - n = \begin{cases} j_2 - j_1 \\ j_1 - j_2 \end{cases} \quad \text{whichever is } \geq 0.$$

Therefore $j_1 + j_2 \geq j \geq |j_1 - j_2|$. The consequence of this is that in the transformations (5.3.9) and (5.3.10) no j-value outside $|j_1 - j_2|, \ldots, j_1 + j_2$ can occur. The CGCs must automatically take care of that.

Combining this result and the one of (5.3.8), we state that

$$\langle j_1 m_1 j_2 m_2 | jm \rangle = 0$$

unless $m_1 + m_2 = m$, $|j_1 - j_2| \le j \le j_1 + j_2$

i.e. $\Delta(j_1 j_2 j) = 1$ (5.3.12)

and, of course, $|m| \le j$, $|m_1| \le j_1$, $|m_2| \le j_2$.

The Δ-symbol is shorthand for $|j_1 - j_2| \le j \le j_1 + j_2$. It says j_1, j_2 and j must be such that they could be the three sides of a triangle:

$$\Delta(a\ b\ c) = \begin{cases} 1 & \text{if } a, b, c \text{ can form a triangle} \\ 0 & \text{otherwise.} \end{cases}$$

This $\Delta(j_1 j_2 j)$ is always understood, although we shall not write it always.

5.3.3 Orthogonality of the Clebsch–Gordan coefficients

The CGCs are elements of a unitary matrix C; the condition $C^\dagger C = CC^\dagger = 1$ leads to certain orthogonality relations (or sum rules) if written in terms of the matrix elements $\langle j_1 m_1 j_2 m_2 | jm \rangle$.

The explicit form of these relations is most easily written down by means of the projection operator onto $\mathcal{H}_{j_1 j_2 m}$ which exists in two forms

$$P_{j_1 j_2 m} = \sum_{\underline{m_1} + m_2 = m} |j_1 \underline{m_1} j_2 m_2 \rangle \langle j_1 \underline{m_1} j_2 m_2 | \quad \text{(uncoupled representation)}$$

(5.3.13)

$$P_{j_1 j_2 m} = \sum_{\underline{j}} |(j_1 j_2) \underline{j} m \rangle \langle (j_1 j_2) \underline{j} m | \qquad \text{(coupled representation).}$$

We have underlined the summation quantum numbers to exhibit them more clearly (we shall use this device on other occasions too). That these two are identical has been discussed in section 5.3.2 above. Inside the subspace $\mathcal{H}_{j_1 j_2 m}$ these operators are just unit operators.

Now the orthogonality relations for the CGCs are immediately written down by introducing the 'unit operators' (5.3.13) into the orthonormality relation of the state vectors ($j_1 j_2 m$ fixed)

$$\langle (j_1 j_2) j' m' | (j_1 j_2) jm \rangle = \delta_{j'j} \delta_{m'm}$$
$$\langle j_1 m'_1 j_2 m'_2 | j_1 m_1 j_2 m_2 \rangle = \delta_{m'_1 m_1} \delta_{m'_2 m_2}.$$

We introduce the first P into the first and the second P into the second orthonormality relation and obtain immediately (omitting the symbol $(j_1 j_2)$ according to our convention (5.3.4)):

$$\sum_{\underline{m_1} + m_2 = m} \langle j'm' | j_1 \underline{m_1} j_2 m_2 \rangle \langle j_1 \underline{m_1} j_2 m_2 | jm \rangle = \delta_{j'j} \delta_{m'm} \Delta(j_1 j_2 j)$$

(5.3.14)

$$\sum_{\underline{j}} \langle j_1 m'_1 j_2 m'_2 | \underline{j} m \rangle \langle \underline{j} m | j_1 m_1 j_2 m_2 \rangle = \delta_{m'_1 m_1} \delta_{m'_2 m_2}$$

(summation variables underlined). If instead of $P_{j_1 j_2 m}$, we introduce the projection $P_{j_1 j_2}$ onto $\mathcal{H}_{j_1 j_2}$:

$$P_{j_1 j_2} = \sum_{m_1 m_2} |j_1 m_1 j_2 m_2\rangle \langle j_1 m_1 j_2 m_2| = \sum_m P_{j_1 j_2 m}$$

$$P_{j_1 j_2} = \sum_{jm} |(j_1 j_2) jm\rangle \langle (j_1 j_2) jm| = \sum_m P_{j_1 j_2 m}$$

(5.3.15)

into the orthogonality relation of the states, we obtain

$$\sum_{\underline{m_1} \, \underline{m_2}} \langle j'm' | j_1 \underline{m_1} j_2 \underline{m_2} \rangle \langle j_1 \underline{m_1} j_2 \underline{m_2} | jm \rangle = \delta_{j'j} \delta_{m'm} \Delta(j_1 j_2 j)$$

$$\sum_{\underline{j} \, \underline{m}} \langle j_1 m'_1 j_2 m'_2 | \underline{jm} \rangle \langle \underline{jm} | j_1 m_1 j_2 m_2 \rangle = \delta_{m'_1 m_1} \delta_{m'_2 m_2}.$$

(5.3.16)

5.3.4 Sketch of the calculation of the Clebsch–Gordan coefficients; phase convention and reality

We draw a figure in which for fixed $j_1 = 2$ and $j_2 = 4$ all 45 states are represented by dots in the $(m_1 m_2)$-plane and in the (jm)-plane for the uncoupled and the coupled set respectively. We observe with its help how the states of $\mathcal{H}_{j_1 j_2}$ are mapped from the uncoupled set $|j_1 m_1 j_2 m_2\rangle$ onto the coupled set $|(j_1 j_2) jm\rangle$ and vice versa.

In figure 5.1 we have indicated which points belong to the same subspace $\mathcal{H}_{j_1 j_2 m}$. Obviously, the unitary transformation C which maps $\mathcal{H}_{j_1 j_2 m}$ onto itself will represent any one given state (e.g. *one* of the circles $\odot$ in figure 5.1(b)) as a linear combination of *all* states of the corresponding subset (e.g. *all* circles $\odot$ in figure 5.1(a)).

We see immediately that there are two subspaces $\mathcal{H}_{j_1 j_2 m}$, namely with $m = j_1 + j_2$ (indicated by $\otimes$) and $m = -j_1 - j_2$ (indicated by $*$ (star)) which consist of one single state each. Here the unitary transformation C consists at most of multiplication by a complex number of modulus 1.

We see that the phase is arbitrary. The usual convention is then to fix it to be $+1$ for the $\otimes$-states. Hence the CGC is $+1$:

$$|(j_1 j_2), j_1 + j_2, j_1 + j_2\rangle = |j_1 j_1 j_2 j_2\rangle$$
$$\langle j_1 j_1 j_2 j_2 | j_1 + j_2, j_1 + j_2 \rangle = 1.$$

(5.3.17)

This convention, however, already fixes uniquely all those CGCs belonging to the states of the column below $\otimes$ in figure 5.1(b), for which $j = j_1 + j_2$. The calculation of these CGCs is achieved by the lowering operators

$$J_- = J_-^{(1)} + J_-^{(2)}$$

(5.3.18)

formed according to the rule $\boldsymbol{J} = \boldsymbol{J}^{(1)} + \boldsymbol{J}^{(2)}$ (see (5.1.1)).

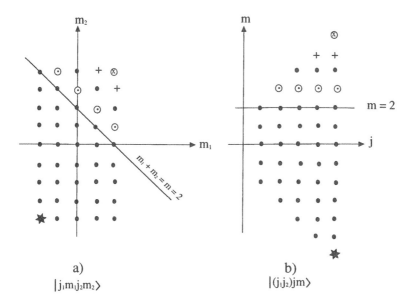

a)
$|j_1 m_1 j_2 m_2\rangle$

b)
$|(j_1 j_2) j m\rangle$

Figure 5.1. States belonging to the same subspace $\mathcal{H}_{j_1 j_2 m}$ in the two sets are drawn with the same symbol for $m = 6, 5, 4, 3, 2, -6$.

Applying J_- to $|(j_1 j_2) j_1 + j_2, j_1 + j_2\rangle$ repeatedly, we descend from this state step by step down the column below $\otimes$ in figure 5.1(b) until we arrive at $*$, namely $|(j_1 j_2) j_1 + j_2, -j_1 - j_2\rangle$.

Applying $J_- = J_-^{(1)} + J_-^{(2)}$ to $|j_1 j_1 j_2 j_2\rangle$ repeatedly, we generate those linear combinations of the states of figure 5.1(a) which make up the states $|(j_1 j_2) j_1 + j_2, m\rangle$.

Thus, applying J_- in figure 5.1(b) three times to $\otimes$ will give $\odot$; applying $J_-^{(1)} + J_-^{(2)}$ in figure 5.1(a) three times to $\otimes$ will give a linear combination of all the states $\odot$ on the diagonal line; in fact exactly that linear combination which is equal to $|(j_1 j_2) j, j - 3\rangle$.

According to $(4.6.12)^2$

$$J_-|(j_1 j_2) j m\rangle = \sqrt{j(j+1) - m(m-1)}|(j_1 j_2) j, m-1\rangle$$

$$J_-^{(1)}|j_1 m_1 j_2 m_2\rangle = \sqrt{j_1(j_1+1) - m_1(m_1-1)}|j_1, m_1 - 1, j_2 m_2\rangle \qquad (5.3.19)$$

$$J_-^{(2)}|j_1 m_1 j_2 m_2\rangle = \sqrt{j_2(j_2+1) - m_2(m_2-1)}|j_1 m_1 j_2, m_2 - 1\rangle.$$

Thus applying $J_-^{(1)} + J_-^{(2)}$ to any state $|j_1 m_1 j_2 m_2\rangle$ generates a superposition of

---

[2] The reader should recall that here $J^{(1)}$ really means $J^{(1)} \otimes 1^{(2)}$ etc. See section 4.3, in particular (4.3.16) and the discussion following it.

two states:

$$[J_-^{(1)} + J_-^{(2)}]|j_1 m_1 j_2 m_2\rangle = \sqrt{j_1(j_1+1) - m_1(m_1-1)}|j_1 m_1 - 1 j_2 m_2\rangle$$
$$+\sqrt{j_2(j_2+1) - m_2(m_2-1)}|j_1 m_1 j_2 m_2 - 1\rangle.$$
$$(5.3.20)$$

As all these superpositions involve real positive coefficients (the square roots in (5.3.20)) it is clear that by repeated application of J_- and of $J_-^{(1)} + J_-^{(2)}$ respectively only sums of products of these real, positive roots can occur. Thus, n-fold application of J_- to (5.3.17) leads to

$$J_-^{(n)}|(j_1 j_2), j_1 + j_2, j_1 + j_2\rangle = (J_-^{(1)} + J_-^{(2)})^n |j_1 j_1 j_2 j_2\rangle \qquad (5.3.21)$$

with real positive coefficients on both sides. On the l.h.s. n square roots multiply; writing $\sqrt{j(j+1) - m(m-1)} = \sqrt{(j+m)(j-m+1)}$, we obtain

$$J_-^{(n)}|(j_1 j_2)jj\rangle = \sqrt{\frac{(2j)!n!}{(2j-n)!}}|(j_1 j_2)j, j-n\rangle.$$

Putting $j - n = m$ yields

$$|(j_1 j_2)jm\rangle = \sqrt{\frac{(j+m)!}{(j-m)!(2j)!}} J_-^{j-m}|(j_1 j_2)jj\rangle. \qquad (5.3.22)$$

In the particular case (5.3.21) we must put $j = j_1 + j_2$ and obtain

$$|(j_1 j_2)j_1 + j_2, m\rangle = \sqrt{\frac{(j_1+j_2+m)!}{(j_1+j_2-m)!(2j_1+2j_2)!}}(J_-^{(1)} + J_-^{(2)})^{j_1+j_2-m}|j_1 j_1 j_2 j_2\rangle.$$
$$(5.3.23)$$

We could now work out the r.h.s. by means of the binomial expansion of $(J_-^{(1)} + J_-^{(2)})^{j_1+j_2-m}$ and (5.3.19). Finally we could compare the coefficients with those of the general formulae

$$|(j_1 j_2)j_1 + j_2, m\rangle = \sum_{m_1+m_2=m} |j_1 m_1 j_2 m_2\rangle\langle j_1 m_1 j_2 m_2 | j_1 + j_2, m\rangle$$

and thereby read off the CGC $\langle j_1 m_1 j_2 m_2 | j_1 + j_2, m\rangle$. We shall not do that now in detail. We only note that from the discussion between (5.3.20) and (5.3.21) it follows that for arbitrary m_1, m_2 and m

$$\langle j_1 m_1 j_2 m_2 | j_1 + j_2, m\rangle \text{ is real and } \geq 0. \qquad (5.3.24)$$

Now we return to figure 5.1. We have just (in principle) calculated CGCs belonging to the states $|(j_1 j_2)j_1 + j_2, m\rangle$ which make up the column farthest to the right in figure 5.1(b). We now do the same for the column next (left) to it. The state on top of it is $|(j_1 j_2), j_1 + j_2 - 1, j_1 + j_2 - 1\rangle$, indicated by $+$. There is another one, also indicated by $+$, which belongs to the $j = j_1 + j_2$ column which has already been settled. Both these $+$ states in figure 5.1(b) are linear combinations of the two $+$ states in figure 5.1(a); and since there are (except for an arbitrary phase) only two orthogonal linear combinations possible (of which one is already used), there remains only one linear combination for the top state $+$ of the second column. The phase, however, is arbitrary and we must again fix it by convention. Once this is done, all other states of the second column are found by repeated application of J_- to the highest state $+$.

Obviously, when we arrive at the top state of the next column in figure 5.1(b), two linear combinations of the three possible ones are already used up and the only remaining one is fixed (up to an arbitrary phase). Its phase will be agreed on by convention and then by repeated application of J_- all states of that column can be calculated.

This situation is found each time when we wish to determine the linear combination of a state on top of a column in figure 5.1(b), i.e. a state of the form $|j_1 j_2 jj\rangle$. All possible mutual orthogonal linear combinations, except one, have been used up, and for this last one we must only fix the phase; then the rest is uniquely determined.

We fix the phases for all the top states $(m = j)$ of figure 5.1(b) by the following phase convention:

In the linear combination

$$|(j_1 j_2)jj\rangle = \sum_{\underline{m_1} + m_2 = j} |j_1 \underline{m_1} j_2 m_2\rangle \langle j_1 \underline{m_1} j_2 m_2 | jj\rangle$$

(5.3.25)

the CGC with the highest $\underline{m_1}$ is real and ≥ 0:

$$\langle j_1 j_1 j_2, j - j_1 | jj\rangle \text{ real} \geq 0.$$

That is, in the linear combinations making up the top states of figure 5.1(b), the states of the rightmost column $(m_1 = j_1)$ of figure 5.1(a) always enter with a positive coefficient.

It is this phase convention which causes the two states $|(j_1 j_2)jm\rangle$ and $|(j_2 j_1)jm\rangle$ to differ by a phase factor. This will be seen in detail in subsection 5.3.5.

This convention—but not our notation!—is used by many authors (see table 5.1).

This convention and the phase convention for the ladder operators (laid down in (5.3.19)) fixes the CGC uniquely. In (5.3.19) only real square roots occur and our convention (5.3.24) does not imply other than real coefficients for the linear combinations making up the states $|j_1 j_2 jj\rangle$. Therefore all CGCs are real in this phase convention:

$$\langle j_1 m_1 j_2 m_2 | jm\rangle = \langle jm | j_1 m_1 j_2 m_2\rangle.$$

(5.3.26)

5.3.5   Calculation of $\langle j_1 m_1 j_2 j - m_1 | jj \rangle$

It is relatively easy to calculate explicitly the coefficient $\langle j_1 m_1 j_2 j - m_1 | jj \rangle$ of (5.3.25), from which the others belonging to the same j are found by application of J_- to the relevant states. We proceed as follows.

We apply $J_+ = J_+^{(1)} + J_+^{(2)}$ to the state $|j_1 j_2 jj\rangle$; it gives zero. Thus

$$J_+ |j_1 j_2 jj\rangle = J_+ \left[\sum_{m_1 + m_2 = j} |j_1 m_1 j_2 m_2\rangle \langle j_1 m_1 j_2 m_2 | jj\rangle \right]$$

$$= \sum_{m_1 + m_2 = j} \sqrt{j_1(j_1 + 1) - m_1(m_1 + 1)} |j_1 m_1 + 1 j_2 m_2\rangle \langle j_1 m_1 j_2 m_2 | jj\rangle$$

$$+ \sum_{m_1 + m_2 = j} \sqrt{j_2(j_2 + 1) - m_2(m_2 + 1)} |j_1 m_1 j_2 m_2 + 1\rangle \langle j_1 m_1 j_2 m_2 | jj\rangle$$

$$= 0.$$

The m are summation variables, therefore we may substitute in the first sum $m_1 - 1$ for m_1 and $m_2 + 1$ for m_2; after this substitution we can compare the coefficients term by term and obtain

$$\sqrt{j_1(j_1 + 1) - m_1(m_1 - 1)} \langle j_1 m_1 - 1 j_2 m_2 + 1 | jj\rangle$$

$$+ \sqrt{j_2(j_2 + 1) - m_2(m_2 + 1)} \langle j_1 m_1 j_2 m_2 | jj\rangle = 0$$

or

$$\langle j_1 m_1 - 1 j_2 m_2 + 1 | jj\rangle = -\sqrt{\frac{j_2(j_2 + 1) - m_2(m_2 + 1)}{j_1(j_1 + 1) - m_1(m_1 - 1)}} \langle j_1 m_1 j_2 m_2 | jj\rangle. \quad (5.3.27)$$

It is clear that by repeated application of this formula we can come from $\langle j_1 j_1 j_2 j - j_1 | jj\rangle$ to any other $\langle j_1 m j_2 j - m | jj\rangle$. For this procedure it is more convenient to rewrite the square root in (5.3.27) by using $j(j+1) - m(m \pm 1) = (j \pm m + 1)(j \mp m)$, hence

$$\langle j_1 m_1 - 1 j_2 m_2 + 1 | jj\rangle = -\sqrt{\frac{(j_2 + m_2 + 1)(j_2 - m_2)}{(j_1 - m_1 + 1)(j_1 + m_1)}} \langle j_1 m_1 j_2 m_2 | jj\rangle. \quad (5.3.28)$$

Applying this formula k times, we arrive at

$$\langle j_1, m_1 - k, \ j_2, m_2 + k | jj\rangle = (-1)^k$$

$$\times \sqrt{\frac{(j_2 + m_2 + k)!(j_2 - m_2)!(j_1 - m_1)!(j_1 + m_1 - k)!}{(j_2 + m_2)!(j_2 - m_2 - k)!(j_1 - m_1 + k)!(j_1 + m_1)!}} \langle j_1 m_1 j_2 m_2 | jj\rangle.$$

We now put $m_1 = j_1, m_2 = j - j_1, k = j_1 - m$ and obtain

$$
\langle j_1 m j_2, j - m | j j \rangle = \sqrt{\frac{(j_1 + j_2 - j)!}{(2j_1)!(j + j_2 - j_1)!}} \left[(-1)^{j_1 - m} \right.
$$

$$
\left. \times \sqrt{\frac{(j + j_2 - m)!(j_1 + m)!}{(j_2 - j + m)!(j_1 - m)!}} \right] \langle j_1 j_1 j_2 j - j_1 | j j \rangle.
$$
(5.3.29)

As—by convention—$\langle j_1 j_1 j_2 j - j_1 | j j \rangle \geq 0$, it follows that

$$
\langle j_1 m j_2, j - m | j j \rangle (-1)^{j_1 - m} \geq 0.
$$
(5.3.30)

Finally we find the value of $\langle j_1 j_1 j_2 j - j_1 | j j \rangle$ by using the orthogonality relation (5.3.14):

$$
\sum_m \langle j_1 m j_2 j - m | j j \rangle^2 = \frac{(j_1 + j_2 - j)!}{(2j_1)!(j + j_2 - j_1)!} \langle j_1 j_1 j_2 j - j_1 | j j \rangle^2
$$

$$
\times \sum_m \frac{(j_1 + m)!(j + j_2 - m)!}{(j_2 + j + m)!(j_1 - m)!} = 1.
$$

It remains to calculate the sum over m. This sum is a special case of the sum

$$
\sum_m \frac{(a + m)!(b - m)!}{(c + m)!(d - m)!}.
$$

This sum is calculated by starting from the addition theorem of binomial coefficients: we expand $(1 + x)^{p+q} = (1 + x)^p (1 + x)^q$ and find

$$
\sum_k \binom{p + q}{k} x^k = \sum_{\alpha, \beta} \binom{p}{\alpha} \binom{q}{\beta} x^{\alpha + \beta}
$$

$$
= \sum_k x^k \sum_r \binom{p}{r} \binom{q}{k - r}.
$$

Hence

$$
\sum_r \binom{p}{r} \binom{q}{k - r} = \binom{p + q}{k}.
$$
(5.3.31)

Further it follows from the definition of the binomial coefficients

$$
\binom{n}{k} = \frac{n(n - 1) \cdots (n - k + 1)}{1 \cdot 2 \cdots k}
$$

that for $n > 0$

$$
\binom{-n}{k} = (-1)^k \binom{n + k - 1}{k}.
$$
(5.3.32)

The sum which we wish to evaluate contains the summation variable m in the numerator and in the denominator, each time once positive, once negative. Thus, if in (5.3.31) we replace $p \to -p$ and $q \to -q$, we obtain a sum of the required type:

$$
\sum_r \binom{p + r - 1}{r} \binom{q + k - r - 1}{k - r} = \binom{p + q + k - 1}{k} \quad (p, q \geq 0)
$$

which, written in factorials, becomes

$$\sum_r \frac{(p-1+r)!(q+k-1-r)!}{r!(k-r)!} = (p-1)!(q-1)! \binom{p+q+k-1}{k}.$$

As r is a summation variable which goes over all values for which the factorials make sense, we may replace r by $c+m$ and sum over m:

$$\sum \frac{(c+p-1+m)!(q+k-c-1-m)!}{(c+m)!(k-c-m)!} = (p-1)!(q-1)! \binom{p+q+k-1}{k}.$$

Now we put $c+p-1=a$; $q+k-c-1=b$; $k-c=d$ and obtain the desired result

$$\sum_m \frac{(a+m)!(b-m)!}{(c+m)!(d-m)!} = \frac{(a+b+1)!(a-c)!(b-d)!}{(a+b+1-c-d)!(c+d)!}. \tag{5.3.33}$$

With the help of the formula just derived, (5.3.33), we obtain

$$\sum_m \frac{(j_1+m)!(j+j_2-m)!}{(j_2-j+m)!(j_1-m)!} = \frac{(j_1+j_2+j+1)!(j_1-j_2+j)!(-j_1+j_2+j)!}{(2j+1)!(j_1+j_2-j)!}$$

and finally (this CGC is by convention positive!)

$$\langle j_1 j_1 j_2, j - j_1 | jj \rangle = \sqrt{\frac{(2j_1)!(2j+1)!}{(j_1+j_2+j+1)!(j_1-j_2+j)!}} \tag{5.3.34}$$

from which by means of (5.3.29) all CGCs of the type $\langle j_1 m_1 j_2 j - m_1 | jj \rangle$ are found:

$$\langle j_1 m_1 j_2, j - m_1 | jj \rangle = (-1)^{j_1 - m_1} \sqrt{\frac{(j+j_2-m_1)!(j_1+m_1)!}{(j_2-j+m_1)!(j_1-m_1)!}} \tag{5.3.35}$$

$$\times \sqrt{\frac{(2j+1)!(j_1+j_2-j)!}{(j_1+j_2+j+1)!(j+j_2-j_1)!(j_1-j_2+j)!}}.$$

From this we could—following the procedure described above—obtain all CGCs by application of J_- to the relevant states. We shall not do this here, because the principle is obvious and the calculation tedious. We shall instead derive a closed formula for the CGCs in subsection 5.3.8 by a completely different method, thereby gaining a new aspect.

5.3.6 Obvious symmetry relations for CGCs

We shall now derive three symmetry relations for the CGCs, which seem rather plausible. We derive them here explicitly because physical intuition leads the way to them. In subsection 5.3.9 below we shall derive the full symmetry group of the CGCs and find the symmetries present among them—but in that derivation their simple geometric meaning does not become apparent. Here we discuss only the following symmetries.

(i)   The first symmetry relation is obtained by interchanging the order in which
        the angular momenta are coupled; that is, we consider the difference
        between $|(j_1 j_2) jm\rangle$ and $|(j_2 j_1) jm\rangle$. One might have expected the two
        CGCs $\langle j_1 m_1 j_2 m_2 | jm \rangle$ and $\langle j_2 m_2 j_1 m_1 | jm \rangle$ to be equal, but our phase
        convention (5.3.25) makes this impossible. This is seen immediately when
        we observe what happens to figure 5.1(a) if the order of the coupling is
        reversed. We display the effect in figure 5.2, where the horizontal axis
        counts the m-values of the system which is taken first, and the vertical axis
        those of the system which is taken as the second one.

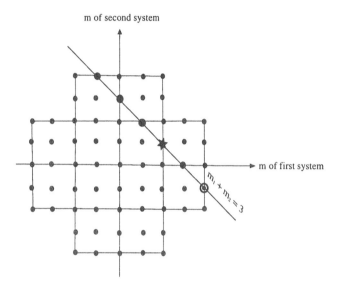

Figure 5.2. Each dot represents a direct product state $|j_1 m_1 j_2 m_2\rangle$. Each frame contains
one complete set of such states.

        The states on the line $m_1 + m_2 = 3$ combine into states $|jm = 3\rangle$; in
        particular: the four states of the upright box are needed to build up a state
        $|(j_1 j_2) jj\rangle$ with $j = 3$, and the four states of the flat box (they are the same
        states in another order) combine into a state $|(j_2 j_1) jj\rangle$. Now our phase
        convention (5.3.25) prescribes that in the first case the state

$$|*\rangle = |2241\rangle$$

enters with a positive coefficient, whereas in the second case

$$|\odot\rangle = |442, -1\rangle$$

is the state with a positive coefficient. These two states do not transform
into each other by reversing the order of coupling (except if $j_1 = j_2$).

Therefore the two states $|(j_1 j_2) jj\rangle$ and $|(j_2 j_1) jj\rangle$ will be different, because in $\mathcal{H}_{j_1 j_2} = \mathcal{H}_{j_2 j_1}$ there is for fixed j and m (inside the allowed region) essentially only one state $|jm\rangle$, of which we have fixed the phase (in an arbitrary but definite way) such that it depends on the order of coupling. This must not disturb us, as obviously a complete orthonormal system remains such if one multiplies each basis state by an arbitrary phase factor. We shall derive below the explicit form of the phase factor.

(ii) Next we consider the relation

$$|(j_1 j_2) jm\rangle = \sum_{m_1 + m_2 = m} |j_1 m_1 j_2 m_2\rangle \langle j_1 m_1 j_2 m_2 | jm\rangle$$

and replace m, m_1, m_2 by $-m, -m_1, -m_2$; then

$$|(j_1 j_2) j, -m\rangle = \sum_{-m_1 - m_2 = m} |j_1, -m_1, j_2, -m_2\rangle \langle j_1, -m_1, j_2, -m_2 | j, -m\rangle$$

which is again true. Now, any state $|jm\rangle$ may be transformed into a state $|j, -m\rangle$ by rotating the space by π about the y (or x)-axis. There will, however, be a change of the phase, because a rotation by π about *any* axis lying in the xy-plane will make $|jm\rangle \rightarrow |j, -m\rangle$ and somehow the new state should bear a label indicating about which axis the space was rotated; the phase of the state $|j, -m\rangle$ is the only place where such a label could be attached. Let then R_y be a rotation by π about the y-axis:

$$R_y |jm\rangle = \varphi_y(jm)|j, -m\rangle.$$

Thus our equation above with the negative m can be transformed into

$$\varphi_y(j, -m)|(j_1 j_2) jm\rangle = \sum \varphi_y(j_1, -m_1) \varphi_y(j_2, -m_2)$$
$$\times |j_1 m_1 j_2 m_2\rangle \langle j_1, -m_1 j_2, -m_2 | j, -m\rangle.$$

Comparing this with the above equation with positive m we see that

$$\varphi_y^*(j, -m)\varphi_y(j_1, -m_1)\varphi_y(j_2, -m_2)\langle j_1, -m_1, j_2, -m_2 | j, -m\rangle \qquad (5.3.36)$$
$$= \langle j_1 m_1 j_2 m_2 | jm\rangle.$$

The whole factor must be real; its exact form will be derived below.

(iii) Considering finally the vector (operator) equation $\boldsymbol{J} = \boldsymbol{J}^{(1)} + \boldsymbol{J}^{(2)}$, we see that we may interchange $\boldsymbol{J}$ with either $\boldsymbol{J}^{(1)}$ or $\boldsymbol{J}^{(2)}$ if the proper signs are taken. We thus might expect that e.g. $\langle j_1 m_1 j_2 m_2 | jm\rangle$ and $\langle j_1 m_1, j, -m | j_2, -m_2\rangle$ would be equal. Again this is not quite true. Of course, after the preceding discussion we are ready to expect a difference in sign. However, even the magnitudes will in general not be the same.

This is seen immediately from the orthogonality relation (5.3.14)

$$\sum_{\underline{m}} \left\{ \sum_{\underline{m_2}} \langle j\underline{m}|j_1 m_1 j_2\underline{m_2}\rangle\langle j_1 m_1 j_2\underline{m_2}|j\underline{m}\rangle \right\} = 2j+1$$

$$\sum_{\underline{m_2}} \left\{ \sum_{\underline{m}} \langle j_2, -\underline{m_2}|j_1 m_1, j, -\underline{m}\rangle\langle j_1 m_1, j, -\underline{m}|j_2, -\underline{m_2}\rangle \right\} = 2j_2+1$$

(5.3.37)

because in both cases the curly bracket is equal to unity. Since we sum over the squares of the CGCs, the possible different signs are eliminated and then both sums go over the same variables. If the two types of CGC were the same, then the sums would be equal—but they are not.

Dividing the upper sum by $2j+1$ and the lower by $2j_2+1$, we find a 1 on both right-hand sides. We might expect therefore that

$$\left|\frac{\langle j_1 m_1 j_2 m_2|jm\rangle}{\sqrt{2j+1}}\right| \quad \text{and} \quad \left|\frac{\langle j_1 m_1 j, -m|j_2, -m_2\rangle}{\sqrt{2j_2+1}}\right|$$

would be equal. That this is true will be shown below, where also the correct sign will be derived.

We turn now to the proof of these three symmetry relations; the reader who is content with the above plausibility arguments may skip these proofs and turn to the results, (5.3.52).

(i) We first prove that by interchanging the order of coupling of systems 1 and 2, we obtain

$$\langle j_2 m_2 j_1 m_1|jm\rangle = (-1)^{j_1+j_2-j}\langle j_1 m_1 j_2 m_2|jm\rangle.$$

Above, in our plausibility consideration we convinced ourselves that because of the phase convention the two CGCs might differ by a phase factor. To calculate this phase factor, we use (5.3.30) (which is a direct consequence of the phase convention (5.3.25)):

$$\langle j_1 m_1 j_2 m_2|jj\rangle\,(-1)^{j_1-m_1} \geq 0.$$

Had we interchanged the order of coupling, the same formula would read

$$\langle j_2 m_2 j_1 m_1|jj\rangle\,(-1)^{j_2-m_2} \geq 0.$$

The direct product states $|j_1 m_1 j_2 m_2\rangle$ and $|j_2 m_2 j_1 m_1\rangle$ in these expressions are the same; only the two states $|(j_1 j_2)jj\rangle$ and $|(j_2 j_1)jj\rangle$ can differ by a phase factor. Thus, as both expressions have equal magnitude and are both non-negative, they are equal. Thus

$$\langle j_1 m_1 j_2 m_2|jj\rangle = (-1)^{j_1+j_2-j}\langle j_2 m_2 j_1 m_1|jj\rangle$$

where $m_1 + m_2 = j$ has been used. With the same arguments (equality of the direct product states and at most a phase factor between the two coupled states) we can immediately draw the conclusion

$$|(j_2 j_1) jj\rangle = (-1)^{j_1+j_2-j}|(j_1 j_2) jj\rangle.$$

Now, by application of J_-, we can generate all $|jm\rangle$ states; the phase factor remains obviously untouched. Hence

$$|(j_2 j_1) jm\rangle = (-1)^{j_1+j_2-j}|(j_1 j_2) jm\rangle \qquad (5.3.38)$$

and from this it follows for the CGCs that

$$\langle j_2 m_2 j_1 m_1 | jm\rangle = (-1)^{j_1+j_2-j} \langle j_1 m_1 j_2 m_2 | jm\rangle. \qquad (5.3.39)$$

(ii) Next we calculate the change in sign which is caused by changing all m values into their negatives. According to (5.3.36) the new sign is

$$\varphi = \varphi_y^*(j, -m)\varphi_y(j_1, -m_1)\varphi_y(j_2, -m_2) \qquad (5.3.40)$$

where $\varphi_y(j, -m)$ is defined for any state $|jm\rangle$ by

$$R_y|jm\rangle = \varphi_y(j, m)|j - m\rangle$$
$$R_y|j - m\rangle = \varphi_y(j, -m)|jm\rangle$$

and R_y is a rotation by π about the y-axis.
For a spin-$\frac{1}{2}$ state we can write down the effect of a rotation R_y immediately by means of (4.8.7), where we put $\beta = \pi$ and obtain

$$[U_a(R_y)] = \begin{pmatrix} 0 & -1 \\ 1 & 0 \end{pmatrix}.$$

Writing $u_+ \equiv \begin{pmatrix} 1 \\ 0 \end{pmatrix}$ and $u_- \equiv \begin{pmatrix} 0 \\ 1 \end{pmatrix}$, we find then

$$R_y u_+ = u_- \qquad R_y u_- = -u_+. \qquad (5.3.41)$$

For arbitrary j and m we use the representation (4.10.8)

$$|jm\rangle = \sqrt{\frac{(2j)!}{(j+m)!(j-m)!}} \left(u_+^{j+m} u_-^{j-m}\right)_\otimes$$

and apply R_y to it (to $|jm\rangle$ on the left-hand side and to each u separately on the right-hand side); using (5.3.41)

$$R_y|jm\rangle = \sqrt{\frac{(2j)!}{(j+m)!(j-m)!}} \left[(R_y u_+)^{j+m}(R_y u_-)^{j-m}\right]_\otimes$$

$$= \sqrt{\frac{(2j)!}{(j+m)!(j-m)!}} (-1)^{j-m} \left(u_-^{j+m} u_+^{j-m}\right)_\otimes$$

or

$$R_y|jm\rangle = (-1)^{j-m}|j-m\rangle. \tag{5.3.42}$$

Thus, by definition, $\varphi_y(jm) = (-1)^{j-m}$, which is always real, because $j - m$ is always an integer. We now obtain with (5.3.40)

$$\varphi = (-1)^{j+m+j_1+m_1+j_2+m_2}.$$

Using $m_1 + m_2 = m$ and adding $j - j = 0$ in the exponent yields

$$\varphi = (-1)^{2j+2m+j_1+j_2-j} = (-1)^{j_1+j_2-j}$$

since $2j + 2m$ is always an even integer. The combined phase factors in (5.3.36) thus have the value $(-1)^{j_1+j_2-j}$ and (5.3.36) becomes

$$\langle j_1, -m_1, j_2, -m_2|j-m\rangle = (-1)^{j_1+j_2-j}\langle j_1 m_1 j_2 m_2|jm\rangle. \tag{5.3.43}$$

(iii) Finally we show what changes follow from interchanging any one of the coupled angular momenta with the resulting one. To this end we consider first the more symmetric problem of coupling three angular momenta $\boldsymbol{J}^{(1)} + \boldsymbol{J}^{(2)} + \boldsymbol{J}^{(3)} = 0$. We then have to calculate the state

$$|(j_1 j_2 j_3)00\rangle.$$

This is achieved by first coupling j_1 and j_2 to j

$$|(j_1 j_2)jm\rangle = \sum |j_1 m_1 j_2 m_2\rangle\langle j_1 m_1 j_2 m_2|jm\rangle \tag{5.3.44}$$

and thereafter coupling j and j_3 to $J = 0$

$$|((j_1 j_2)j j_3)00\rangle = \sum |(j_1 j_2)jm j_3 m_3\rangle\langle jm j_3 m_3|00\rangle. \tag{5.3.45}$$

Here the symbol $|(j_1 j_2)jm j_3 m_3\rangle$ is an abbreviation for the direct product state $|(j_1 j_2)jm\rangle\otimes|j_3 m_3\rangle$. Inserting here the expression (5.3.44) for $|(j_1 j_2)jm\rangle$ gives a sum over threefold direct product states $|j_1 m_1\rangle\otimes|j_2 m_2\rangle\otimes|j_3 m_3\rangle \equiv |j_1 m_1 j_2 m_2 j_3 m_3\rangle$. Hence the combination of (5.3.44) and (5.3.45) yields

$$|((j_1 j_2)j j_3)00\rangle = \sum |j_1 m_1 j_2 m_2 j_3 m_3\rangle\langle j_1 m_1 j_2 m_2|jm\rangle\langle jm j_3 m_3|00\rangle \tag{5.3.46}$$

where the sum goes over all m such that $m_1 + m_2 = m$ and $m_3 + m = 0$. The last CGC vanishes unless $j = j_3$ and $m = -m_3$. Its value follows immediately from (5.3.35):

$$\langle j_3, -m_3, j_3 m_3|00\rangle = \frac{(-1)^{j_3+m_3}}{\sqrt{2j_3 + 1}} \tag{5.3.47}$$

and this is the CGC to be inserted into (5.3.46), where of course jm is to be replaced by j_3 and $-m_3$. Hence we can write $|((j_1 j_2) j_3) \, 00\rangle$ instead of $|((j_1 j_2) j j_3) \, 00\rangle$ and find

$$|((j_1 j_2) j_3) \, 00\rangle = \sum |j_1 m_1 j_2 m_2 j_3 m_3\rangle \langle j_1 m_1 j_2 m_2 | j_3 - m_3 \rangle \frac{(-1)^{j_3 + m_3}}{\sqrt{2 j_3 + 1}} .$$

$$(5.3.48)$$

Now we couple the three states in a different order and remember that $|j_1 m_1 j_2 m_2 j_3 m_3\rangle$ is invariant under the permutation of the order of $1, 2, 3$. Hence

$$|((j_2 j_3) j_1) \, 00\rangle = \sum |j_1 m_1 j_2 m_2 j_3 m_3\rangle \langle j_2 m_2 j_3 m_3 | j_1 - m_1 \rangle \frac{(-1)^{j_1 + m_1}}{\sqrt{2 j_1 + 1}} .$$

$$(5.3.49)$$

We now claim that the two states $|((j_1 j_2) j_3) \, 00\rangle$ and $|((j_2 j_3) j_1) \, 00\rangle$ differ at most by a phase. Both states belong to the direct product space $\mathcal{H}_{j_1 j_2 j_3}$ spanned by the vectors $|j_1 m_1 j_2 m_2 j_3 m_3\rangle$. We must only show that there cannot exist more than one state $|J = 0, \ M = 0\rangle$ in $\mathcal{H}_{j_1 j_2 j_3}$. We couple j_1 and j_2 to j with the possible values $|j_1 - j_2| \le j \le j_1 + j_2$. Coupling any one of these intermediate j with j_3 will lead to a total J with the possible values $|j - j_3| \le J \le j + j_3$. The value $J = 0$ is present in only one of these sequences, namely in the sequence obtained with $j = j_3$. As j itself occurs only once in the sequence of possible j values, it follows that only one state $|J = 0, \ M = 0\rangle$ exists in $\mathcal{H}_{j_1 j_2 j_3}$. A different order of coupling can at most lead to another phase. (Note that the above conclusions hold only for $J = 0, \ M = 0$; for $J \ne 0$ there will in general be several independent states $|JM\rangle$ for given J and M, see section 5.4).

We thus obtain

$$|((j_2 j_3) j_1) \, 00\rangle = \varphi \, ((j_2 j_3) j_1) \, |((j_1 j_2) j_3) \, 00\rangle .$$

Comparing coefficients in (5.3.48) and (5.3.49) yields

$$\langle j_2 m_2 j_3 m_3 | j_1, -m_1 \rangle = \varphi(j_2 j_3 j_1) \, (-1)^{j_1 + j_3 + m_1 + m_3} \sqrt{\frac{2 j_1 + 1}{2 j_3 + 1}} \qquad (5.3.50)$$

$$\times \langle j_1 m_1 j_2 m_2 | j_3, -m_3 \rangle .$$

We determine $\varphi \, ((j_2 j_3) j_1)$ by making the CGC on the right-hand side positive (see (5.3.25)). We put

$$m_1 = j_1$$
$$m_3 = -j_3$$
$$m_2 = j_3 - j_1 .$$

Then, collecting the positive numbers on the r.h.s. into one symbol P we have

$$\langle j_2, j_3 - j_1, j_3, -j_3 | j_1, -j_1 \rangle = P \, \varphi \, ((j_2 j_3) j_1) \, (-1)^{2j_1}.$$

On the l.h.s. we change all m-values into their negatives and then exchange 2 with 3. A glance at (5.3.39) and (5.3.43) shows that both operations give rise to the same sign $(-1)^{j_2+j_3-j_1}$, whose square is one. But after these operations we have on the left-hand side $\langle j_3 j_3 j_2, j_1 - j_3 | j_1 j_1 \rangle$ which is positive (see (5.3.25)). Hence

$$\varphi(j_2 j_3 j_1) = (-1)^{2j_1} = (-1)^{-2j_1}.$$

If we insert that into (5.3.30) and remember $m_1 + m_3 = -m_2$, we obtain

$$\langle j_2 m_2 j_3 m_3 | j_1, -m_1 \rangle = (-1)^{j_3 - j_1 - m_2} \sqrt{\frac{2j_1 + 1}{2j_3 + 1}} \langle j_1 m_1 j_2 m_2 | j_3 - m_3 \rangle. \quad (5.3.51)$$

This formula gives us the result of the cyclic permutations; (5.3.39) adds to them the remaining three. Thus, with these two equations we can calculate any permutation of the arguments of the CGCs.

We collect the results of this section (the equations are numbered as they originally appeared):

$$\langle jm, j, -m | 00 \rangle = \frac{(-1)^{j-m}}{\sqrt{2j+1}} \qquad\qquad\qquad [5.3.47]$$

$$\langle j_2 m_2 j_1 m_1 | jm \rangle = (-1)^{j_1+j_2-j} \langle j_1 m_1 j_2 m_2 | jm \rangle \qquad [5.3.39]$$

$$|(j_2 j_1) jm \rangle = (-1)^{j_1+j_2-j} |(j_1 j_2) jm \rangle \qquad\qquad [5.3.38]$$

$$\langle j_2 m_2 jm | j_1 - m_1 \rangle = (-1)^{j-j_1-m_2} \sqrt{\frac{2j_1+1}{2j+1}} \langle j_1 m_1 j_2 m_2 | j - m \rangle \quad [5.3.51]$$

$$\langle j_1, -m_1, j_2, -m_2 | j - m \rangle = (-1)^{j_1+j_2-j} \langle j_1 m_1 j_2 m_2 | jm \rangle. \qquad [5.3.43]$$

$$(5.3.52)$$

From this, further relations can be derived by repeated application and specialization of the values. For instance, from the last relation we read off

$$\langle l_1 0 l_2 0 | l 0 \rangle = 0 \qquad \text{if } l_1 + l_2 + l \text{ is odd } (l_i \text{ integers}). \qquad (5.3.53)$$

5.3.7 Wigner's $3j$-symbol and Racah's $V(j_1 j_2 j_3 | m_1 m_2 m_3)$-symbol

Wigner introduced another symbol instead of the CGC, which shows more symmetry. It is defined by[3]

$$\langle j_1 m_1 j_2 m_2 | j_3, -m_3 \rangle = \sqrt{2j_3 + 1}(-1)^{j_1 - j_2 - m_3} \begin{pmatrix} j_1 & j_2 & j_3 \\ m_1 & m_2 & m_3 \end{pmatrix}. \quad (5.3.54)$$

---

[3] The reason for this definition is that (5.3.48) becomes simple in this new notation: it is now (5.3.62).

From the properties of CGCs it follows that

$$\begin{pmatrix} j_1 & j_2 & j_3 \\ m_1 & m_2 & m_3 \end{pmatrix} = 0 \qquad \text{unless } m_1 + m_2 + m_3 = 0; \ \Delta(j_1 j_2 j_3) = 0.$$

Introducing this into (5.3.51), we find

$$\sqrt{2j_1 + 1}(-1)^{j_2-j_3-m_1} \begin{pmatrix} j_2 & j_3 & j_1 \\ m_2 & m_3 & m_1 \end{pmatrix}$$

$$= \sqrt{\frac{2j_1+1}{2j_3+1}}(-1)^{j_3-j_1-m_2} \sqrt{2j_3+1}(-1)^{j_1-j_2-m_3} \begin{pmatrix} j_1 & j_2 & j_3 \\ m_1 & m_2 & m_3 \end{pmatrix}$$

$$= \sqrt{2j_1+1}(-1)^{j_3-j_2+m_1} \begin{pmatrix} j_1 & j_2 & j_3 \\ m_1 & m_2 & m_3 \end{pmatrix}.$$

Hence (because $j_3 - j_2 + m_1 = $ integer)

$$\begin{pmatrix} j_1 & j_2 & j_3 \\ m_1 & m_2 & m_3 \end{pmatrix} = \begin{pmatrix} j_2 & j_3 & j_1 \\ m_2 & m_3 & m_1 \end{pmatrix} = \begin{pmatrix} j_3 & j_1 & j_2 \\ m_3 & m_1 & m_2 \end{pmatrix}. \qquad (5.3.55)$$

It is a matter of a few minutes of simple algebra to derive from (5.3.39) and (5.3.43) the following further formulae:

$$\begin{pmatrix} j_2 & j_1 & j_3 \\ m_2 & m_1 & m_3 \end{pmatrix} = (-1)^{j_1+j_2+j_3} \begin{pmatrix} j_1 & j_2 & j_3 \\ m_1 & m_2 & m_3 \end{pmatrix}$$

$$= \begin{pmatrix} j_1 & j_2 & j_3 \\ -m_1 & -m_2 & -m_3 \end{pmatrix} \qquad (5.3.56)$$

where the first corresponds to (5.3.39) and the second to (5.3.43), whereas (5.3.17) and (5.3.24) become

$$\begin{pmatrix} j_1 & j_2 & (j_1 + j_2) \\ j_1 & j_2 & -(j_1 + j_2) \end{pmatrix} = \frac{(-1)^{2j_1}}{\sqrt{2(j_1 + j_2) + 1}}$$

$$(-1)^{j_1-j_2+j_3} \begin{pmatrix} j_1 & j_2 & j_3 \\ j_1 & m_2 & -j_3 \end{pmatrix} \geq 0. \qquad (5.3.57)$$

The orthogonality relations (5.3.16) appear now as

$$\sum_{\underline{m_1 m_2}} \begin{pmatrix} j_1 & j_2 & j_3 \\ \underline{m_1} & \underline{m_2} & m_3 \end{pmatrix} \begin{pmatrix} j_1 & j_2 & j_3' \\ \underline{m_1} & \underline{m_2} & m_3' \end{pmatrix} = \delta_{j_3 j_3'} \delta_{m_3 m_3'} \frac{\Delta(j_1 j_2 j_3)}{2j_3 + 1}$$

$$\sum_{\underline{j_3 m_3}} (2j_3 + 1) \begin{pmatrix} j_1 & j_2 & j_3 \\ m_1 & m_2 & \underline{m_3} \end{pmatrix} \begin{pmatrix} j_1 & j_2 & j_3 \\ m_1' & m_2' & \underline{m_3} \end{pmatrix} = \delta_{m_1 m_1'} \delta_{m_2 m_2'}. \qquad (5.3.58)$$

Summing the first one over m_3 after putting $m_3 = m_3'$ and $j_3 = j_3'$ yields

$$\sum_{\underline{m_1 m_2 m_3}} \begin{pmatrix} j_1 & j_2 & j_3 \\ \underline{m_1} & \underline{m_2} & \underline{m_3} \end{pmatrix} \begin{pmatrix} j_1 & j_2 & j_3 \\ \underline{m_1} & \underline{m_2} & \underline{m_3} \end{pmatrix} = 1. \qquad (5.3.59)$$

Note that the sums in (5.3.58) are in fact only sums over one variable and in (5.3.59) only over two variables, because $m_1 + m_2 + m_3 = 0$. Furthermore, by means of the symmetry properties (5.3.55) and (5.3.56), we may permute arbitrarily the columns of the $3j$-symbols in the sums of (5.3.58) without any consequence other than that the summation goes over another variable (after changing their names). Hence

$$
(2j_3 + 1) \sum_{\underline{m}_1 \underline{m}_2} \begin{pmatrix} j_1 & j_2 & j_3 \\ \underline{m}_1 & \underline{m}_2 & m_3 \end{pmatrix} \begin{pmatrix} j_1 & j_2 & j_3 \\ \underline{m}_1 & \underline{m}_2 & m_3' \end{pmatrix}
$$
$$
= (2j_1 + 1) \sum_{\underline{m}_2 \underline{m}_3} \begin{pmatrix} j_1 & j_2 & j_3 \\ m_1' & \underline{m}_2 & \underline{m}_3 \end{pmatrix} \begin{pmatrix} j_1 & j_2 & j_3 \\ m_1 & \underline{m}_2 & \underline{m}_3 \end{pmatrix} = \ldots
$$

$$
\sum_{\underline{j}_3, \underline{m}_3} (2j_3 + 1) \begin{pmatrix} j_1 & j_2 & \underline{j}_3 \\ m_1 & m_2 & \underline{m}_3 \end{pmatrix} \begin{pmatrix} j_1 & j_2 & \underline{j}_3 \\ m_1' & m_2' & \underline{m}_3 \end{pmatrix}
$$
$$
= \sum_{\underline{j}_2, \underline{m}_2} (2j_2 + 1) \begin{pmatrix} j_1 & \underline{j}_2 & j_3 \\ m_1 & \underline{m}_2 & m_3 \end{pmatrix} \begin{pmatrix} j_1 & \underline{j}_2 & j_3 \\ m_1{}' & \underline{m}_2 & m_3' \end{pmatrix} = \ldots .
$$

$$(5.3.60)$$

Racah has introduced still another symbol by

$$
V(j_1 j_2 j_3 | m_1 m_2 m_3) = (-1)^{j_1 - j_2 - j_3} \begin{pmatrix} j_1 & j_2 & j_3 \\ m_1 & m_2 & m_3 \end{pmatrix}
$$
$$
= \frac{(-1)^{j_3 - m_3}}{\sqrt{2j_3 + 1}} \langle j_1 m_1 j_2 m_2 | j_3 - m_3 \rangle.
$$

The properties of the V-coefficients are similar to those of Wigner's, but they are a little less symmetric.

5.3.8 Racah's formula for the CGCs

In this subsection we shall derive a closed formula for the CGCs; this formula is due to Racah, although he derived it differently. We shall use the Wigner notation ($3j$-symbol) for the CGCs and follow the usual custom of replacing the letters $j_1 j_2 j_3$ by abc and $m_1 m_2 m_3$ by $\alpha\beta\gamma$.

We start our consideration by remembering that any $|jm\rangle$ can be represented by a suitable direct product of

$$
u_+ = |\tfrac{1}{2}, \tfrac{1}{2}\rangle = \begin{pmatrix} 1 \\ 0 \end{pmatrix} \quad \text{and} \quad u_- = |\tfrac{1}{2}, -\tfrac{1}{2}\rangle = \begin{pmatrix} 0 \\ 1 \end{pmatrix}
$$

states. This was fully discussed in section 4.10. Thus, according to (4.10.8) we may build up three states

$$|a\alpha\rangle_u = \sqrt{\frac{(2a)!}{(a+\alpha)!(a-\alpha)!}} \left(u_+^{a+\alpha} u_-^{a-\alpha}\right)_\otimes$$

$$|b\beta\rangle_v = \sqrt{\frac{(2b)!}{(b+\beta)!(b-\beta)!}} \left(v_+^{b+\beta} v_-^{b-\beta}\right)_\otimes \qquad (5.3.61)$$

$$|c\gamma\rangle_w = \sqrt{\frac{(2c)!}{(c+\gamma)!(c-\gamma)!}} \left(w_+^{c+\gamma} w_-^{c-\gamma}\right)_\otimes$$

where u, v, w serve to distinguish three physical systems.

Next we remember (5.3.48), which allows us to couple three such states to a state with total angular momentum zero, i.e. *to a state which is invariant under rotations*; we rewrite (5.3.48) using the Wigner $3j$-notation and obtain

$$|\,((ab)c)\,00\rangle = (-1)^{a-b+c} \sum_{\alpha+\beta+\gamma=0} |(a\alpha)_u(b\beta)_v(c\gamma)_w\rangle \begin{pmatrix} a & b & c \\ \alpha & \beta & \gamma \end{pmatrix}. \quad (5.3.62)$$

If we insert (5.3.61) into (5.3.62), a sum over terms like

$$\left(u_+^{a+\alpha} u_-^{a-\alpha}\right)_\otimes \otimes \left(v_+^{b+\beta} v_-^{b-\beta}\right)_\otimes \otimes \left(w_+^{c+\gamma} w_-^{c-\gamma}\right)_\otimes \qquad (5.3.63)$$

results, and *this sum is invariant under rotations*.

The main idea is now to build up an invariant state from factors $u_\pm$, $v_\pm$ and $w_\pm$ in a different way and not containing CGCs, then to rearrange it such that it takes the form of the sum (5.3.62), (5.3.63) and compare coefficients. As the coefficients will contain the CGCs, we obtain in this way a formula for them.

How can we build up an invariant state? First we look back at (4.8.8) which describes the transformation properties of spinors. As one sees at once, this matrix has determinant unity (it is the original representation of $SU(2)$) and therefore leaves the 'determinant' $(u_+v_- - v_+u_-)$ and similar ones invariant[4]. These expressions are therefore also invariant under the corresponding rotations of the three-dimensional space. Hence we can build up invariant states by taking direct products of any number of determinants:

$$\begin{aligned} \delta_u &\equiv (v_+w_- - w_+v_-) \\ \delta_v &\equiv (w_+u_- - u_+w_-) \qquad (5.3.64) \\ \delta_w &\equiv (u_+v_- - v_+u_-). \end{aligned}$$

We only have to do it in such a way that the total number of factors u, v and w is equal to $2a$, $2b$ and $2c$ respectively—then this state is equal to (up to phase and normalization) the state $|\,((ab)c)\,00\rangle_{uvw}$.

---

[4] It is recommended that the reader checks this statement by an explicit calculation. Hint: use the standard $U = \begin{pmatrix} a & b \\ -b^* & a^* \end{pmatrix}$, with $aa^* + bb^* = 1$ for simplicity.

Thus we put

$$| ((ab)c) \, 00)_{uvw} = N(pqr) \left(\delta_u^p \delta_v^q \delta_w^r \right)_\otimes \qquad (5.3.65)$$

where the numbers p, q, r and the factor $N(pqr)$ which contains both normalization and phase will be determined later.

We use the binomial expansion for the powers of the determinants and collect all $u_\pm$, all $v_\pm$ and all $w_\pm$ together. The result is

$$N(pqr) \left(\delta_u^p \delta_v^q \delta_w^r \right)_\otimes = N(pqr) \sum_{p'q'r'} (-1)^{p'+q'+r'} \begin{pmatrix} p \\ p' \end{pmatrix} \begin{pmatrix} q \\ q' \end{pmatrix} \begin{pmatrix} r \\ r' \end{pmatrix}$$
$$\times \left[(u_+^{r+q'-r'} u_-^{q+r'-q'})_\otimes \otimes (v_+^{p+r'-p'} v_-^{r+p'-r'})_\otimes \otimes (w_+^{q+p'-q'} w_-^{p+q'-p'})_\otimes \right].$$
$$(5.3.66)$$

We now only have to compare exponents and to insert some normalization factors in order to arrive at a sum of type (5.3.62), (5.3.63). In fact, considering (5.3.61) we see that the sum of the two exponents must equal $2j$ ($= 2a, 2b, 2c$) and their difference $2m$ ($= 2\alpha, 2\beta, 2\gamma$). We thus replace, e.g., according to (5.3.61)

$$(u_+^{r+q'-r'} u_-^{q+r'-q'})_\otimes = \sqrt{\frac{(r+q'-r')!(q+r'-q')!}{(r+q)!}}$$
$$\times \left| \frac{r+q}{2}, \frac{1}{2}(r-q+2q'-2r') \right\rangle_u$$

where now $\frac{1}{2}(r+q) = a$ and $\frac{1}{2}(r-q+2q'-2r') = \alpha$ will do what we want. We thus have to put

$$
\begin{array}{ll}
r + q = 2a & r - q + 2q' - 2r' = 2\alpha \\
r + p = 2b & p - r + 2r' - 2p' = 2\beta \qquad (5.3.67) \\
p + q = 2c & q - p + 2p' - 2q' = 2\gamma.
\end{array}
$$

The three equations for p, q, r are solved at once:

$$
\begin{array}{l}
p = -a + b + c \\
q = a - b + c \qquad (5.3.68) \\
r = a + b - c
\end{array}
$$

but we cannot solve the other three for $p'q'r'$ because the three equations are not independent: their sum gives $0 = 0$, as it should ($\alpha + \beta + \gamma$ must be zero!). Since we wish, however, to change from the summation variables p', q', r' to α, β, γ which appear in the sum (5.3.62), we have to keep one of the three—say r'—and solve for the other two. We call r' from now on ρ, and obtain from (5.3.67)

$$
\begin{array}{ll}
r' = \rho & p - p' = b + \beta - \rho \\
q' = \alpha - b + c + \rho & q - q' = a - \alpha - \rho \qquad (5.3.69) \\
p' = -\beta + c - a + \rho & r - r' = a + b - c - \rho.
\end{array}
$$

From here on α, β and ρ are the summation variables. The factor $(-1)^{p'+q'+r'}$ is calculated with the following congruence mod 2 (note that $2(j \pm m) \equiv 0$ mod 2):

$$p' + q' + r' \equiv \rho + \alpha - \beta + 2c - b - a + (2a - 2\alpha) \bmod 2$$
$$\equiv \rho - \alpha - \beta + 2c - b + a - (2c + 2\gamma) \bmod 2$$
$$\equiv \rho - \underbrace{(\alpha + \beta + \gamma)}_{=0} + a - b - \gamma \bmod 2.$$

Hence

$$(-1)^{p'+q'+r'} = (-1)^{a-b-\gamma}(-1)^{\rho}. \tag{5.3.70}$$

With these substitutions the sum (5.3.66) becomes (we use pqr along with abc, but consider them only as an abbreviation, see (5.3.68)), writing out all binomial coefficients:

$$|((ab)c)\,00\rangle_{uvw} = N \sum_{\alpha+\beta+\gamma=0} X(\alpha, \beta, \gamma)\,|(a\alpha)_u (b\beta)_v (c\gamma)_w\rangle$$

$$N = N(pqr)\frac{p!q!r!}{\sqrt{(2a)!(2b)!(2c)!}}$$

$$X(\alpha, \beta, \gamma) = (-1)^{a-b-\gamma}\sqrt{(a+\alpha)!(a-\alpha)!(b+\beta)!(b-\beta)!(c+\gamma)!(c-\gamma)!}$$

$$\times \sum_{\rho} (-1)^{\rho} \frac{1}{\left[\begin{array}{l} \rho!(a+b-c-\rho)!(\alpha-b+c+\rho)!(a-\alpha-\rho)! \\ (-\beta+c-a+\rho)!(b+\beta-\rho)! \end{array}\right]}.$$

$$\tag{5.3.71}$$

Comparison with (5.3.62) shows that

$$(-1)^{a-b+c}\begin{pmatrix} a & b & c \\ \alpha & \beta & \gamma \end{pmatrix} = NX(\alpha, \beta, \gamma)$$

where N and $X(\alpha, \beta, \gamma)$ are defined in (5.3.71). Explicitly

$$\begin{pmatrix} a & b & c \\ \alpha & \beta & \gamma \end{pmatrix} = (-1)^{c+\gamma}N(pqr)\frac{p!q!r!\sqrt{\left[\begin{array}{l}(a+\alpha)!(a-\alpha)!(b+\beta)! \\ (b-\beta)!(c+\gamma)!(c-\gamma)!\end{array}\right]}}{\sqrt{(2a)!(2b)!(2c)!}}$$

$$\times \sum_{\rho}(-1)^{\rho}\frac{1}{\left[\begin{array}{l}\rho!(a+b-c-\rho)!(\alpha-b+c+\rho)! \\ (a-\alpha-\rho)!(-\beta+c-a+\rho)!(b+\beta-\rho)!\end{array}\right]}.$$

$$\tag{5.3.72}$$

It remains only to evaluate the sign and magnitude of $N(pqr)$. As N does not depend on α, β, γ, ρ, this can be done at once by evaluating the r.h.s. of (5.3.72) in the case $a = \alpha$ where in the sum over ρ both $(\rho)!$ and $(-\rho)!$ appear; thus necessarily $\rho = 0$ (all sums extend over such values of the variables, that the factorials make sense).

Furthermore, we may put $\gamma = -c$ and write with (5.3.54)

$$\begin{pmatrix} a & b & c \\ a & c-a & -c \end{pmatrix} = \frac{(-1)^{a-b+c}}{\sqrt{2c+1}} \langle aab(c-a)|cc \rangle.$$

The last CGC has been calculated earlier in (5.3.34) which gives in the Wigner notation

$$\begin{pmatrix} a & b & c \\ a & c-a & -c \end{pmatrix} = (-1)^{a-b+c} \sqrt{\frac{(2a)!(2c)!}{(a+b+c+1)!(a-b+c)!}}. \qquad (5.3.73)$$

For this same CGC we obtain now from (5.3.72)

$$\begin{pmatrix} a & b & c \\ a & c-a & -c \end{pmatrix} = N(pqr) \frac{\sqrt{(-a+b+c)!(a+b-c)!}}{\sqrt{(2b)!}} \qquad (5.3.74)$$

so that

$$N(pqr) = (-1)^{a-b+c} \sqrt{\frac{(2a)!(2b)!(2c)!}{(a+b+c+1)!(-a+b+c)!(a-b+c)!(a+b-c)!}}. \qquad (5.3.75)$$

Inserting this into (5.3.72) yields Racah's formula

$$\begin{pmatrix} a & b & c \\ \alpha & \beta & \gamma \end{pmatrix} = (-1)^{a-b-\gamma} \sqrt{\frac{(-a+b+c)!(a-b+c)!(a+b-c)!}{(a+b+c+1)!}}$$

$$\times \sqrt{(a+\alpha)!(a-\alpha)!(b+\beta)!(b-\beta)!(c+\gamma)!(c-\gamma)!}$$

$$\times \sum_\rho \frac{(-1)^\rho}{\left[\begin{array}{c} \rho!(a+b-c-\rho)!(\alpha-b+c+\rho)! \\ (a-\alpha-\rho)!(-a-\beta+c+\rho)!(b+\beta-\rho)! \end{array} \right]}$$

$$(5.3.76)$$

with the conditions that $\alpha + \beta + \gamma = 0$ and $\Delta(abc) = 1$. The sum over ρ extends only over such integer values where the factorials have non-negative argument.

It could be argued that, in order to determine the sign and magnitude of $N(pqr)$, we have used a formula for the special CGC $\langle aab(c-a)|cc \rangle$ which was calculated earlier by means of the usual technique involving recursion formulae obtained by application of $J_\pm$. So our present derivation would not be independent of these techniques. It is left to the reader as an exercise to prove that we can determine $N(pqr)$ with little more effort also without the help of $\langle aab(c-a)|cc \rangle$. One uses the orthogonality of CGCs as expressed in (5.3.58) and (5.3.60) and writes it down in terms of (5.3.72), thereby obtaining an expression for $N^2(pqr)$. Here again, one uses the trick to put $\alpha = a$ which implies that the (this time two) sums over ρ and ρ' have only one single term.

The ensuing sum over factorials can be evaluated by means of formula (5.3.33) and $N^2(pqr)$ is found. Its sign is then determined from the requirement that

$$(-1)^{a-b+c} \begin{pmatrix} a & b & c \\ a & c-a & -c \end{pmatrix} > 0$$

(phase convention (5.3.25), (5.3.57)).

The general Racah formula is rather complicated and not too useful for the calculation of CGCs, except if one uses computers. But in many important cases it reduces to much simpler expressions. In general, one needs the CGCs for rather small values of its arguments and then it is easier to write down some explicit formulae. We shall collect such formulae at the end of this chapter in subsection 5.3.10.

5.3.9 Regge's symmetry of CGCs

The first closed formula (similar to Racah's) for the CGCs was given by Wigner in 1931 (Wigner (1931, 1959)). The rather obvious symmetry relations

$$\begin{pmatrix} a & b & c \\ \alpha & \beta & \gamma \end{pmatrix} = \begin{pmatrix} b & c & a \\ \beta & \gamma & \alpha \end{pmatrix} = \begin{pmatrix} c & a & b \\ \gamma & \alpha & \beta \end{pmatrix} = -\begin{pmatrix} b & a & c \\ \beta & \alpha & \gamma \end{pmatrix}$$

(which were derived above in subsection 5.3.6) were known all the time. It took twenty-seven years before some further, much less obvious symmetries of the CGCs were discovered by Regge (1958). We shall discuss them now. We write down once more (5.3.65)

$$\left(\delta_u^p \delta_v^q \delta_w^r \right)_\otimes = \frac{1}{N} | \left((ab)c \right) 00 \rangle_{uvw}$$

and insert $| ((ab)c) 00 \rangle_{uvw}$ from (5.3.62) and (5.3.63) and N from (5.3.75) with the result

$$\left(\delta_u^p \delta_v^q \delta_w^r \right)_\otimes = \sqrt{(a+b+c+1)!(-a+b+c)!(a-b+c)!(a+b-c)!}$$

$$\times \sum_{\alpha+\beta+\gamma=0} \begin{pmatrix} a & b & c \\ \alpha & \beta & \gamma \end{pmatrix} \frac{\left(u_+^{a+\alpha} u_-^{a-\alpha} \right)_\otimes \left(v_+^{b+\beta} v_-^{b-\beta} \right)_\otimes \left(w_+^{c+\gamma} w_-^{c-\gamma} \right)_\otimes}{\sqrt{\begin{bmatrix} (a+\alpha)!(a-\alpha)!(b+\beta)!(b-\beta)! \\ (c+\gamma)!(c-\gamma)! \end{bmatrix}}}$$

$$p \equiv -a+b+c \qquad q \equiv a-b+c \qquad r \equiv a+b-c$$
$$p+q+r = a+b+c \equiv k.$$

$$(5.3.77)$$

Consider now the determinant

$$\begin{vmatrix} u_0 & v_0 & w_0 \\ u_+ & v_+ & w_+ \\ u_- & v_- & w_- \end{vmatrix}_\otimes = u_0(v_+w_- - w_+v_-)_\otimes + v_0(w_+u_- - u_+w_-)_\otimes$$

$$+ w_0(u_+v_- - v_+u_-)_\otimes$$

$$= (u_0\delta_u + v_0\delta_v + w_0\delta_w)_\otimes$$

and its kth power $(k = a + b + c)$

$$\frac{1}{k!} \begin{vmatrix} u_0 & v_0 & w_0 \\ u_+ & v_+ & w_+ \\ u_- & v_- & w_- \end{vmatrix}_\otimes^k = \sum_{p+q+r=k} \frac{u_0^p v_0^q w_0^r}{p!q!r!} \left(\delta_u^p \delta_v^q \delta_w^r \right)_\otimes . \tag{5.3.78}$$

(For the definition of δ_u, δ_v and δ_w see (5.3.64).) The new quantities $u_0 \, v_0 \, w_0$ may be considered to be parameters (just numbers). We encounter here our familiar invariant state $\left(\delta_u^p \delta_v^q \delta_w^r \right)_\otimes$, which we now insert from (5.3.77) into (5.3.78) with the result

$$\frac{1}{k!} \begin{vmatrix} u_0 & v_0 & w_0 \\ u_+ & v_+ & w_+ \\ u_- & v_- & w_- \end{vmatrix}_\otimes^k = \sqrt{(k+1)!} \sum_{\substack{\alpha+\beta+\gamma=0 \\ p+q+r=k}} \begin{pmatrix} a & b & c \\ \alpha & \beta & \gamma \end{pmatrix}$$

$$\times \frac{\left(u_0^p u_+^{a+\alpha} u_-^{a-\alpha} \right) \left(v_0^q v_+^{b+\beta} v_-^{b-\beta} \right) \left(w_0^r w_+^{c+\gamma} w_-^{c-\gamma} \right)_\otimes}{\sqrt{p!(a+\alpha)!(a-\alpha)!q!(b+\beta)!(b-\beta)!r!(c+\gamma)!(c-\gamma)!}} . \tag{5.3.79}$$

Now the well known symmetries of the determinant on the l.h.s. imply that the r.h.s. has the same symmetries. Thus even (odd) permutations of columns leave the determinant invariant (multiply it by -1). On the r.h.s. this means corresponding permutations of u, v, w. In order to restore the sum one only has to rename the summation variables and to perform corresponding permutations of the columns of $\begin{pmatrix} a & b & c \\ \alpha & \beta & \gamma \end{pmatrix}$, because the square root in the denominator is invariant. According to what happens to the determinant, we conclude that $\begin{pmatrix} a & b & c \\ \alpha & \beta & \gamma \end{pmatrix}$ multiplies by $(-1)^{a+b+c}$ under odd and stays invariant under even permutations of its columns. Interchanging the last two rows of the determinant multiplies it by -1. On the r.h.s. the corresponding operation leads to $\alpha \to -\alpha, \beta \to -\beta, \gamma \to -\gamma$. Thus

$$\begin{pmatrix} a & b & c \\ -\alpha & -\beta & -\gamma \end{pmatrix} = (-1)^{a+b+c} \begin{pmatrix} a & b & c \\ \alpha & \beta & \gamma \end{pmatrix} .$$

These two types of symmetry are our old well known ones of (5.3.55) and (5.3.56). But there are now many more: arbitrary permutations of rows and arbitrary permutations of columns and reflection on the diagonal. These symmetries sum up to 72 different operations: $(3! = 6$ column permutations) $\times$ (6 row permutations) $\times$ (2 reflections on the main diagonal), whereas the old symmetries comprise only 12 operations.

In order to bring out more clearly the complete symmetry group of CGC, Regge introduced a new notation:

$$\begin{pmatrix} a & b & c \\ \alpha & \beta & \gamma \end{pmatrix} \equiv \begin{bmatrix} -a+b+c & a-b+c & a+b-c \\ a+\alpha & b+\beta & c+\gamma \\ a-\alpha & b-\beta & c-\gamma \end{bmatrix} . \tag{5.3.80}$$

The Regge symbol is highly redundant with its nine entries for five independent variables—but between the Wigner $3j$-symbol and the Regge symbol there is a one-to-one correspondence. With this notation (5.3.79) becomes

$$\frac{1}{k!}\begin{vmatrix} u_0 & v_0 & w_0 \\ u_+ & v_+ & w_+ \\ u_- & v_- & w_- \end{vmatrix}_\otimes^k = \sqrt{(k+1)!}\sum_{\substack{\alpha+\beta+\gamma=0 \\ p+q+r=k}}\begin{bmatrix} p & q & r \\ a+\alpha & b+\beta & c+\gamma \\ a-\alpha & b-\beta & c-\gamma \end{bmatrix}$$

$$\times\frac{\left(u_0^p u_+^{a+\alpha} u_-^{a-\alpha}\right)\left(v_0^q v_+^{b+\beta} v_-^{b-\beta}\right)\left(w_0^r w_+^{c+\gamma} w_-^{c-\gamma}\right)_\otimes}{\sqrt{p!(a+\alpha)!(a-\alpha)!q!(b+\beta)!(b-\beta)!r!(c+\gamma)!(c-\gamma)!}} \qquad (5.3.81)$$

$$p = -a+b+c \qquad q = a-b+c \qquad r = a+b-c.$$

In this form each of the symmetry operations on the determinant implies exactly the same operation on the Regge symbol:

$$\begin{bmatrix} -a+b+c & a-b+c & a+b-c \\ a+\alpha & b+\beta & c+\gamma \\ a-\alpha & b-\beta & c-\gamma \end{bmatrix} \equiv \begin{pmatrix} a & b & c \\ \alpha & \beta & \gamma \end{pmatrix} \qquad (5.3.82)$$

is invariant under the following operations on the Regge symbol and the corresponding ones on the $3j$-symbol:

(i) reflection of the Regge symbol on the main diagonal;

(ii) even permutation of rows and/or columns of the Regge symbol;

(iii) simultaneous multiplication by $(-1)^{a+b+c}$ and odd permutations of rows and/or columns of the Regge symbol.

5.3.10 Collection of formulae for the CGCs; a table of special values

We collect here some formulae which either have been derived in the text or are found by specialization of formulae given in the text. We shall, however, not prove here these more special formulae. Table 5.2 contains the values of $\langle j_1 m_1 00|jm\rangle$, $\langle j_1 m_1 \frac{1}{2} m_2|jm\rangle$ and $\langle j_1 m_1 1 m_2|jm\rangle$ in Wigner's $3j$-notation. The following notations are in current use:

$$j_1 \leftrightarrow a$$
$$m_1 \leftrightarrow \alpha$$
$$j_2 \leftrightarrow b$$
$$m_2 \leftrightarrow \beta$$
$$j_3 \leftrightarrow j \to c$$
$$m_3 \leftrightarrow m \to \gamma.$$

The CGC is defined as the scalar product of two states

$$|(ab)c\gamma\rangle \equiv |(j_1 j_2)jm\rangle \qquad \text{'coupled state'}$$

$$|a\alpha b\beta\rangle \equiv |j_1 m_1\rangle \otimes |j_2 m_2\rangle \qquad \text{'uncoupled direct product'}.$$

The scalar product is written with $(j_1 j_2)$ omitted:

$$\langle a\alpha b\beta | c\gamma \rangle = \langle c\gamma | a\alpha b\beta \rangle \quad \text{reality} \qquad\qquad ((5.3.26))$$

$$= 0 \qquad \text{unless } \alpha + \beta = \gamma \qquad ((5.3.12))$$

$$\text{and } \Delta(abc) = 1$$

where $\Delta(abc) = 1$, if a, b, c could be the three sides of a triangle; otherwise $\Delta = 0$.

Other useful notations—Wigner and Regge

$$\begin{pmatrix} a & b & c \\ \alpha & \beta & \gamma \end{pmatrix} = \begin{bmatrix} -a+b+c & a-b+c & a+b-c \\ a+\alpha & b+\beta & c+\gamma \\ a-\alpha & b-\beta & c-\gamma \end{bmatrix}$$

$$\uparrow \qquad\qquad\qquad\qquad \uparrow$$
$$\text{Wigner} \qquad\qquad\qquad \text{Regge}$$

$$= \frac{(-1)^{a-b-\gamma}}{\sqrt{2c+1}} \langle a\alpha b\beta | c - \gamma \rangle.$$

Orthogonality

$$(2c+1) \sum_{\alpha,\beta} \begin{pmatrix} a & b & c \\ \alpha & \beta & \gamma \end{pmatrix} \begin{pmatrix} a & b & c' \\ \alpha & \beta & \gamma' \end{pmatrix} = \delta_{cc'} \delta_{\gamma\gamma'}$$

$$\sum_{c,\gamma} (2c+1) \begin{pmatrix} a & b & c \\ \alpha & \beta & \gamma \end{pmatrix} \begin{pmatrix} a & b & c \\ \alpha' & \beta' & \gamma \end{pmatrix} = \delta_{\alpha\alpha'} \delta_{\beta\beta'}$$

$$((5.3.16), (5.3.58), (5.3.60)).$$

$$\begin{bmatrix} \text{remains true} \\ \text{under any} \\ \text{permutation of} \\ \begin{pmatrix} a \\ \alpha \end{pmatrix}, \begin{pmatrix} b \\ \beta \end{pmatrix}, \begin{pmatrix} c \\ \gamma \end{pmatrix} \end{bmatrix}$$

Symmetry

$$\begin{bmatrix} -a+b+c & a-b+c & a+b-c \\ a+\alpha & b+\beta & c+\gamma \\ a-\alpha & b-\beta & c-\gamma \end{bmatrix} \quad \text{is invariant under} \qquad ((5.3.82))$$

(i)   even permutations of rows and/or columns;
(ii) reflection on the main diagonal;
(iii) simultaneous multiplication by $(-1)^{a+b+c}$ *and* odd permutations of rows and/or columns.

Phase convention

$$\langle a a b\beta | cc \rangle = \begin{pmatrix} a & b & c \\ a & \beta & -c \end{pmatrix} \sqrt{2c+1}\,(-1)^{a-b+c} \geq 0. \qquad ((5.3.25), (5.3.57))$$

Computation of CGCs: Racah's general formula (5.3.76)

$$\begin{pmatrix} a & b & c \\ \alpha & \beta & \gamma \end{pmatrix} = (-1)^{a-b-\gamma} \sqrt{\frac{(-a+b+c)!(a-b+c)!(a+b-c)!}{(a+b+c+1)!}}$$

$$\times \sqrt{(a+\alpha)!(a-\alpha)!(b+\beta)!(b-\beta)!(c+\gamma)!(c-\gamma)!}$$

$$\times \sum_\rho \frac{(-1)^\rho}{\left[\begin{array}{c} \rho!(a+b-c-\rho)!(\alpha-b+c+\rho)!(a-\alpha-\rho)! \\ (-a-\beta+c+\rho)!(b+\beta-\rho)! \end{array} \right]}.$$

Special formulae (5.3.73), (5.3.34) and (5.3.35)

$$\begin{pmatrix} a & b & c \\ a & c-a & -c \end{pmatrix} = \frac{(-1)^{a-b+c}}{\sqrt{2c+1}} \langle aab, c-a|cc\rangle$$

$$= (-1)^{a-b+c} \sqrt{\frac{(2a)!(2c)!}{(a+b+c+1)!(a-b+c)!}}$$

$$\begin{pmatrix} a & b & c \\ \alpha & c-\alpha & -c \end{pmatrix} = \frac{(-1)^{a-b+c}}{\sqrt{2c+1}} \langle a\alpha b, c-\alpha|cc\rangle$$

$$= (-1)^{\alpha-b+c} \sqrt{\frac{(c+b-\alpha)!(a+\alpha)!}{(b-c+\alpha)!(a-\alpha)!}}$$

$$\times \sqrt{\frac{(2c)!(a+b-c)!}{(a+b+c+1)!(-a+b+c)!(a-b+c)!}}.$$

Putting $c = 0$ requires $b = a$ and we obtain ((5.3.47))

$$\begin{pmatrix} a & a & 0 \\ \alpha & -\alpha & 0 \end{pmatrix} = (-1)^{\alpha-a} \frac{1}{\sqrt{2a+1}} = \langle a\alpha a, -\alpha|00\rangle.$$

$$\begin{pmatrix} a & b & c \\ 0 & 0 & 0 \end{pmatrix} = \begin{cases} 0, & \text{if } a+b+c = J = \text{odd number} \\[2mm] (-1)^{J/2} \sqrt{\dfrac{(J-2a)!(J-2b)!(J-2c)!}{(J+1)!}} \\[2mm] \times \dfrac{(J/2)!}{(J/2-a)!(J/2-b)!(J/2-c)!}, \\ \text{if } J = a+b+c = \text{even number } (a,b,c \text{ integers!}) \\[2mm] (\text{can be derived from Racah's formula}). \end{cases}$$

Further special formulae follow from these by means of the symmetry of the CGCs (see table 5.2).

Table 5.2 allows one to compute any CGC in which at least one j-value is 0, 1/2 or 1; of course the symmetry properties have to be used. We shall not go

Table 5.2. Some special formulae for $3j$-symbols.

$$\begin{pmatrix} j & j & 0 \\ m & -m & 0 \end{pmatrix} = (-1)^{j-m} \frac{1}{\sqrt{2j+1}}$$

$$\begin{pmatrix} j+1/2 & j & 1/2 \\ m & -m-1/2 & 1/2 \end{pmatrix} = (-1)^{j-m-1/2} \sqrt{\frac{(j-m+1/2)}{(2j+2)(2j+1)}}$$

$$\begin{pmatrix} j+1 & j & 1 \\ m & -m-1 & 1 \end{pmatrix} = (-1)^{j-m-1} \sqrt{\frac{(j-m)(j-m+1)}{(2j+3)(2j+2)(2j+1)}}$$

$$\begin{pmatrix} j+1 & j & 1 \\ m & -m & 0 \end{pmatrix} = (-1)^{j-m-1} \sqrt{\frac{(j+m+1)(j-m+1)}{(2j+3)(2j+1)(j+1)}}$$

$$\begin{pmatrix} j & j & 1 \\ m & -m-1 & 1 \end{pmatrix} = (-1)^{j-m} \sqrt{\frac{(j-m)(j+m+1)}{2(2j+1)(j+1)j}}$$

$$\begin{pmatrix} j & j & 1 \\ m & -m & 0 \end{pmatrix} = (-1)^{j-m} \frac{m}{\sqrt{(2j+1)(j+1)j}}$$

into any further detail. A table of special formulae of the above type, when one of the angular momenta is $3/2$ or 2, is found in Edmonds (1957). For a survey on numerical tables see section 5.6 at the end of the present chapter.

5.4 Combining three angular momenta; recoupling coefficients

5.4.1 General remarks; statement of the problem

The general problem was discussed in sections 5.1 and 5.2, where it was found that the coupling of any number of angular momenta was reducible to repeated coupling of two angular momenta. In what follows, we shall see that explicitly.

Let us return to the summary on page 126 in section 5.2. Applying it to the case of three subsystems, we see that there are four different sets of states characterized by the quantum numbers of some of the operators

$$\begin{array}{cccccccccc} J^2 & J_z & J^{(1)^2} & J^{(2)^2} & J^{(3)^2} & J_z^{(1)} & J_z^{(2)} & J_z^{(3)} & J^{(23)^2} & J^{(31)^2} & J^{(12)^2} \\ j & m & j_1 & j_2 & j_3 & m_1 & m_2 & m_3 & j_{23} & j_{31} & j_{12} \end{array}$$

(first line, possible operators; $J^{(ik)^2}$ is the squared total angular momentum of systems i and k coupled together; second line, corresponding quantum numbers).

The invariant subspace $\mathcal{H}_{j_1 j_2 j_3}$ is spanned by

$$
\begin{array}{ll}
|j_1 m_1 j_2 m_2 j_3 m_3\rangle & \text{or by} \\
|((j_1 j_2) j_{12} j_3) jm\rangle & \text{or by} \\
|((j_2 j_3) j_{23} j_1) jm\rangle & \text{or by} \\
|((j_3 j_1) j_{31} j_2) jm\rangle;
\end{array}
\tag{5.4.1}
$$

in each case by $(2j_1 + 1)(2j_2 + 1)(2j_3 + 1)$ orthonormal states. As the operators $\boldsymbol{J}^{(12)^2}$, $\boldsymbol{J}^{(23)^2}$ and $\boldsymbol{J}^{(31)^2}$ do not commute with each other, it follows that the corresponding bases of $\mathcal{H}_{j_1 j_2 j_3}$ cannot be identical, but as each of them spans the whole of $\mathcal{H}_{j_1 j_2 j_3}$, they must be related to each other and to the uncoupled (direct product) basis by unitary transformations. The above notation in which $j_1 j_2 j_3$ are cyclically permuted would seem the most natural. Common use prefers, however, to consider (instead of the third and fourth states of the above list) the states $|(j_1(j_2 j_3) j_{23}) jm\rangle$ and $|(j_2(j_3 j_1) j_{31}) jm\rangle$, respectively, which differ from those of (5.4.1) only by a phase (see (5.3.38)). Furthermore it is clear that we need not consider all the unitary transformations between the four bases; it suffices to know the transformation between

$$|j_1 m_1 j_2 m_2 j_3 m_3\rangle \longleftrightarrow |((j_1 j_2) j_{12} j_3) jm\rangle$$

and

$$|((j_1 j_2) j_{12} j_3) jm\rangle \longleftrightarrow |(j_1(j_2 j_3) j_{23}) jm\rangle$$

because the others are obtained from these two by a suitable rearrangement. We therefore study the following unitary transformations:

$$
\begin{array}{l}
|j_1 m_1 j_2 m_2 j_3 m_3\rangle \xleftrightarrow{\;U^{(12)}\;} |((j_1 j_2) j_{12} j_3) jm\rangle \\[2mm]
|j_1 m_1 j_2 m_2 j_3 m_3\rangle \xleftrightarrow{\;U^{(23)}\;} |(j_1(j_2 j_3) j_{23}) jm\rangle \\[2mm]
|((j_1 j_2) j_{12} j_3) jm\rangle \xleftrightarrow{\;V\;} |(j_1(j_2 j_3) j_{23}) jm\rangle.
\end{array}
\tag{5.4.2}
$$

V bears a suggestive name: recoupling transformation. For the two U-transformations, it is obvious how they are found by twice applying the coupling rule for two angular momenta, but then we also automatically have the V-transformation—so there is no problem; we shall carry out the trivial calculation later on and define the $6j$-symbol, which is essentially the V-matrix element.

We shall first stop a moment to ask ourselves why there is a recoupling transformation V at all and on what quantum numbers it depends and what is the dimension of the space in which it works.

If we couple two angular momenta j_1 and j_2, then the resulting j can have values between $|j_1 - j_2|$ and $j_1 + j_2$ and each of the possible values appears once and only once; thus a state $|(j_1 j_2) jm\rangle$ is defined uniquely except for a phase, which is fixed by convention. When we couple three angular momenta, then $j_1 j_2 j_3$ may compose in different ways to form one resultant j:

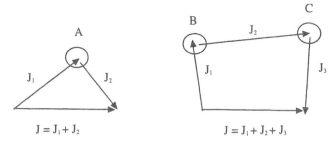

$$J = J_1 + J_2 \qquad\qquad\qquad J = J_1 + J_2 + J_3$$

In these two diagrams the point A can rotate about J, that corresponds to a phase factor for the otherwise uniquely fixed triangle. The figure in which three angular momenta are added has additional degrees of freedom: B and C can—apart from rotating the figure about J—change their relative position and thereby lead to different shapes of the figure; different shapes correspond to independent states. There is one exception, namely when one of the four numbers j_1, j_2, j_3, j is zero; in that case the second diagram reduces to the first one and we conclude that (except for an arbitrary phase) there is only one way to couple three angular momenta to a $j = 0$ state, a result which we derived earlier (between (5.3.49) and (5.3.50)).

Thus our above qualitative diagrams show that now j and m do not define a state but a whole subspace, and that the vectors in that subspace must be labelled by a further quantum number: either j_{12} or j_{23} or j_{31}. Each of these labellings defines another orthonormal basis and the unitary transformation V transforms these into each other. One might think that j_{12}, j_{23} and j_{31} have different ranges of possible values, e.g. $|j_1 - j_2| \le j_{12} \le j_1 + j_2|$, $j_2 - j_3| \le j_{23} \le j_2 + j_3$, but the condition that each of these must combine with the remaining j_k to a given j reduces all of them to the same range. We leave it as an exercise to the reader to prove that the *number* of possible values is the same for j_{23}, j_{31} and j_{12}. One only has to inspect the various possibilities of the four inequalities

$$\begin{aligned} |j_1 - j_2| &\le j_{12} \le j_1 + j_2 & |j_2 - j_3| &\le j_{23} \le j_2 + j_3 \\ |j - j_3| &\le j_{12} \le j + j_3 & |j - j_1| &\le j_{23} \le j + j_1 \end{aligned} \qquad (5.4.3)$$

(assume $j_1 \le j_2 \le j_3$ and discuss the four possibilities $j \le j_1$, $j_1 \le j \le j_2$, $j_2 \le j \le j_3$ and $j_3 \le j$; show that if any one of the four values j_1, j_2, j_3, j equals zero, the subspace $\mathcal{H}_{j_1 j_2 j_3 jm}$ is one dimensional).

Next we observe that *the unitary transformation V does not depend on m*; that is: it is the same for all subspaces $\mathcal{H}_{j_1 j_2 j_3 jm}$ once j_1, j_2, j_3 and j are fixed. This follows immediately from its definition (5.4.2):

$$\begin{aligned} |((j_1 j_2) j_{12} j_3) jm\rangle &= \sum_{j_{23}} |(j_1 (j_2 j_3) j_{23}) jm\rangle \langle (j_1 (j_2 j_3) j_{23}) jm| \\ &\quad \times |((j_1 j_2) j_{12} j_3) jm\rangle \\ &= \sum_{j_{23}} |(j_1 (j_2 j_3) j_{23}) jm\rangle V^{(j)}_{j_{23} j_{12}}. \end{aligned} \qquad (5.4.4)$$

Whatever the values of all other quantum numbers are, the states appearing in (5.4.4) are of the form $|\gamma jm\rangle$ (γ labels all other quantum numbers whose operators in fact commute with $J_x J_y J_z$—see (5.1.4)) and thus obey the usual rules

$$J_\pm|\gamma jm\rangle = \sqrt{j(j+1) - m(m \pm 1)}|\gamma j \, m \pm 1\rangle.$$

Therefore, applying J_+ or J_- a suitable number of times to (5.4.4), one obtains the transformation law for all m ($-j \le m \le j$) and clearly the numerical coefficients $V^{(j)}_{j_{23}j_{12}}$ remain untouched.

The matrix elements of the unitary transformation $V^{(j)}$ are not only interesting insofar as they transform between two different coupling schemes; the very fact that they do not depend on m is an interesting feature in itself: it means that these matrix elements are invariant under rotations (in contradistinction to the m-dependent CGCs) and therefore are likely to come up in any calculation where the result is expressed in a coordinate-independent way. That does not necessarily mean that by formulating one's problems by using invariant equations as much as possible, these transformation coefficients will automatically appear (quite apart from the fact that there are additional invariants if there are more than three subsystems); what may—and frequently will—happen, however, is that sums over products of CGCs are obtained which can be written in terms of these invariant coefficients and thereby considerably simplified.

In the next subsection we shall define the $6j$-symbol (which is almost the same as a $V^{(j)}$-matrix element) and the Racah coefficient (which differs from it by a phase). In the last subsection we shall collect the most important formulae for the $6j$-symbol without deriving them. We shall not derive them because after having defined and expressed the coupling coefficients in terms of CGCs all the listed formulae are obtained by straightforward[5] elementary algebra, which probably every reader of this book would skip anyway.

5.4.2 The $6j$-symbol and the Racah coefficients

We start by working out the unitary transformations $U^{(12)}$ and $U^{(23)}$ defined by (5.4.2). We define $U^{(12)}$ by

$$|((j_1 j_2)j_{12}j_3)jm\rangle = \sum_{m_i} |j_1 m_1 j_2 m_2 j_3 m_3\rangle U^{(12)}_{m_1 m_2 m_3|j_{12}jm} \tag{5.4.5}$$

and $U^{(23)}$ by

$$|(j_1(j_2 j_3)j_{23})jm\rangle = \sum_{m_i} |j_1 m_1 j_2 m_2 j_3 m_3\rangle U^{(23)}_{m_1 m_2 m_3|j_{23}jm}. \tag{5.4.6}$$

The summation over all three m_i is formal, because $m_1 + m_2 + m_3 = m$; we may define the U-matrix elements to be zero if that condition is violated (they

---

[5] Do not take this literally: 'straightforward' means only that no new ideas are needed; clever substitutions, extensive use of known symmetry and orthogonality relations, all these, in the right combination, will lead to the collected formulae.

are indeed automatically zero in that case). From the two expressions (5.4.5) and (5.4.6) and the definition of the V-coefficients by (5.4.4) we obtain directly the relation between V and the two U.

$$
\begin{aligned}
V^{(j)}_{j_{23}j_{12}} &= \langle (j_1(j_2j_3)j_{23})jm|((j_1j_2)j_{12}j_3)jm\rangle \\
&= \sum_{m_i} U^{(23)}_{j_{23}jm|m_1m_2m_3} U^{(12)}_{m_1m_2m_3)|j_{12}jm}.
\end{aligned}
\tag{5.4.7}
$$

As the U are obtained by twice using CGC coupling, the V-matrix elements are a sum over products of four CGCs.

We find actually

$$
|((j_1j_2)j_{12}j_3)jm\rangle = \sum_{m_3} |j_{12}m_{12}j_3m_3\rangle\langle j_{12}m_{12}j_3m_3|jm\rangle
$$

$$
|(j_1j_2)j_{12}m_{12}\rangle = \sum |j_1m_1j_2m_2\rangle\langle j_1m_1j_2m_2|j_{12}m_{12}\rangle.
$$

Inserting the second into the first yields

$$
\begin{aligned}
|((j_1j_2)j_{12}j_3)jm\rangle = \sum_{m_i} |j_1m_1j_2m_2j_3m_3\rangle\langle j_1m_1j_2m_2|j_{12}m_{12}\rangle \\
\times \langle j_{12}m_{12}j_3m_3|jm\rangle.
\end{aligned}
\tag{5.4.8}
$$

Similarly

$$
\begin{aligned}
|(j_1(j_2j_3)j_{23})jm\rangle = \sum_{m_i} |j_1m_1j_2m_2j_3m_3\rangle\langle j_2m_2j_3m_3|j_{23}m_{23}\rangle \\
\times \langle j_1m_1j_{23}m_{23}|jm\rangle.
\end{aligned}
\tag{5.4.9}
$$

These two equations may be used to read off immediately the matrix elements of $U^{(12)}$ and $U^{(23)}$ respectively. Equation (5.4.7) then gives for the recoupling coefficient (we suppress m, because the coefficient is m independent as we know already):

$$
\begin{aligned}
V^{(j)}_{j_{23}j_{12}} &= \langle (j_1(j_2j_3)j_{23})j|((j_1j_2)j_{12}j_3)j\rangle \\
&= \sum_{\text{all } m_i} [\langle j_2m_2j_3m_3|j_{23}m_{23}\rangle\langle j_1m_1j_{23}m_{23}|jm\rangle \\
&\quad \times \langle j_1m_1j_2m_2|j_{12}m_{12}\rangle\langle j_{12}m_{12}j_3m_3|jm\rangle].
\end{aligned}
\tag{5.4.10}
$$

As this quantity does not depend on m, we may sum over m and at the same time divide by $2j+1$; furthermore we may, without any consequences, sum also over m_{12} and over m_{23}, because the relevant CGCs are zero unless $m_{12} = m_1 + m_2$ and $m_{23} = m_2 + m_3$. If we do all that and finally replace the CGC by the $3j$-symbol

$$
\langle a\alpha b\beta|c\gamma\rangle = (-1)^{a-b+\gamma}\sqrt{2c+1}\begin{pmatrix} a & b & c \\ \alpha & \beta & -\gamma \end{pmatrix}
$$

we obtain

$$V_{j_{23}j_{12}}^{(j)} = \sqrt{(2j_{23}+1)(2j_{12}+1)}$$

$$\times \sum_{\substack{m_1 m_2 m_3 m \\ m_{12} m_{23}}} (-1)^{2j_1-2j_3+j_{12}-j_{23}+2m+m_{12}+m_{23}}$$

$$\times \begin{pmatrix} j_2 & j_3 & j_{23} \\ m_2 & m_3 & -m_{23} \end{pmatrix} \begin{pmatrix} j_1 & j_{23} & j \\ m_1 & m_{23} & -m \end{pmatrix}$$

$$\times \begin{pmatrix} j_1 & j_2 & j_{12} \\ m_1 & m_2 & -m_{12} \end{pmatrix} \begin{pmatrix} j_{12} & j_3 & j \\ m_{12} & m_3 & -m \end{pmatrix}.$$

One now defines the $6j$-symbol by

$$\begin{Bmatrix} j_1 & j_2 & j_{12} \\ j_3 & j & j_{23} \end{Bmatrix} \equiv \frac{(-1)^{j_1+j_2+j_3+j}}{\sqrt{(2j_{23}+1)(2j_{12}+1)}} V_{j_{23}j_{12}}^{(j)}. \tag{5.4.11}$$

By an obvious substitution of new names for the variables and by some simple manipulations in the exponent of -1, one arrives at the formula

$$\begin{Bmatrix} j_1 & j_2 & j_3 \\ l_1 & l_2 & l_3 \end{Bmatrix} = (-1)^{l_1+l_2+l_3} \sum_{\substack{\text{all } m_i \\ \text{all } \mu_i}} (-1)^{\mu_1+\mu_2+\mu_3} \left[\begin{pmatrix} j_1 & j_2 & j_3 \\ m_1 & m_2 & m_3 \end{pmatrix} \right.$$

$$\left. \times \begin{pmatrix} j_1 & l_2 & l_3 \\ m_1 & \mu_2 & -\mu_3 \end{pmatrix} \begin{pmatrix} l_1 & j_2 & l_3 \\ -\mu_1 & m_2 & \mu_3 \end{pmatrix} \begin{pmatrix} l_1 & l_2 & j_3 \\ \mu_1 & -\mu_2 & m_3 \end{pmatrix} \right] \tag{5.4.12}$$

$$\equiv (-1)^{j_1+j_2+l_1+l_2} W(j_1 j_2 l_2 l_1; j_3 l_3).$$

The symbol $W(j_1 j_2 l_2 l_1; j_3 l_3)$ is the Racah W-coefficient. Other notations are in current use; in the more recent literature the $6j$-symbol is gaining steadily more territory owing to its simple symmetry properties. The other notations are given in the next subsection, where we list the most important formulae. For derivations refer to Edmonds (1957), Biedenharn and van Dam (1956) and Brink and Satchler (1968).

5.4.3   Collection of formulae for recoupling coefficients

Notation

$$V_{j_{23}j_{12}}^{(j)} = \langle j_1(j_2 j_3) j_{23}) j | ((j_1 j_2) j_{12} j_3) j \rangle$$

$$= \sqrt{(2j_{23}+1)(2j_{12}+1)}(-1)^{j_1+j_2+j_3+j} \begin{Bmatrix} j_1 & j_2 & j_{12} \\ j_3 & j & j_{23} \end{Bmatrix} \tag{5.4.13}$$

(Wigner in Biedenharn and van Dam (1956)).

Other notations are

$$\left\{ \begin{matrix} j_1 & j_2 & j_3 \\ l_1 & l_2 & l_3 \end{matrix} \right\} = (-1)^{j_1+j_2+l_1+l_2} W(j_1 j_2 l_2 l_1; j_3 l_3)$$

(Racah (1942); reprinted in Biedenharn and van Dam (1956))

$$= \frac{(-1)^{j_1+j_2+l_1+l_2}}{\sqrt{(2j_3+1)(2l_3+1)}} U(j_1 j_2 l_2 l_1; j_3 l_3)$$

(Jahn (1951))

$$= \frac{(-1)^{l_1+l_2+\frac{1}{2}(j_2-j_1+j_3)} \cdot Z(j_1 l_2 j_2 l_1; l_3 j_3)}{\left(\begin{matrix} j_1 & j_2 & j_3 \\ 0 & 0 & 0 \end{matrix} \right) \sqrt{(2j_1+1)(2j_2+1)(2j_3+1)(2l_1+1)(2l_2+1)}} \quad (5.4.14)$$

(Biedenharn *et al* (1952), reprinted in Biedenharn and van Dam (1956)).

Symmetries

$$\left\{ \begin{matrix} j_1 & j_2 & j_3 \\ l_1 & l_2 & l_3 \end{matrix} \right\} = \left\{ \begin{matrix} j_1 & l_2 & l_3 \\ l_1 & j_2 & j_3 \end{matrix} \right\} = \left\{ \begin{matrix} l_1 & j_2 & l_3 \\ j_1 & l_2 & j_3 \end{matrix} \right\}$$
$$= \left\{ \begin{matrix} l_1 & l_2 & j_3 \\ j_1 & j_2 & l_3 \end{matrix} \right\}. \quad (5.4.15)$$

Each of these is, moreover, invariant under any permutation of its columns:

$$\left\{ \begin{matrix} j_1 & j_2 & j_3 \\ l_1 & l_2 & l_3 \end{matrix} \right\} = \left\{ \begin{matrix} j_i & j_k & j_m \\ l_i & l_k & l_m \end{matrix} \right\} \quad \begin{matrix} (i,k,m \text{ is any permutation} \\ \text{of } 1, 2, 3); \end{matrix} \quad (5.4.16)$$

$$\left\{ \begin{matrix} j_1 & j_2 & j_3 \\ l_1 & l_2 & l_3 \end{matrix} \right\} = 0 \quad \begin{matrix} \text{(unless one can draw a tetrahedron with sides} \\ \text{of lengths } j_1, j_2, j_3, l_1, l_2, l_3 \text{ and such} \\ \text{that the sum of the lengths of the sides of} \\ \text{each triangular face sum to an integer):} \end{matrix} \quad (5.4.17)$$

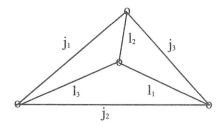

$$
\begin{Bmatrix} j_1 & j_2 & j_3 \\ l_1 & l_2 & l_3 \end{Bmatrix} = \begin{Bmatrix} j_1 & \frac{1}{2}(j_2+j_3-l_2+l_3) & \frac{1}{2}(j_2+j_3+l_2-l_3) \\ l_1 & \frac{1}{2}(-j_2+j_3+l_2+l_3) & \frac{1}{2}(j_2-j_3+l_2+l_3) \end{Bmatrix}
$$

$$
= \begin{Bmatrix} \frac{1}{2}(j_1+j_3-l_1+l_3) & j_2 & \frac{1}{2}(j_1+j_3+l_1-l_3) \\ \frac{1}{2}(-j_1+j_3+l_1+l_3) & l_2 & \frac{1}{2}(j_1-j_3+l_1+l_3) \end{Bmatrix}
$$

$$
= \begin{Bmatrix} \frac{1}{2}(j_1+j_2-l_1+l_2) & \frac{1}{2}(j_1+j_2+l_1-l_2) & j_3 \\ \frac{1}{2}(-j_1+j_2+l_1+l_2) & \frac{1}{2}(j_1-j_2+l_1+l_2) & l_3 \end{Bmatrix}
$$

$$
= \begin{Bmatrix} \frac{1}{2}(j_1+j_2-l_1+l_2) & \frac{1}{2}(j_2+j_3-l_2+l_3) & \frac{1}{2}(j_1+j_3+l_1-l_3) \\ \frac{1}{2}(-j_1+j_2+l_1+l_2) & \frac{1}{2}(-j_2+j_3+l_2+l_3) & \frac{1}{2}(j_1-j_3+l_1+l_3) \end{Bmatrix}
$$

$$
= \begin{Bmatrix} \frac{1}{2}(j_1+j_2+l_1-l_2) & \frac{1}{2}(j_2+j_3+l_2-l_3) & \frac{1}{2}(j_1+j_3-l_1+l_3) \\ \frac{1}{2}(j_1-j_2+l_1+l_2) & \frac{1}{2}(j_2-j_3+l_2+l_3) & \frac{1}{2}(-j_1+j_3+l_1+l_3) \end{Bmatrix}.
$$

$$(5.4.18)$$

The first symmetry in (5.4.18) is due to Regge (1959), reprinted in Biedenharn and van Dam (1956); the other symmetries are 'old'. Note that under Regge symmetry the sums of the numbers in each row and each column stay invariant.

Closed formula

We define an expression $F(abc)$, which occurs already in the general formula for $3j$-symbols (see (5.3.76)), by

$$
F(abc) \equiv \sqrt{\frac{(-a+b+c)!(a-b+c)!(a+b-c)!}{(a+b+c+1)!}}. \tag{5.4.19}
$$

With this abbreviation we have Racah's formula

$$
\begin{Bmatrix} a & b & c \\ d & e & f \end{Bmatrix} = F(abc)\,F(aef)\,F(dbf)\,F(dec)
$$
$$
\times \sum_\rho \frac{(-1)^\rho(\rho+1)!}{\begin{bmatrix} (a+b+d+e-\rho)!(b+c+e+f-\rho)!(a+c+d+f-\rho)! \\ \times(\rho-a-b-c)!(\rho-a-e-f)!(\rho-b-d-f)!(\rho-d-e-c)! \end{bmatrix}} \tag{5.4.20}
$$

(Racah (1942); reprinted in Biedenharn and van Dam (1956)).

Orthogonality; sum rules

$$
\sum_l (2l+1)(2j+1) \begin{Bmatrix} j_1 & j_2 & j \\ j_1 & j_2 & l \end{Bmatrix} \begin{Bmatrix} j_1 & j_2 & j' \\ j_1 & j_2 & l \end{Bmatrix} = \delta_{jj'}
$$

i.e. $M_{jl} \equiv \sqrt{(2l+1)(2j+1)} \begin{Bmatrix} j_1 & j_2 & j \\ j_1 & j_2 & l \end{Bmatrix}$ is a real orthogonal matrix.

$$(5.4.21)$$

(These relations follow from the unitarity of the recoupling transformation; see (5.4.13).)

$$\sum_l (-1)^{j+l+j_3}(2l+1)\begin{Bmatrix} j_1 & j_2 & j_3 \\ j_1 & j_2 & \underline{l} \end{Bmatrix}\begin{Bmatrix} j_1 & l_1 & j \\ j_2 & l_2 & \underline{l} \end{Bmatrix} = \begin{Bmatrix} j_1 & j_2 & j_3 \\ l_2 & l_1 & j \end{Bmatrix} \quad (5.4.22)$$

(Racah; see reference below (5.4.20));

$$\sum_k (-1)^{s+k}(2k+1)\begin{Bmatrix} l_1 & j_2 & l_3 \\ l'_3 & l'_2 & \underline{k} \end{Bmatrix}\begin{Bmatrix} j_2 & j_3 & j_1 \\ l'_1 & l'_3 & \underline{k} \end{Bmatrix}\begin{Bmatrix} l_1 & j_3 & l_2 \\ l'_1 & l'_2 & \underline{k} \end{Bmatrix}$$

$$= \begin{Bmatrix} j_1 & j_2 & j_3 \\ l_1 & l_2 & l_3 \end{Bmatrix}\begin{Bmatrix} l_3 & j_1 & l_2 \\ l'_1 & l'_2 & l'_3 \end{Bmatrix} \quad (5.4.23)$$

$$s = \sum_i j_i + \sum_i l_i + \sum_i l'_i$$

(Biedenharn (1953), Elliot (1953); both reprinted in Biedenharn and van Dam (1956)).

Special formulae

Special formulae for $6j$-symbols are given in table 5.3.

Table 5.3. $J = j_1 + j_2 + j_3$.

$\begin{Bmatrix} j_1 & j_2 & j_3 \\ 0 & j_3 & j_2 \end{Bmatrix}$	$= \dfrac{(-1)^J}{\sqrt{(2j_2+1)(2j_3+1)}}$
$\begin{Bmatrix} j_1 & j_2 & j_3 \\ \frac{1}{2} & j_3-\frac{1}{2} & j_2+\frac{1}{2} \end{Bmatrix}$	$= (-1)^J\sqrt{\dfrac{(J-2j_2)(J-2j_3+1)}{(2j_2+1)(2j_2+2)2j_3(2j_3+1)}}$
$\begin{Bmatrix} j_1 & j_2 & j_3 \\ \frac{1}{2} & j_3-\frac{1}{2} & j_2-\frac{1}{2} \end{Bmatrix}$	$= (-1)^J\sqrt{\dfrac{(J+1)(J-2j_1)}{2j_2(2j_2+1)2j_3(2j_3+1)}}$
$\begin{Bmatrix} j_1 & j_2 & j_3 \\ 1 & j_3-1 & j_2-1 \end{Bmatrix}$	$= (-1)^J\sqrt{\dfrac{J(J+1)(J-2j_1-1)(J-2j_1)}{(2j_2-1)2j_2(2j_2+1)(2j_3-1)2j_3(2j_3+1)}}$
$\begin{Bmatrix} j_1 & j_2 & j_3 \\ 1 & j_3-1 & j_2 \end{Bmatrix}$	$= (-1)^J\sqrt{\dfrac{2(J+1)(J-2j_1)(J-2j_2)(J-2j_3+1)}{2j_2(2j_2+1)(2j_2+2)(2j_3-1)2j_3(2j_3+1)}}$
$\begin{Bmatrix} j_1 & j_2 & j_3 \\ 1 & j_3-1 & j_2+1 \end{Bmatrix}$	$= (-1)^J\sqrt{\dfrac{(J-2j_2-1)(J-2j_2)(J-2j_3+1)(J-2j_3+2)}{(2j_2+1)(2j_2+2)(2j_3+3)(2j_3-1)2j_3(2j_3+1)}}$
$\begin{Bmatrix} j_1 & j_2 & j_3 \\ 1 & j_3 & j_2 \end{Bmatrix}$	$= (-1)^{J+1}\dfrac{2[j_2(j_2+1)+j_3(j_3+1)-j_1(j_1+1)]}{\sqrt{2j_2(2j_2+1)(2j_2+2)2j_3(2j_3+1)(2j_3+2)}}$

Together with the symmetry relation ((5.4.15), (5.4.16)) these formulae are sufficient to calculate any $6j$-symbol with one entry being 0, $\frac{1}{2}$ or 1. By means of the Regge symmetry some other $6j$-symbols can be calculated (see (5.4.18)).

5.5 Combining more than three angular momenta

Here we give only a few remarks and references. Evidently the number of different ways of coupling $J^{(1)} + J^{(2)} + \cdots + J^{(n)}$ to a resultant J, i.e. into a $|jm\rangle$ state, grows rapidly with n, the number of coupled systems. The principle is clear after our general discussion in sections 5.1 and 5.2 (see in particular the summary on page 126) and its application to the case of three angular momenta. Although the recoupling coefficients for four angular momenta (the '$9j$-symbol' appears in the problem of transforming from LS- to jj-coupling) and for five momenta (the '$12j$-symbol' is one particular recoupling coefficient) are of practical importance, we do not consider them here. Information may be found for instance in

(i)   Rotenberg *et al* (1959),
(ii) Biedenharn and van Dam (1956),
(iii) Edmonds (1957) and
(iv) de-Shalit and Talmi (1963).

5.6 Numerical tables and important references on addition of angular momenta

We list here some numerical tables and books in which further information may be found. Short special tables and formulae are scattered in the literature on atomic and nuclear spectroscopy; it is plainly impossible to give here a complete list of them.

General references

(i)   Biedenharn and van Dam (1956)
(ii) Rose (1957)
(iii) Edmonds (1957)
(iv) de-Shalit and Talmi (1963)
(v)   Varshalovich *et al* (1988).

Numerical tables

(i)   Rotenberg *et al* (1959) (contains bibliography of tables)
(ii) Ishidzu *et al* (1960)
(iii) Nikiforov *et al* (1965)
(iv) see also Varshalovich *et al* (1988).

We finally note that a powerful and elegant graphical representation for CGCs exists which can be used in the calculations of coupling of angular momenta. We shall not discuss such a technique here. As references we list

(i)   Brink and Satchler (1968)
(ii) Varshalovich *et al* (1988).

6

REPRESENTATIONS OF THE ROTATION GROUP

For what follows it is useful to have in mind the discussions in chapters 3 and 4. All states $|\gamma jm\rangle$ are assumed to be standard ones.

6.1 Active and passive interpretation; definition of $D^{(j)}_{m'm}$; the invariant subspaces $\mathcal{H}_j$

We shall define the representations $D^{(j)}$ of the rotation group by means of the abstract state vectors $|jm\rangle$ and not—as many authors do, and as we could do as well—by means of wave functions. That implies that the rotations, for which we define the representations, are active rotations.

On the other hand, we could also discuss everything in terms of wave functions and then we would have the choice between active and passive rotations. In that case we would find unitary representations

$$U_a(\alpha, \beta, \gamma) = U(R_a(\alpha, \beta, \gamma))$$
$$U_p(\alpha, \beta, \gamma) = U(R_p(\alpha, \beta, \gamma)). \tag{6.1.1}$$

In section 3.2.2, we found that for a sequence of rotations the three-dimensional matrices M have the property (see (3.2.14))

$$M_a(\alpha, \beta, \gamma)M_p(\alpha, \beta, \gamma) = M_p(\alpha, \beta, \gamma)M_a(\alpha, \beta, \gamma) = 1. \tag{6.1.2}$$

This, as will be shown now, is also true for the unitary representations of these rotations. We have not yet, however, defined what $U_p(\eta)$ or $U_p(\alpha, \beta, \gamma)$ means. As said above, this is not possible without using the concept of wave functions. We shall consider both the active and passive interpretations for a given axis and angle or rotation

$$\eta \equiv \eta n \text{ (angle } \eta \text{ and axis } n \text{ of rotation)}. \tag{6.1.3}$$

Let P be a point in space. If we introduce a frame in reference, P will have the coordinates $x \equiv x, y, z$.

An *active rotation* means P, together with all its physical properties (field at P) is transferred to a new position P':

$$P \rightarrow P' = R_a(\eta)P \tag{6.1.4}$$

with coordinates

$$x'_a = M_a(\eta)x. \tag{6.1.5}$$

170

A *passive rotation* means the coordinates of the point P are referring to a rotated frame of reference; its new coordinates become

$$x'_a = M_p(\eta)x \tag{6.1.6}$$

$$M_a(\eta)\,M_p(\eta) = 1. \tag{6.1.7}$$

This and what follows is illustrated in figure 6.1, where we assume for simplicity a rotation by η about z.

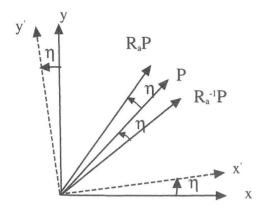

Figure 6.1. Active and passive rotations.

We consider the change of the wave function in active and passive rotations respectively.

In the *active interpretation* the physical system—characterized in space by the wave function—is bodily rotated. The state of the rotated system is $U_a(\eta)|\psi\rangle$ and its wave function at the point P is the same as that of $|\psi\rangle$ at the point $R_a^{-1}P$, because it has been bodily transferred from there to P. Hence

$$\psi'_a(x) \equiv \langle x|U_a(\eta)|\psi\rangle = \psi(M_a^{-1}(\eta)x). \tag{6.1.8}$$

In the *passive interpretation* the frame of reference is rotated, whereas P and the local physical state at P remain unchanged. We must require that there is a description $\psi'_p(x')$ which uses the new coordinates but describes the old situation; it is defined by

$$\psi'_p(x'_p) = \psi'_p(M_p(\eta)x) \equiv \psi(x) \tag{6.1.9}$$

which can be rewritten as

$$\psi'_p(x) \equiv \psi(M_p^{-1}(\eta)x). \tag{6.1.10}$$

We now *define* a unitary transformation $U_p(\eta)$ by

$$\psi'_p(x) \equiv \langle x|U_p(\eta)|\psi\rangle = \psi(M_p^{-1}(\eta)x). \tag{6.1.11}$$

From (6.1.6) and (6.1.10) we read off (for the same η)

$$\psi(M_a^{-1}M_p^{-1}x) = \langle x|U_aU_p|\psi\rangle = \psi(x) = \langle x|\psi\rangle$$

$$\psi(M_p^{-1}M_a^{-1}x) = \langle x|U_pU_a|\psi\rangle = \psi(x) = \langle x|\psi\rangle.$$

Hence

$$U_p(\eta) \quad \equiv U_a^{-1}(\eta) \quad = U_a(-\eta)$$
$$U_p(\alpha, \beta, \gamma) \equiv U_a^{-1}(\alpha, \beta, \gamma) = U_a(-\gamma, -\beta, -\alpha). \tag{6.1.12}$$

Note that for a sequence of rotations the order is reversed in the two interpretations! In the active interpretation the sequence $R_a = R_{a,2}R_{a,1}$ is represented by the unitary matrix $U_a = U_{a,2}U_{a,1}$. In the passive interpretation the sequence $R_p = R_{p,2}R_{p,1}$ is represented by the unitary matrix $U_p = U_{p,1}U_{p,2}$ since only then does $U_p = U_a^{-1}$. This has to be kept in mind if one compares formulae in different books: some authors use the passive interpretation and find $U(R_2R_1) = U(R_1)U(R_2)$, whereas others use the active point of view and obtain the same formulae as we do.

From now on (if not stated otherwise) we mean always the active interpretation. Thus from (3.4.2) and (3.4.5) we have

$$U_a(\eta) = e^{-i\eta\cdot J}$$
$$U_a(\alpha, \beta, \gamma) = e^{-i\alpha J_z}\, e^{-i\beta J_y}\, e^{-i\gamma J_z}. \tag{6.1.13}$$

We define the $(2j + 1)$-dimensional unitary representation $D^{(j)}(\alpha, \beta, \gamma)$ of the active rotation $R_a(\alpha, \beta, \gamma)$ by the transformation of the eigenstates of angular momentum $|jm\rangle$

$$U_a(\alpha, \beta, \gamma)|jm\rangle = \sum_{m'}|jm'\rangle\langle jm'|U_a(\alpha, \beta, \gamma)|jm\rangle$$

$$\equiv \sum_{m'}|jm'\rangle D_{m'm}^{(j)}(\alpha, \beta, \gamma) \tag{6.1.14}$$

$$D_{m'm}^{(j)}(\alpha, \beta, \gamma) \equiv \langle jm'|e^{-i\alpha J_z}\, e^{-i\beta J_y}\, e^{-i\gamma J_z}|jm\rangle.$$

Since we know that J_x, J_y, J_z all commute with J^2, it is clear that the operator $U_a(\alpha, \beta, \gamma)$ also commutes with J^2 and therefore has no matrix elements between states of different j. Therefore the states $|jm\rangle$ of the subspace $\mathcal{H}_j$ transform among each other: the subspace $\mathcal{H}_j$ is invariant under rotations in the three-dimensional space. In other words: if the states $|\gamma jm\rangle$ ($\gamma =$ all remaining quantum numbers) are used as the basis of the Hilbert space, then the unitary representation $U_a(\alpha, \beta, \gamma)$ of the rotation group splits up into an infinite set of submatrices $D^{(j)}$ with dimensions $2j + 1$, which transform only states inside the corresponding invariant subspaces $\mathcal{H}_j$. Since J_y and J_x do not commute with J_z (whose quantum number is m) and $\mathcal{H}_j$ is spanned by the $2j + 1$ states $|jm\rangle$, it is clear that no further splitting into smaller invariant subspaces is possible. We call the subspaces $\mathcal{H}_j$ irreducible subspaces and the

$(2j + 1)$-dimensional matrices $D^{(j)}$ irreducible representations of the rotation group.

$$\mathcal{H} = \sum_j \oplus \mathcal{H}_j$$

$$U_a(\alpha, \beta, \gamma) = \sum_j \oplus D^{(j)}(\alpha, \beta, \gamma)$$

$$(6.1.15)$$

6.2   The explicit form of $D_{m'm}^{(j)}(\alpha, \beta, \gamma)$

By definition, (6.1.14), the $D_{m'm}^{(j)}$ are the matrix elements of U_a between the states $\langle jm'|$ and $|jm\rangle$. Since these states are eigenstates of J_z, we obtain

$$D_{m'm}^{(j)}(\alpha, \beta, \gamma) = \langle jm'|e^{-i\alpha J_z} e^{-i\beta J_y} e^{-i\gamma J_z}|jm\rangle$$
$$= e^{-i(\alpha m' + \gamma m)}\langle jm'|e^{-i\beta J_y}|jm\rangle. \qquad (6.2.1)$$

That leaves us with the problem of calculating

$$d_{m'm}^{(j)}(\beta) \equiv \langle jm'|e^{-i\beta J_y}|jm\rangle. \qquad (6.2.2)$$

In all that follows we shall use this convention: if any matrix $M_{m'm}$ is written explicitly, m' and m will assume their maximum values $m' = m = j$ in the upper left corner, thus

$$M = \begin{pmatrix} M_{jj} & M_{jj-1} & \cdots \\ M_{j-1j} & M_{j-1j-1} & \cdots \\ \vdots & \vdots & & M_{-j-j} \end{pmatrix}. \qquad (6.2.3)$$

6.2.1   The spin-$\frac{1}{2}$ case

For the spin-$\frac{1}{2}$ case the problem has already been solved when we considered the spin-$\frac{1}{2}$ state with components $\pm\frac{1}{2}$ in a given direction ϑ, φ. Indeed, what does (6.2.2) require? It requires the calculation of the state

$$e^{-i\beta J_y}|\tfrac{1}{2}, m\rangle = |\tfrac{1}{2}, m\rangle_{\vartheta=\beta; \varphi=0}$$

with $m = \pm\frac{1}{2}$; namely that state which results from rotating the state $|\frac{1}{2}, m\rangle$ by β about the y-axis. Hence it is the eigenstate $|\frac{1}{2}m\rangle$ in the direction $\vartheta = \beta$; $\varphi = 0$. This state has been calculated in (4.8.11):

$$e^{-i\beta J_y}|\tfrac{1}{2}, \tfrac{1}{2}\rangle = |\tfrac{1}{2}, \tfrac{1}{2}\rangle_{\beta/2,0} = \begin{pmatrix} \cos\dfrac{\beta}{2} \\ \sin\dfrac{\beta}{2} \end{pmatrix}$$

$$= \cos\frac{\beta}{2}|\tfrac{1}{2}, \tfrac{1}{2}\rangle + \sin\frac{\beta}{2}|\tfrac{1}{2}, -\tfrac{1}{2}\rangle$$

$$e^{-i\beta J_y}|\tfrac{1}{2}, -\tfrac{1}{2}\rangle = |\tfrac{1}{2}, -\tfrac{1}{2}\rangle_{\beta/2,0} = \begin{pmatrix} -\sin\dfrac{\beta}{2} \\ \cos\dfrac{\beta}{2} \end{pmatrix} \qquad (6.2.4)$$

$$= -\sin\frac{\beta}{2}|\tfrac{1}{2}, \tfrac{1}{2}\rangle + \cos\frac{\beta}{2}|\tfrac{1}{2}, -\tfrac{1}{2}\rangle.$$

With this, we obtain immediately from (6.2.2)

$$d^{(1/2)}(\beta) = \begin{pmatrix} \cos\dfrac{\beta}{2} & -\sin\dfrac{\beta}{2} \\ \sin\dfrac{\beta}{2} & \cos\dfrac{\beta}{2} \end{pmatrix} \qquad (6.2.5)$$

and with (6.2.1) we find again our old result (4.8.8):

$$D^{(1/2)}(\alpha, \beta, \gamma) = \begin{pmatrix} e^{-i\alpha/2}\cos\dfrac{\beta}{2}e^{-i\gamma/2} & -e^{-i\alpha/2}\sin\dfrac{\beta}{2}e^{i\gamma/2} \\ e^{i\alpha/2}\sin\dfrac{\beta}{2}e^{-i\gamma/2} & e^{i\alpha/2}\cos\dfrac{\beta}{2}e^{i\gamma/2} \end{pmatrix}. \qquad (6.2.6)$$

We shall calculate the same again in another way. From (4.8.5) we have for $n = (\sin\vartheta\cos\varphi, \sin\vartheta\sin\varphi, \cos\vartheta)$

$$(\boldsymbol{n}\cdot\boldsymbol{J})_{1/2} = \frac{1}{2}\begin{pmatrix} \cos\vartheta & e^{-i\varphi}\sin\vartheta \\ e^{i\varphi}\sin\vartheta & -\cos\vartheta \end{pmatrix} = \frac{1}{2}\begin{pmatrix} n_3 & n_1 - in_2 \\ n_1 + in_2 & -n_3 \end{pmatrix}. \quad (6.2.7)$$

In order to represent $R_a(\eta)$ we must calculate

$$\left(e^{-i\eta\boldsymbol{n}\cdot\boldsymbol{J}}\right)_{\frac{1}{2}} = \sum_{k=0}^{\infty} \frac{(-i\eta)^k}{k!}(\boldsymbol{n}\cdot\boldsymbol{J})_{\frac{1}{2}}^k. \qquad (6.2.8)$$

From (6.2.7) we find

$$(\boldsymbol{n}\cdot\boldsymbol{J})_{\frac{1}{2}}^2 = \tfrac{1}{4}\times 1.$$

Thus even powers $(2k)$ of $(n \cdot J)_{\frac{1}{2}}$ are equal to $\left(\frac{1}{2}\right)^{2k}$ and odd powers $(2k+1)$ are equal to $(n \cdot J)_{\frac{1}{2}} \left(\frac{1}{2}\right)^{2k+1}$. Therefore

$$\left(e^{-i\eta n \cdot J}\right)_{\frac{1}{2}} = \cos\frac{\eta}{2} - i\left(\begin{matrix} n_3 & n_1 - in_2 \\ n_1 + in_2 & -n_3 \end{matrix}\right)\sin\frac{\eta}{2}.$$

In particular

$$\left(e^{-i\alpha J_z}\right)_{\frac{1}{2}} = \left(\begin{matrix} e^{-i\alpha/2} & 0 \\ 0 & e^{i\alpha/2} \end{matrix}\right) \tag{6.2.9}$$

$$\left(e^{-i\beta J_y}\right)_{\frac{1}{2}} = \left(\begin{matrix} \cos\dfrac{\beta}{2} & -\sin\dfrac{\beta}{2} \\ \sin\dfrac{\beta}{2} & \cos\dfrac{\beta}{2} \end{matrix}\right). \tag{6.2.10}$$

Carrying out the matrix product $e^{-i\alpha J_z} e^{-i\beta J_y} e^{-i\gamma J_z}$ yields (6.2.6).

6.2.2 The general case

We know how a spin-$\frac{1}{2}$ state transforms under a rotation; we have just derived it. Can we take advantage of that knowledge to derive the transformation law for a $j > \frac{1}{2}$ state? Indeed, we may do that by using the fact that we can compose several spin-$\frac{1}{2}$ states to one total angular momentum j state. In (4.10.8) the most general normalized state $|jm\rangle$ was given as a totally symmetrized direct product of $2j$ spin-$\frac{1}{2}$ states. The symmetrization concerns only the labels (i) which we might attach to $2j$ systems, while the number of u_+ states and that of u_- states remains $j+m$ and $j-m$ respectively. Therefore the $(2j)!$ permutations leave the transformation properties untouched, which implies that for our present purpose the symmetrization is an unnecessary luxury, which we omit. We thus evaluate (6.2.2)

$$d_{m'm}^{(j)}(\beta) = \langle jm'|e^{-i\beta J_y}|jm\rangle \tag{6.2.11}$$

between (simplified) states according to (4.10.8)

$$|jm\rangle = C_{jm}u_+^{j+m}u_-^{j-m}; \quad C_{jm} = \sqrt{\frac{(2j)!}{(j+m)!(j-m)!}}. \tag{6.2.12}$$

There are $2j$ systems of spin $\frac{1}{2}$ which for a moment we may again imagine to be labelled $(u)^{(i)}; i = 1, \ldots, 2j$; then, since the different $J_y^{(i)}$ commute:

$$e^{-i\beta J_y} = e^{-i\beta \sum_i J_y^{(i)}} = \prod_{i=1}^{2i} e^{-i\beta J_y^{(i)}} \tag{6.2.13}$$

where $J_y^{(i)}$ acts only on the ith spin state and is the unit operator for all others. Thus there is exactly one operator per spin state $u^{(i)}$. How it acts on $u^{(i)}$ we know from (6.2.4):

$$e^{-i\beta J_y^{(i)}} u_+^{(i)} = \cos\frac{\beta}{2} u_+^{(i)} + \sin\frac{\beta}{2} u_-^{(i)}$$

$$e^{-i\beta J_y^{(i)}} u_-^{(i)} = -\sin\frac{\beta}{2} u_+^{(i)} + \cos\frac{\beta}{2} u_-^{(i)}.$$

(6.2.14)

Hence (forgetting from now on the labelling) with (6.2.12)

$$e^{-i\beta J_y}|jm\rangle = C_{jm} \left(\cos\frac{\beta}{2} u_+ + \sin\frac{\beta}{2} u_-\right)^{j+m} \left(-\sin\frac{\beta}{2} u_+ + \cos\frac{\beta}{2} u_-\right)^{j-m}.$$

(6.2.15)

We multiply out (remember that $j\pm m$ is always an integer ≥ 0)

$$e^{-i\beta J_y}|jm\rangle = C_{jm} \sum_{\nu\mu} \binom{j+m}{\nu}\binom{j-m}{\mu}(-1)^{\mu}$$

$$\times \left(\cos\frac{\beta}{2}\right)^{\nu+j-m-\mu}\left(\sin\frac{\beta}{2}\right)^{j+m-\nu+\mu} u_+^{\nu+\mu} u_-^{2j-\nu-\mu}.$$

We transform the sum indices:

$$\nu + \mu \equiv j + m' \qquad\qquad \mu = j + m' - \nu$$

$$\text{or}$$

$$\nu \equiv \nu \qquad\qquad\qquad \nu = \nu$$

and obtain in the sum the product

$$u_+^{j+m'} u_-^{j-m'} = \frac{1}{C_{jm'}}|jm'\rangle.$$

Therefore

$$e^{-i\beta J_y}|jm\rangle = \sum_{m'}|jm'\rangle \left\{ \frac{C_{jm}}{C_{jm'}} \sum_{\nu} \binom{j+m}{\nu}\binom{j-m}{j+m'-\nu} \right.$$

$$\times (-1)^{j+m'-\nu}\left(\cos\frac{\beta}{2}\right)^{2\nu-m-m'}\left(\sin\frac{\beta}{2}\right)^{2j+m+m'-2\nu} \Bigg\}$$

$$= \sum_{m'}|jm'\rangle d_{m'm}^{(j)}(\beta)$$

(see (6.2.2)). Inserting the normalization constants C_{jm} from (6.2.12) gives the explicit form

$$
d_{m'm}^{(j)}(\beta) = \sqrt{\frac{(j+m')!(j-m')!}{(j+m)!(j-m)!}} \left(\cos\frac{\beta}{2} \right)^{-m-m'} \left(\sin\frac{\beta}{2} \right)^{2j+m+m'}
$$

$$
\times (-1)^{j+m'} \sum_{\nu} \left[(-1)^{\nu} \binom{j+m}{\nu} \binom{j-m}{j+m'-\nu} \right.
$$

$$
\left. \times \left(\cos\frac{\beta}{2} \right)^{2\nu} \left(\sin\frac{\beta}{2} \right)^{-2\nu} \right]
$$

$$
D_{m'm}^{(j)}(\alpha, \beta, \gamma) = e^{-i(\alpha m' + \gamma m)} d_{m'm}^{(j)}(\beta).
$$

(6.2.16)

The sum over ν goes over all integer values $\nu \geq 0$ for which the binomial coefficients do not vanish.

Example. For $j = 1$ one finds

$$
d^{(1)}(\beta) = \begin{pmatrix} \frac{1}{2}(1+\cos\beta) & -\frac{1}{\sqrt{2}}\sin\beta & \frac{1}{2}(1-\cos\beta) \\ \frac{1}{\sqrt{2}}\sin\beta & \cos\beta & -\frac{1}{\sqrt{2}}\sin\beta \\ \frac{1}{2}(1-\cos\beta) & \frac{1}{\sqrt{2}}\sin\beta & \frac{1}{2}(1+\cos\beta) \end{pmatrix}.
$$

(6.2.17)

If one compares (6.2.16) to other formulae in the literature, one has to keep in mind that some authors give $D_{m'm}^{(j)}(-\gamma, -\beta, -\alpha)$, which refers to the passive interpretation $R_p(\alpha, \beta, \gamma)$, and call it $D_{m'm}^{(j)}(\alpha, \beta, \gamma)$; it also happens that in the same book sometimes the active and sometimes the passive interpretation is used. Apart from that real difference, the above formula (6.2.16) can be written in various other equivalent forms.

6.3 General properties of $D^{(j)}$

In what follows we shall investigate the general properties of the $D^{(j)}$-matrices. We expect close relations to the eigenfunctions of angular momentum and to the coupling coefficients (CGCs). The unitarity relation $[D^{\dagger}(\alpha, \beta, \gamma)]_{mm'}^{(j)} \equiv D_{m'm}^{(j)*}(\alpha, \beta, \gamma) = [D^{-1}(\alpha, \beta, \gamma)]_{mm'}^{(j)} = D_{mm'}^{(j)}(-\gamma, -\beta, -\alpha)$ will often be used.

6.3.1 Relation to the Clebsch–Gordan coefficients

We consider various matrix elements of the unitary transformation representing a rotation $R_a(\alpha, \beta, \gamma)$ or $R_a(\eta)$; we shall omit the arguments α, β, γ and η; in products of D-matrices all of them have the same arguments unless stated otherwise.

The relations of the D matrices to the CGCs are obtained if we use mixed matrix elements, i.e. between coupled and uncoupled states. Let R_a be an active rotation, U_a the induced unitary transformation. Then

$$U_a|jm'\rangle = \sum_m |jm\rangle\langle jm|U_a|jm'\rangle = \sum_m |jm\rangle D_{mm'}^{(j)}$$

$$\langle jm|U_a = \sum_{m'} \langle jm|U_a|jm'\rangle\langle jm'| = \sum_{m'} D_{mm'}^{(j)}\langle jm'|.$$

(6.3.1)

Now consider the matrix element $\langle j_1 m_1 j_2 m_2|U_a|(j_1 j_2)jm'\rangle$, let U act once to the left and once to the right and equate these two expressions:

to the right

$$\langle j_1 m_1 j_2 m_2|U_a|(j_1 j_2)jm'\rangle = \sum_\mu \langle j_1 m_1 j_2 m_2|j\mu\rangle D_{\mu m'}^{(j)}$$

to the left

$$\langle j_1 m_1 j_2 m_2|U_a|(j_1 j_2)jm'\rangle = \sum_{m_1'm_2'} D_{m_1 m_1'}^{(j_1)} D_{m_2 m_2'}^{(j_2)} \langle j_1 m_1' j_2 m_2'|jm'\rangle.$$

By equating these two expressions we obtain two equations; namely, we may use the orthogonality relations of the CGCs in order to eliminate these on either the left- or the right-hand side of

$$\sum_\mu \langle j_1 m_1 j_2 m_2|j\mu\rangle D_{\mu m'}^{(j)} = \sum_{m_1'm_2'} D_{m_1 m_1'}^{(j_1)} D_{m_2 m_2'}^{(j_2)} \langle j_1 m_1' j_2 m_2'|jm'\rangle.$$

(6.3.2)

Using the first of equations (5.3.16), i.e. multiplying by $\langle j_1 m_1 j_2 m_2|jm\rangle$ and summing over m_1 and m_2, we obtain

$$D_{mm'}^{(j)} = \sum_{\substack{m_1,m_2 \\ m_1',m_2'}} \langle jm|j_1 m_1 j_2 m_2\rangle D_{m_1 m_1'}^{(j_1)} D_{m_2 m_2'}^{(j_2)} \langle j_1 m_1' j_2 m_2'\rangle|jm'\rangle$$

(6.3.3)

(only terms with $m_1 + m_2 = m$ and $m_1' + m_2' = m$ contribute). Using the second of equations (5.3.16), i.e. multiplying by $\langle jm'|j_1\mu_1 j_2\mu_2\rangle$ and summing over j and m', we obtain

$$D_{m_1\mu_1}^{(j_1)} D_{m_2\mu_2}^{(j_2)} = \sum_{\mu jm'} \langle j_1 m_1 j_2 m_2|j\mu\rangle D_{\mu m'}^{(j)} \langle jm'|j_1\mu_1 j_2\mu_2\rangle.$$

We change the subscripts $\mu_{1,2} \to m_{1,2}'$; $\mu \to m$ and find

$$D_{m_1 m_1'}^{(j_1)} D_{m_2 m_2'}^{(j_2)} = \sum_j \langle j_1 m_1 j_2 m_2|jm\rangle D_{m_1+m_2,m_1'+m_2'}^{(j)} \langle jm'|j_1 m_1' j_2 m_2'\rangle.$$

(6.3.4)

Finally we consider the matrix element (the CGC) $\langle j_1 m_1 j_2 m_2 | (j_1 j_2) j m \rangle$, where we introduce $1 = U_a U_a^\dagger$

$$\langle j_1 m_1 j_2 m_2 | U_a U_a^\dagger | j m \rangle = \langle j_1 m_1 j_2 m_2 | j m \rangle$$

$$= \sum_{m_1' m_2' m'} D^{(j_1)}_{m_1 m_1'} D^{(j_2)}_{m_2 m_2'} D^{(j)*}_{mm'} \langle j_1 m_1' j_2 m_2' | j m' \rangle.$$

Thus

$$\langle j_1 m_1 j_2 m_2 | j m \rangle = \sum_{m_1' m_2' m'} D^{(j_1)}_{m_1 m_1'} D^{(j_2)}_{m_2 m_2'} D^{(j)*}_{m_1+m_2, m_1'+m_2'} \langle j_1 m_1' j_2 m_2' | j m' \rangle. \quad (6.3.5)$$

Multiplying by $\langle j_1 m_1 j_2 m_2 | j' \mu' \rangle$ and summing over m_1 and m_2 yields, according to the first equation of (5.3.16)

$$\delta_{j'j} \delta_{\mu' m} = \sum_{\substack{m_1 m_2 \\ m_1' m_2' m'}} \left[\langle j_1 m_1 j_2 m_2 | j' \mu' \rangle D^{(j_1)}_{m_1 m_1'} D^{(j_2)}_{m_2 m_2'} D^{(j)*}_{m_1+m_2, m_1'+m_2'} \right.$$

$$\left. \times \langle j_1 m_1' j_2 m_2' | j m' \rangle \right]. \quad (6.3.6)$$

6.3.2 Significance of the relation to the CGCs

What do these equations, in particular (6.3.3), mean? They mean what in group theory is called transforming a direct product of two representations into a form where it splits up into irreducible parts. We shall look into the details.

Consider a matrix M with a particular structure; namely, let it be built up of submatrices along the diagonal, like this:

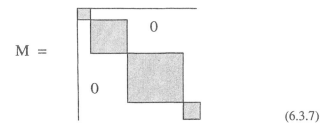

$$M = \qquad (6.3.7)$$

Transform M by a unitary matrix U. Then $M' = U^\dagger M U$ will no longer have the box structure of M. Now, any matrix of the type M', which can be transformed into the form M (box structure), is called reducible; if not, it is called irreducible. If the boxes showing up in M are irreducible, then one says that M is split into its irreducible parts.

Now, how is it with the D-matrices? Let us consider the coupling of two angular momenta, j_1 and j_2, which we hold fixed in the following discussion. By fixing j_1 and j_2 we have singled out a subspace $\mathcal{H}_{j_1 j_2}$ of the total Hilbert space $\mathcal{H}$. The basis states in $\mathcal{H}_{j_1 j_2}$ can be chosen to be

$$\{|j_1 m_1 j_2 m_2\rangle\} \qquad \text{or} \qquad \{|(j_1 j_2) jm\rangle\}$$

with variables

(6.3.8)

$$-j_1 \leq m_1 \leq j_1 \qquad\qquad -j \leq m \leq j$$
$$\text{or}$$
$$-j_2 \leq m_2 \leq j_2 \qquad\qquad |j_1 - j_2| \leq j \leq j_1 + j_2.$$

The dimension of $\mathcal{H}_{j_1 j_2}$—in either representation—is $(2j_1+1)(2j_2+1)$. Consider now how these states transform under a rotation: the coupled states transform with $D^{(j)}$, namely

$$|(j_1 j_2) jm'\rangle = \sum_m |(j_1 j_2) jm\rangle D^{(j)}_{mm'} \qquad (6.3.9)$$

that is, they transform with the irreducible representation $D^{(j)}$. There are, however, several different j values. We may combine all the $D^{(j)}$ matrices in a 'direct sum' $D^{(c)}$ where the superscript c indicates that $D^{(c)}$ transforms the 'coupled' states)

$$D^{(c)}_{cc'} \equiv \sum_{j=|j_1-j_2|}^{j_1+j_2} \oplus D^{(j)}_{mm'} \qquad (6.3.10)$$

i.e. in one single matrix—just by lining them up along the diagonal—like this (we take the example $j_1 = 1$; $j_2 = 3$; $2 \leq j \leq 4$):

$$D^{(c)} = (D^{(2)} \oplus D^{(3)} \oplus D^{(4)})_{cc'} \qquad (6.3.11)$$

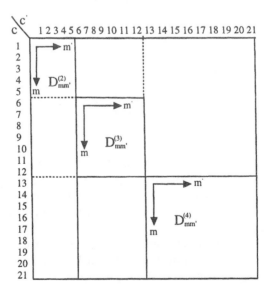

This matrix transforms the subspace $\mathcal{H}_{j_1 j_2}$ if the set $\{|(j_1 j_2)jm\rangle\}$ is chosen as a basis. It is fully split up into its irreducible parts $D^{(j)}$ $(j = |j_1 - j_2|, \ldots, j_1 + j_2)$.

Next, we consider the other basis in $\mathcal{H}_{j_1 j_2}$, that one which is provided by the uncoupled states: $\{|j_1 m_1 j_2 m_2\rangle\}$. These states we have called 'direct product' states $|j_1 m_1 j_2 m_2\rangle \equiv |j_1 m_1\rangle \otimes |j_2 m_2\rangle$. Here each factor transforms with its own D-matrix. Hence together with what is called the direct product of D-matrices

$$|j_1 m_1' j_2 m_2'\rangle = \sum_{m_1 m_2} |j_1 m_1 j_2 m_2\rangle D^{(j_1)}_{m_1 m_1'} D^{(j_2)}_{m_2 m_2'}. \qquad (6.3.12)$$

The 'direct product' $D^{(u)}$ transforms the 'uncoupled' states:

$$D^{(u)}_{uu'} \equiv \left(D^{(j_1)}_{m_1 m_1'}\right) \otimes \left(D^{(j_2)}_{m_2 m_2'}\right) = \left(D^{(j_1)} \otimes D^{(j_2)}\right)_{m_1 m_1' m_2 m_2'} \qquad (6.3.13)$$

which is a matrix with four subscripts. It could be written in a four-dimensional matrix scheme, but it need not be. Indeed, we can combine two-dimensional schemes in an interlacing way to write down (6.3.13) like this (we take the same example $j_1 = 1$; $j_2 = 3$):

$$D^{(u)} = D^{(1)} \otimes D^{(3)} = \left(D^{(1)} \otimes D^{(3)}\right)_{uu'} \qquad (6.3.14)$$

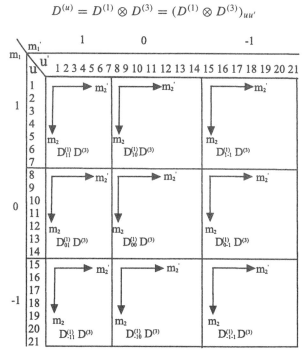

This matrix transforms the subspace $\mathcal{H}_{j_1 j_2}$ if the set $\{|j_1 m_1 j_2 m_2\rangle\}$ is chosen as basis. It is not split up into irreducible parts, but it can be 'reduced'.

Now the situation is this: we have in the subspace $\mathcal{H}_{j_1 j_2}$ of dimension $(2j_1 + 1)(2j_2 + 1)$ two orthonormal bases, each containing $(2j_1 + 1)(2j_2 + 1)$ basis states. Let us denote the coupled states by $|c_i\rangle \equiv |(j_1 j_2)jm\rangle$, where c_i means 'coupled' and i goes from 1 to $(2j_1 + 1)(2j_2 + 1)$, and the uncoupled states by $|u_i\rangle \equiv |j_1 m_1 j_2 m_2\rangle$, where u_i means 'uncoupled' and i has the same range as before. Then the transformation under rotations is given by

$$\begin{aligned}
|c_i\rangle' &= \sum_k |c_k\rangle\langle c_k|c_i\rangle' \\
&= \sum_k D_{ki}^{(c)}|c_k\rangle \equiv D^{(c)}|c_i\rangle \\
|u_i\rangle' &= \sum_k |u_k\rangle\langle u_k|u_i\rangle' \\
&= \sum_k D_{ki}^{(u)}|u_k\rangle \equiv D^{(u)}|u_i\rangle.
\end{aligned}$$

$$(6.3.15)$$

Between the coupled states $|c_i\rangle$ and the uncoupled ones $|u_i\rangle$ is the unitary transformation by the CGCs—we call it T—which should be the same before and after a rotation:

$$\begin{aligned}
|c_i\rangle &= \sum_k |u_k\rangle\langle u_k|c_i\rangle = \sum_k T_{ki}|u_k\rangle \equiv T|u_i\rangle \\
|c_i\rangle' &= \sum_k |u_k\rangle'\langle u_k|'c_i\rangle' = \sum_k T_{ki}|u_k'\rangle \equiv T|u_i\rangle'.
\end{aligned}$$

$$(6.3.16)$$

Here $T_{ki} = \langle u_k|c_i\rangle = \langle j_1 m_1 j_2 m_2|jm\rangle$ is a CGC. From (6.3.15) and (6.3.16) we read off

$$|c_i\rangle' = T|u_i\rangle' = TD^{(u)}|u_i\rangle = TD^{(u)}T^{-1}|c_i\rangle$$

$$|c_i\rangle' = D^{(c)}|c_i\rangle.$$

$$(6.3.17)$$

Hence

$$D^{(c)} = TD^{(u)}T^{-1}$$

$$D^{(u)} = T^{-1}D^{(c)}T.$$

$$(6.3.18)$$

The first of these equations is equivalent to (6.3.3) and the second to (6.3.4). They become identical with those equations if in (6.3.3) the 'direct sum' over j is understood. We have thus found the following very plausible result:

> The same unitary transformation which relates the coupled and uncoupled states to each other, and whose matrix elements are the CGCs—this same unitary transformation, if applied to the direct product $D^{(j_1)} \otimes D^{(j_2)} \equiv D^{(u)}$, transforms it into the direct sum $\sum \oplus D^{(c)}$ of the irreducible representations $D^{(j)}$ with $|j_1 - j_2| \le j \le j_1 + j_2$.

$$(6.3.19)$$

6.3.3   Relation to the eigenfunctions of angular momentum

Remark. It should be mentioned that the $D_{mm'}^{(l)}$ are the eigenfunctions of the quantum mechanical symmetric top. This fact has found applications in nuclear physics. We shall not discuss it here. See, however, Edmonds (1957) for further details.

6.3.3.1   Transformation of eigenfunctions under rotations

The eigenfunctions of angular momentum are defined as the wave functions

$$\psi_{jm}(x) = \langle x | jm \rangle \tag{6.3.20}$$

where x are the coordinates of a point P on the unit sphere, i.e.

$$x \equiv (\sin \vartheta \, \cos \varphi, \, \sin \vartheta \, \sin \varphi, \, \cos \vartheta). \tag{6.3.21}$$

As we have already found in various discussions, for the last time in (6.1.10), an *active* rotation $R_a(\alpha, \beta, \gamma)$ on the state carries the wave function from P to $M_a P$ and from $M_a^{-1} P$ to the point P, thus

$$\psi'_{jm,a}(x) \equiv \langle x | U_a | jm \rangle = \psi_{jm}(M_a^{-1}x) = \psi_{jm}(M_p x). \tag{6.3.22}$$

$M_p x = x'_p$ are the new coordinates of the point P after a *passive* rotation $R_p(\alpha, \beta, \gamma)$:

$$\psi_{jm}(x'_p) = \langle x | U_a | jm \rangle. \tag{6.3.23}$$

Now

$$U_a | jm \rangle = \sum_{m'} | jm' \rangle D_{mm'}^{(j)}(\alpha, \beta, \gamma). \tag{6.3.24}$$

Hence (remember x lies on the unit sphere)

$$\psi_{jm}(x'_p) = \sum_{m'} \psi_{jm'}(x) D_{m'm}^{(j)}(\alpha, \beta, \gamma) = \psi'_{jm,a}(x) \tag{6.3.25}$$

where x'_p are the new coordinates of the old point P after the rotation $R_p(\alpha, \beta, \gamma)$ of the coordinate system, and $D_{m'm}^{(j)}(\alpha, \beta, \gamma)$ is the representation of $R_a(\alpha, \beta, \gamma)$. For $j = l = $ integer, $\psi_{lm} = Y_{lm}$.

   This transformation may be written also by taking the other interpretation on the left-hand side; namely, remembering that

$$x'_p = M_p x = M_a^{-1} x$$

and taking the inverse of the whole transformation, we obtain, using $D^\dagger(\alpha, \beta, \gamma) = D^{-1}(\alpha, \beta, \gamma)$

$$\begin{aligned}
\psi_{jm}(x'_a) &= \sum_{m'} \psi_{jm'}(x) [D^\dagger(\alpha, \beta, \gamma)]_{m'm}^{(j)} \\
&= \sum D_{mm'}^{(j)*}(\alpha, \beta, \gamma) \psi_{jm'}(x)
\end{aligned} \tag{6.3.26}$$

where x'_a denote the coordinates of the new point $P' = R_a(\alpha, \beta, \gamma)P$ in the old coordinate system and $D^{(j)}_{m'm}(\alpha, \beta, \gamma)$ is as in (6.3.25). For $j = l =$ integer, $\psi_{lm} = Y_{lm}$.

6.3.3.2 The $D^{(l)}_{m0}$ are spherical harmonics

We may represent the wave function at the point x' on the unit sphere by a suitable rotation of the wave function at the point $(0, 0, 1)$. This is accomplished by (6.3.26). In that equation we put $x = (0, 0, 1)$, i.e. $\vartheta = 0, \varphi = 0$.

The point x'_a is the point into which $x = (0, 0, 1)$ is carried by the active rotation $R_a(\alpha, \beta, \gamma)$. If x'_a is given, we only have to choose the correct $R_a(\alpha, \beta, \gamma)$.

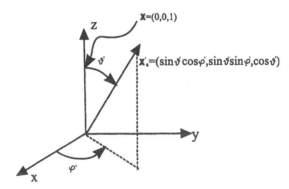

Figure 6.2. The wave function at point x'_a is obtained by rotating the wave function at point x.

Figure 6.2 shows that we must first rotate by ϑ' about the y-axis and then by φ' about the z-axis. We remember that

$$R_a(\alpha, \beta, \gamma) \equiv R_a(\alpha e_3) R_a(\beta e_2) R_a(\gamma e_3)$$

(see (3.2.4)); hence $\gamma = 0$, $\beta = \vartheta'$ and $\alpha = \varphi'$. We then obtain from (6.3.26)

$$\psi_{jm}(\vartheta', \varphi') = \sum_{m'} D^{(j)*}_{mm'}(\varphi', \vartheta', 0)\psi_{jm'}(0, 0). \tag{6.3.27}$$

The $\psi_{jm'}(0, 0)$ are simply constants. The equation states that once these constants are determined, $\psi_{jm}(\vartheta', \varphi')$ is the wavefunction for the $|jm\rangle$ state—whether j is an integer or not. This relation makes it possible to define wave functions for half-integer angular momenta; these wave functions are essentially the $D^{(j)}(\vartheta, \varphi)$ with half-integer j. This possibility does not, however, lead very far, since the half-integer angular momentum has its origin in the spin part. Spin, however, is locally attached to the 'elementary particles' and should not be described by a wave function in space, which gives the probability amplitude for the orbital part of the angular momentum. Thus the full significance of (6.3.27) will show up by taking $j = l =$ integer. Then

$$\psi_{jm'}(\vartheta = \varphi = 0) = Y_{lm'}(0, 0)$$

and (6.3.27) reads

$$Y_{lm}(\vartheta', \varphi') = \sum_{m'} D_{mm'}^{(j)*}(\varphi', \vartheta', 0) Y_{jm'}(0, 0). \tag{6.3.28}$$

First we put $\vartheta' = 0$ and obtain with (6.2.16) and (6.2.2)

$$D_{mm'}^{(j)*}(\varphi', 0, 0) = e^{im\varphi'} \delta_{mm'}$$

$$Y_{lm}(0, \varphi') = e^{im\varphi'} Y_{lm}(0, 0). \tag{6.3.29}$$

The Y_{lm} are one-valued functions (see (4.7.21)); hence $Y_{lm}(0, \varphi')$ must be independent of φ'; that is, according to (6.3.28), only possible if $Y_{lm}(0, 0)$ vanishes for $m \neq 0$. Hence

$$Y_{lm}(0, \varphi) = \begin{cases} 0 & \text{for } m \neq 0 \\ Y_{l0}(0, 0) = \sqrt{\dfrac{2l+1}{4\pi}} \quad (4.7.30) & \text{for } m = 0 \end{cases} \tag{6.3.30}$$

which also can be read off from (4.7.27). With this, (6.3.28) gives

$$Y_{lm}(\vartheta, \varphi) = \sqrt{\frac{2l+1}{4\pi}} D_{m0}^{(l)*}(\varphi, \vartheta, \gamma)$$

$$D_{m0}^{(l)}(\varphi, \vartheta, \gamma) = \sqrt{\frac{4\pi}{2l+1}} Y_{lm}^*(\vartheta, \varphi) \quad \text{(independent of } \gamma\text{)}. \tag{6.3.31}$$

Since D is unitary, we obtain from $D^\dagger = D^{-1}$

$$D_{0m}^{(l)*}(0, -\vartheta, -\varphi) = D_{m0}^l(\varphi, \vartheta, 0) = \sqrt{\frac{4\pi}{2l+1}} Y_{lm}^*(\vartheta, \varphi).$$

We change $-\vartheta, -\varphi$ into ϑ, φ:

$$D_{0m}^{(l)*}(0, \vartheta, \varphi) = \sqrt{\frac{4\pi}{2l+1}} Y_{lm}^*(-\vartheta, -\varphi) \equiv \sqrt{\frac{4\pi}{2l+1}} Y_{lm}^*(\vartheta, \pi - \varphi);$$

by means of (4.7.27) and (4.7.28)

$$Y_{lm}(\vartheta, \varphi) = \text{constant } e^{im\varphi} P_l^m(\cos \vartheta)$$

which gives

$$Y_{lm}(\vartheta, \pi - \varphi) = \text{constant } (-1)^m e^{-im\varphi} P_l^m(\cos \vartheta) = (-1)^m Y_{lm}^*(\vartheta, \varphi).$$

Hence

$$D^{(l)}_{0m}(\alpha, \vartheta, \varphi) = \sqrt{\frac{4\pi}{2l+1}}(-1)^m Y^*_{lm}(\vartheta, \varphi), \quad \text{independent of } \alpha. \qquad (6.3.32)$$

Finally with $m = 0$ (see (4.7.27) and (4.7.28))

$$D^{(l)}_{00}(\alpha, \vartheta, \gamma) = \sqrt{\frac{4\pi}{2l+1}}Y^*_{l0}(\vartheta) = P_l(\cos\vartheta). \qquad (6.3.33)$$

6.3.3.3 The addition theorem and the composition rule of spherical harmonics; integral over three spherical harmonics

From (6.3.25) and (6.3.26) it follows in either interpretation that

$$\sum_m \varphi_{jm}(\boldsymbol{x}_1)\psi^*_{jm}(\boldsymbol{x}_2) = \sum_m \varphi_{jm}(\boldsymbol{x}'_1)\psi^*_{jm}(\boldsymbol{x}'_2) \qquad (6.3.34)$$

whenever $\boldsymbol{x}_1 \leftrightarrow \boldsymbol{x}'_1$ and $\boldsymbol{x}_2 \leftrightarrow \boldsymbol{x}'_2$ are connected by a rotation. The situation is shown in figure 6.3.

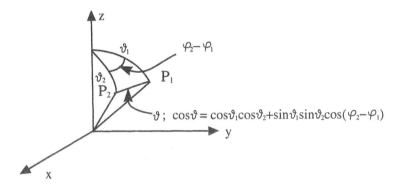

Figure 6.3. The relationship between the spherical coordinates of points P_1 and P_2.

We now take advantage of the rotational invariance of (6.3.34), by rotating such that $P_1 \to P'_1 = (0, 0, 1)$ and that $P_2 \to P'_2$ comes to lie in the xz-plane, i.e. $P'_2 = (\sin\vartheta, 0, \cos\vartheta)$. The angle ϑ is, according to the figure, the angle between P_1 and P_2 (and between P'_1 and P'_2). We specialize to $j = l$ (integer); in that case $\psi_{lm} \to Y_{lm}$; in particular (see (6.3.30))

$$Y_{lm}(P'_1) = Y_{lm}(0, \varphi'_1) = \delta_{m0}\sqrt{\frac{2l+1}{4\pi}}$$

so that (6.3.34) becomes

$$Y^*_{l0}(\vartheta, 0)\sqrt{\frac{2l+1}{4\pi}} = \sum_m Y_{lm}(\vartheta, \varphi)Y^*_{lm}(\vartheta_2, \varphi_2)$$

or, with (see (4.7.27) and (4.7.28))

$$Y_{l0}^*(\vartheta, 0) = Y_{l0}(\vartheta, 0) = \sqrt{\frac{2l+1}{4\pi}} P_l(\cos\vartheta)$$

$$P_l(\cos\vartheta) = \frac{4\pi}{2l+1} \sum_m Y_{lm}(\vartheta_1, \varphi_1) Y_{lm}^*(\vartheta_2, \varphi_2)$$

(6.3.35)

(compare figure 6.3). With $\varphi_1 = \varphi_2 = \varphi$ this angle drops out and $\vartheta = \vartheta_2 - \vartheta_1$. Then we obtain from (4.7.27)

$$P_l(\cos(\vartheta_2 - \vartheta_1)) = \sum_m \frac{(l-m)!}{(l+m)!} P_l^m(\cos\vartheta_1) P_l^m(\cos\vartheta_2). \qquad (6.3.36)$$

These two equations are known as the addition theorems of spherical harmonics and Legendre polynomials. Two further useful formulae for spherical harmonics are found by specializing to $m_1' = m_2' = m' = 0$ in the combination formula for D-matrices, (6.3.4). The D-matrices become then spherical harmonics (see (6.3.31)). We obtain immediately the composition rule for spherical harmonics

$$Y_{l_1 m_1}(\vartheta, \varphi) Y_{l_2 m_2}(\vartheta, \varphi) = \sqrt{\frac{(2l_1+1)(2l_2+1)}{4\pi}} \sum_{lm} \langle l_1 m_1 l_2 m_2 | lm \rangle$$

(6.3.37)

$$\times \sqrt{\frac{1}{2l+1}} Y_{lm}(\vartheta, \varphi) \langle l0 | l_1 0 l_2 0 \rangle.$$

The CGCs take care of $m = m_1 + m_2$ and of the 'conservation of parity': namely, both sides should transform equally under reflections. Since $Y_{lm}(-x^\circ) = (-1)^l Y_{lm}(x^\circ)$ we must require that on the right-hand side only such l contribute, that $(-1)^{l_1+l_2} = (-1)^l$; thus $l_1 + l_2 + l$ must be even. The CGC $\langle l0 | l_1 0 l_2 0 \rangle$ does that automatically (see (5.3.53)). Note that the argument of all spherical harmonics is the same. This is different from the formula for combining two angular momentum eigenfunctions into a new one:

$$|j_1 m_1 j_2 m_2\rangle = \sum |(j_1 j_2) jm\rangle \langle jm | j_1 m_1 j_2 m_2\rangle$$

which, translated to wave functions with integer j reads

$$Y_{l_1 m_1}(\vartheta_1, \varphi_1) Y_{l_2 m_2}(\vartheta_2, \varphi_2) = \sum \langle l_1 m_1 l_2 m_2 | lm \rangle \psi_{lm}(\vartheta_1, \varphi_1, \vartheta_2, \varphi_2)$$

where $\psi_{lm}(\vartheta_1, \varphi_1, \vartheta_2, \varphi_2)$ is no longer a spherical harmonic.

From (6.3.37) we easily obtain the integral over three spherical harmonics. Multiplying (6.3.37) by $Y_{l_3 m_3}^*(\vartheta, \varphi)$ and integrating over the whole sphere gives (because of (4.7.10))

$$\int Y_{lm} Y_{l_3 m_3}^* \sin\vartheta \, d\vartheta \, d\varphi = \delta_{ll_3} \delta_{mm_3}$$

and we obtain the desired result:

$$\int Y_{l_1 m_1}(\vartheta, \varphi) Y_{l_2 m_2}(\vartheta, \varphi) Y^*_{l_3 m_3}(\vartheta, \varphi) \sin\vartheta \, d\vartheta \, d\varphi$$

$$= \sqrt{\frac{(2l_1 + 1)(2l_2 + 1)}{4\pi(2l_3 + 1)}} \langle l_1 m_1 l_2 m_2 | l_3 m_3 \rangle \langle l_1 0 l_2 0 | l_3 0 \rangle. \tag{6.3.38}$$

The $*$ may be transferred by means of (4.7.29).

6.3.4   Orthogonality relations and integrals over D-matrices

The D-matrices, as representatives of the unitary transformations $U_a(\alpha, \beta, \gamma)$, have the property

$$D^{(j)\dagger} D^{(j)} = D^{(j)} D^{(j)\dagger} = 1$$

that is

$$\sum_{m'} D^{(j)*}_{m'm} D^{(j)}_{m'n} = \sum_{m'} D^{(j)}_{m'm} D^{(j)*}_{nm'} = \delta_{mn}. \tag{6.3.39}$$

If one puts $m = n = 0$, one obtains

$$\sum_m Y_{lm}(\vartheta, \varphi) Y^*_{lm}(\vartheta, \varphi) = \frac{2l + 1}{4\pi} \tag{6.3.40}$$

which is a special case of the more general (6.3.35).

There is another orthogonality relation, namely with respect to integrations over all three Euler angles. We derive that now. Let us integrate (6.3.4) over all angles α, β, γ:

$$\int D^{(j_1)}_{m_1 m'_1} D^{(j_2)}_{m_2 m'_2} \, d\Omega = \sum_j \langle j_1 m_1 j_2 m_2 | j m \rangle \langle j m' | j_1 m'_1 j_2 m'_2 \rangle$$

$$\times \int D^{(j)}_{m_1 + m_2 m'_1 + m'_2} \, d\Omega \tag{6.3.41}$$

where

$$\int d\Omega = \int_0^{2\pi} d\alpha \int_0^\pi \sin\beta \, d\beta \int_0^{2\pi} d\gamma.$$

On the r.h.s. the integration over $d\alpha$ and $d\gamma$ can be done at once, because (see (6.2.1) and (6.2.16))

$$D^{(j)}_{m_1 + m_2 m'_1 + m'_2} = e^{-i[(m_1 + m_2)\alpha + (m'_1 + m'_2)\gamma]} d^{(j)}_{m_1 + m_2, m'_1 + m'_2}(\beta). \tag{6.3.42}$$

The integral over this vanishes unless $m_2 = -m_1$; $m'_2 = -m'_1$ and we thus obtain

$$\int D^{(j_1)}_{m_1 m'_1} D^{(j_2)}_{m_2 m'_2} \, d\Omega = 4\pi^2 \delta_{m_1, -m_2} \delta_{m'_1, -m'_2} \tag{6.3.43}$$

$$\times \sum_j \left[\langle j_1 m_1 j_2 - m_1 | j 0 \rangle \langle j 0 | j_1 m'_1 j_2 - m'_1 \rangle \int d^{(j)}_{00}(\beta) \sin\beta \, d\beta \right].$$

Now j must be an integer l, since $m = 0$. Hence with (6.3.33) and table 4.2:

$$d_{00}^{(l)}(\beta) = \sqrt{\frac{4\pi}{2l+1}} Y_{l0}^*(\beta, 0) = \frac{4\pi}{\sqrt{2l+1}} Y_{l0}^*(\beta, 0) Y_{00}(\beta, 0). \qquad (6.3.44)$$

Therefore

$$\int d_{00}^{(l)}(\beta) \sin \beta \, d\beta \, \frac{d\varphi}{2\pi} = \frac{4\pi}{2\pi\sqrt{2l+1}} \underbrace{\int Y_{l0}^*(\beta, 0) Y_{00}(\beta, 0) \sin \beta \, d\beta \, d\varphi}_{\delta_{l0}}$$

$$= 2\delta_{l0}.$$

We thus obtain from (6.3.43)

$$\int D_{m_1 m_1'}^{(j_1)} D_{m_2 m_2'}^{(j_2)} \, d\Omega = 8\pi^2 \delta_{m_1,-m_2} \delta_{m_1',-m_2'} \qquad (6.3.45)$$

$$\times \langle j_1 m_1 j_2 - m_1 | 00 \rangle \langle j_1 m_1' j_2 - m_1' | 00 \rangle.$$

The CGCs vanish, unless $j_1 = j_2$ (triangle condition). We change now $m_1 \to -m_1$ and $m_1' \to -m_1'$ and obtain

$$\int D_{-m_1-m_1'}^{(j_1)} D_{m_2 m_2'}^{(j_2)} \, d\Omega$$

$$= 8\pi^2 \delta_{j_1 j_2} \delta_{m_1, m_2} \delta_{m_1'. m_2'} \langle j_1, -m_1 j_1 m_1 | 00 \rangle \langle j_1, -m_1' j_1 m_1' | 00 \rangle.$$

From (5.3.47), namely

$$\langle j, -mjm | 00 \rangle = \frac{(-1)^{j+m}}{\sqrt{2j+1}}$$

it follows that the product of the two CGCs is equal to

$$\frac{(-1)^{2j_1+m_1+m_1'}}{2j_1+1}.$$

With a slightly changed notation we obtain finally

$$(-1)^{2j+m+n} \left(\frac{2j+1}{8\pi^2}\right) \int D_{-m,-n}^{(j)} D_{m'n'}^{(j')} \, d\Omega = \delta_{jj'} \delta_{mm'} \delta_{nn'} \qquad (6.3.46)$$

where

$$\int d\Omega \equiv \int_0^{2\pi} d\alpha \int_0^{2\pi} d\gamma \int_0^{\pi} \sin \beta \, d\beta.$$

Using a result from subsection 6.3.7, namely (6.3.68)

$$D_{-m,-n}^{(j)} = (-1)^{2j+m+n} D_{mn}^{(j)*}$$

we obtain the orthogonality relation of D-matrices:

$$\frac{2j+1}{8\pi^2} \int D^{(j)*}_{mn} D^{(j')}_{m'n'} \, d\Omega = \delta_{jj'}\delta_{mm'}\delta_{nn'}. \tag{6.3.47}$$

This result enables us to carry out the integral over three D-matrix elements. Equation (6.3.4) allows us to combine two D-factors into one:

$$\int d\Omega \, D^{(j_1)}_{m_1 m'_1} D^{(j_2)}_{m_2 m'_2} D^{(j_3)*}_{m_3 m'_3}$$

$$= \int d\Omega \sum_j \langle j_1 m_1 j_2 m_2 | j m \rangle D^{(j)}_{m_1+m_2, m'_1+m'_2} D^{(j_3)*}_{m_3 m'_3} \langle j m' | j_1 m'_1 j_2 m'_2 \rangle.$$

Now the integral can be read off from (6.3.47) and inserted; the result is

$$\int d\Omega \, D^{(j_1)}_{m_1 m'_1} D^{(j_2)}_{m_2 m'_2} D^{(j_3)*}_{m_3 m'_3}$$

$$= \frac{8\pi^2}{2j_3+1} \langle j_1 m_1 j_2 m_2 | j_3 m_3 \rangle \langle j_1 m'_1 j_2 m'_2 | j_3 m'_3 \rangle. \tag{6.3.48}$$

If the last two results are specialized to integer j values and $n = n' = 0$ and $m'_1 = m'_2 = m'_3 = 0$, respectively, then (6.3.47) gives the orthogonality and (6.3.48) the '$3Y$-formula' (6.3.38) of the spherical harmonics.

The last equation, (6.3.48), can be used to calculate CGCs: by putting $m_1 = m'_1$, $m_2 = m'_2$ and $m_3 = m'_3$, one has

$$\langle j_1 m_1 j_2 m_2 | j_3 m_3 \rangle^2 = \frac{2j_3+1}{8\pi^2} \int d\Omega \, D^{(j_1)}_{m_1 m_1} D^{(j_2)}_{m_2 m_2} D^{(j_3)*}_{m_3 m_3} \tag{6.3.49}$$

which determines the CGCs up to the sign. It should be noted that in the non-vanishing integrals (6.3.47), (6.3.48) and (6.3.49) in fact only the $\sin\beta \, d\beta$ integration remains, while the $d\alpha \, d\gamma$ gives just a factor $4\pi^2$.

6.3.5   A projection formula

We shall derive here a formula by which one can project out the jm-part of an arbitrary state. The orthogonality relation (6.3.47) will be the direct origin of the projection formula.

Consider a state $|\gamma' j' m'\rangle$ and apply to it an active rotation $R_a(\omega)$ with $\omega = (\alpha, \beta, \gamma)$. Then

$$R_a(\omega)|\gamma' j' m'\rangle = \sum_{n'} |\gamma' j' n'\rangle D^{(j')}_{n'm'}(\omega).$$

We now use the orthogonality relation (6.3.47) in order to eliminate the sum on the r.h.s.

$$\frac{2j+1}{8\pi^2} \int d\omega\, D_{nm}^{(j)*}(\omega) R_a(\omega) |\gamma' j' m'\rangle$$

$$= \delta_{jj'} \delta_{mm'} \sum_n \delta_{nn'} |\gamma' j' n'\rangle \qquad (6.3.50)$$

$$= \delta_{jj'} \delta_{mm'} |\gamma' j' n\rangle$$

where $d\omega = d\alpha\, d\gamma \sin\beta\, d\beta$; $0 \le (\alpha, \gamma) < 2\pi$; $0 \le \beta < \pi$. That is, $|\gamma' j' m'\rangle$ is transformed into $|\gamma' j' n\rangle$ if $j' = j$, $m' = m$; otherwise it is annihilated. Hence, if a general state

$$|\psi\rangle = \sum_{\gamma' j' m'} |\gamma' j' m'\rangle \langle \gamma' j' m' |\psi\rangle \qquad (6.3.51)$$

is given, then applying (6.3.50) to it yields the projection formula (attention: $d\omega \equiv d\alpha\, d\gamma \sin\beta\, d\beta$)

$$\frac{2j+1}{8\pi^2} \int d\omega\, D_{nm}^{(j)*}(\omega) R_a(\omega) |\psi\rangle = \sum_{\gamma'} |\gamma' jn\rangle \langle \gamma' jm |\psi\rangle. \qquad (6.3.52)$$

By putting $n = m$ one obtains the projection operator

$$P_{jm} = \frac{2j+1}{8\pi^2} \int d\omega\, D_{mm}^{(j)*}(\omega) R_a(\omega) = \sum_{\gamma'} |\gamma' jm\rangle \langle \gamma' jm| \qquad (6.3.53)$$

and by multiplying (6.3.52) by $\langle \gamma jn|$ one finds

$$\langle \gamma jn| \frac{2j+1}{8\pi^2} \int d\omega\, D_{nm}^{(j)*}(\omega) R_a(\omega) |\psi\rangle = \langle \gamma jm |\psi\rangle. \qquad (6.3.54)$$

The usefulness of this formula depends, of course, on what we know about the effect of a rotation of the state $|\psi\rangle$.

6.3.6 Completeness relation for the D-matrices

Starting from (6.3.53) we shall now prove three different completeness relations for the matrix of the irreducible representations $D^{(j)}(\alpha, \beta, \gamma)$.

Obviously, if we sum the projection operator P_{jm} over j and m, we must obtain the unit operator; hence, for whatever state $|\psi\rangle$, we find

$$\int d\alpha\, d\cos\beta\, d\gamma \left\{ \sum_{jm} \frac{2j+1}{8\pi^2} D_{mm}^{(j)*}(\alpha, \beta, \gamma) \right\} R_a(\alpha, \beta, \gamma) |\psi\rangle = |\psi\rangle. \quad (6.3.55)$$

Let us for a moment call the curly bracket $f(\alpha, \beta, \gamma)$, and the rotated state $R_a(\alpha, \beta, \gamma) |\psi\rangle \equiv |\psi_{(\alpha,\beta,\gamma)}\rangle$. Then the last equation says in that new notation

$$\int d\alpha\, d\cos\beta\, d\gamma\, f(\alpha, \beta, \gamma) |\psi_{(\alpha,\beta,\gamma)}\rangle = |\psi\rangle.$$

As $|\psi\rangle$ is completely arbitrary, this is only possible if $f(\alpha, \beta, \gamma) = \delta(\alpha)\delta(\cos\beta - 1)\delta(\gamma)$ (where $\cos\beta = 1$ means to imply $\beta = 0$ in order to cover also the case of half-integer j representations). Thus the expression inside the curly brackets is equal to these δ-functions. This is already essentially our desired result. In a somewhat symbolic notation we can write it as

$$\sum_{jm} \frac{2j+1}{8\pi^2} D_{mm}^{(j)*}(R) = \delta(R - E) \tag{6.3.56}$$

(where $R \equiv (\alpha, \beta, \gamma)$; $E \equiv (0, 0, 0)$, $\delta(R - E) \equiv \delta(\alpha)\delta(\cos\beta - 1)\delta(\gamma)$ and $dR \equiv d\alpha\, d\cos\beta\, d\gamma$). We shall, however, write this result in a somewhat more general form which we derive now. Returning to (6.3.55) we notice that when an integral over R goes through the whole group (i.e. all values $0 \leq \alpha < 2\pi$, $0 \leq \gamma < 2\pi$ and $0 \leq \beta < \pi$) then RR_0^{-1} with fixed R_0 does the same. Hence

$$\int dR \left\{ \sum_{jm} \frac{2j+1}{8\pi^2} D_{mm}^{(j)*}(R) \right\} R|\psi\rangle$$

$$= \int dR \left\{ \sum_{jm} \frac{2j+1}{8\pi^2} D_{mm}^{(j)*}(RR_0^{-1}) \right\} RR_0^{-1}|\psi\rangle = |\psi\rangle \tag{6.3.57}$$

which, by the same reasoning as above, implies

$$\sum_{jm} \frac{2j+1}{8\pi^2} D_{mm'}^{(j)*}(RR_0^{-1}) = \sum_{jmm'} \frac{2j+1}{8\pi^2} D_{mm'}^{(j)*}(R) D_{m'm}^{(j)*}(R_0^{-1})$$

$$= \sum_{jmm'} \frac{2j+1}{8\pi^2} D_{mm'}^{(j)*}(R) D_{mm'}^{(j)}(R_0) = \delta(R - R_0) \tag{6.3.58}$$

which, for R_0 or R equal to E, reduces to (6.3.56). This completeness relation is the counterpart to the orthogonality relation (6.3.47).

Next we derive two further relations in the following way: we multiply the last line of (6.3.58) by $e^{-n(\gamma - \gamma_0)}$ and integrate over γ_0. Here n will be any integer or half integer such that $j - n$ is integer and $|n| \leq j$ (i.e. n is one of the possible values of m, m'). On the r.h.s., with the use of

$$\delta(R - R_0) \equiv \delta(\alpha - \alpha_0)\delta(\cos\beta - \cos\beta_0)\delta(\gamma - \gamma_0)$$

$$\delta(\alpha - \alpha_0)\delta(\cos\beta - \cos\beta_0) \int d\gamma_0\, e^{-in(\gamma - \gamma_0)}\delta(\gamma - \gamma_0)$$

$$= \delta(\alpha - \alpha_0)\delta(\cos\beta - \cos\beta_0)$$

whereas on the l.h.s., by means of

$$D_{mm'}^{(j)}(\alpha, \beta, \gamma) = e^{-i(m\alpha + m'\gamma)} d_{mm'}^{j}(\beta)$$

(see (6.2.16)) we obtain

$$\sum_{jmm'} \frac{2j+1}{8\pi^2} e^{im(\alpha-\alpha_0)} d^j_{mm'}(\beta) d^j_{mm'}(\beta_0) \int_0^{2\pi} e^{im'(\gamma-\gamma_0)} e^{-in(\gamma-\gamma_0)} d\gamma_0.$$

The integral is equal to

$$e^{i\gamma(m'-n)} \int_0^{2\pi} d\gamma_0 e^{-i\gamma_0(m'-n)} = 2\pi \delta_{m'n}$$

because the difference $m' - n$ is an integer by the choice of n. Introducing this in place of the integral and summing over m' yields with arbitrary γ

$$\sum_{jm} \frac{2j+1}{4\pi} e^{im(\alpha-\alpha_0)} d^{(j)}_{mn}(\beta) d^{(j)}_{mn}(\beta_0) e^{i\gamma n} e^{-i\gamma n}.$$

The $e^{\pm in\gamma}$ factors may be recombined with the rest to $D^{(j)}$-matrix elements. The net result is then—independent of n and of γ

$$\sum_{jm} \frac{2j+1}{4\pi} D^{(j)*}_{mn}(\alpha, \beta, \gamma) D^{(j)}_{mn}(\alpha_0, \beta_0, \gamma) = \delta(\alpha-\alpha_0)\delta(\cos\beta-\cos\beta_0). \quad (6.3.59)$$

Next we multiply by $e^{-i\lambda(\alpha-\alpha_0)}$ (where $|\lambda| \leq j$; $j - \lambda =$ integer) and integrate over α_0. By exactly the same reasoning as above we arrive at

$$\sum_j \frac{2j+1}{2} d^{(j)}_{mn}(\beta) d^{(j)}_{mn}(\beta_0) = \delta(\cos\beta - \cos\beta_0) \qquad (6.3.60)$$

independently of the values of m and n.

The collected results are the three completeness relations:

(i) $\displaystyle\sum_{jmm'} \frac{2j+1}{8\pi^2} D^{(j)*}_{mm'}(\alpha, \beta, \gamma) D^{(j)}_{mm'}(\alpha_0, \beta_0, \gamma_0)$

$\qquad = \delta(\alpha - \alpha_0)\delta(\cos\beta - \cos\beta_0)\delta(\gamma - \gamma_0)$

(ii) $\displaystyle\sum_{jm} \frac{2j+1}{4\pi} D^{(j)*}_{mn}(\alpha, \beta, \gamma) D^{(j)}_{mn}(\alpha_0, \beta_0, \gamma) = \delta(\alpha - \alpha_0)\delta(\cos\beta - \cos\beta_0)$

$\qquad$ (independent of n and γ; indeed γ drops out)

(iii) $\displaystyle\sum_j \frac{2j+1}{2} D^{(j)*}_{mn}(\alpha, \beta, \gamma) D^{(j)}_{mn}(\alpha, \beta_0, \gamma) = \delta(\cos\beta - \cos\beta_0)$

$\qquad$ (independent of m, n, α and γ; indeed α and γ drop out).

$$(6.3.61)$$

6.3.7 Symmetry properties of the D-matrices

The symmetries of the D-matrices arise from two of their properties: they are unitary and they have a simple exponential behaviour in two arguments (see (6.2.1)):

$$D_{m'm}^{(j)*}(\alpha, \beta, \gamma) = D_{mm'}^{(j)}(-\gamma, -\beta, -\alpha) \quad \text{unitarity}$$

$$D_{m'm}^{(j)*}(\alpha, \beta, \gamma) = e^{-i(m'\alpha+m\gamma)} d_{m'm}^{(j)}(\beta) \quad \text{with } d_{m'm}^{(j)} \text{ real.}$$

(6.3.62)

For the $d_{m'm}^{(j)}(\beta)$, unitarity and reality imply

$$d_{m'm}^{(j)}(\beta) = d_{mm'}^{(j)}(-\beta).$$

(6.3.63)

Interchanging α and γ leads therefore to

$$
\begin{aligned}
D_{m'm}^{(j)}(\gamma, \beta, \alpha) &= e^{-i(m'\gamma+m\alpha)} \, d_{m'm}^{(j)}(\beta) \\
&= e^{-i(m\alpha+m'\gamma)} \, d_{mm'}^{(j)}(-\beta) \\
&= D_{mm'}^{(j)}(\alpha, -\beta, \gamma) \\
&= D_{m'm}^{(j)*}(-\gamma, \beta, -\alpha).
\end{aligned}
$$

(6.3.64)

Another relation is found as follows (see (6.2.2)):

$$
\begin{aligned}
d_{m'm}^{(j)}(\beta) &= \langle jm'|e^{-i\beta J_y}|jm\rangle \\
&= \langle jm'|e^{+i\pi J_x} e^{+i\beta J_y} e^{-i\pi J_x}|jm\rangle
\end{aligned}
$$

because rotating by π about x, then by β about y and again by π about x, is the same as rotating by $-\beta$ about y. Now obviously

$$e^{-i\pi J_x}|jm\rangle = \varphi_x(j, m)|j, -m\rangle$$

where $\varphi_x(jm)$ is a phase. The phase factors cancel[1] and we have

$$d_{m'm}^{(j)}(\beta) = d_{-m'-m}^{(j)}(-\beta) = d_{-m-m'}^{(j)}(\beta).$$

(6.3.65)

A glance at the explicit form $d_{m'm}^{(j)}(\beta)$ (6.2.16) shows that

$$d_{m'm}^{(j)}(-\beta) = (-1)^{2j+m'+m} d_{m'm}^{(j)}(\beta) = (-1)^{m'-m} d_{m'm}^{(j)}(\beta)$$

(6.3.66)

(the two sign factors are equal because they differ by $(-1)^{2j+2m} = +1$; $j\pm m$ is always integer). Combining (6.3.65) and (6.3.66) we find

$$d_{m'm}^{(j)}(\beta) = (-1)^{2j+m'+m} d_{-m'-m}^{(j)}(\beta).$$

(6.3.67)

---

[1] The phase factor $\varphi_x(j, m)$ may be calculated along the lines in which the corresponding phase factor for $R_y(\pi)$ was found (between (5.3.40) and (5.3.42)). One finds $\varphi_x(j, m) = (-1)^{-j}$ independent of m.

All these relations are combined in the following formula, where we use from top to bottom (6.3.64)–(6.3.66) and from left to right unitarity:

$$D^{(j)}_{m'm}(\alpha, \beta, \gamma) = D^{(j)*}_{mm'}(-\gamma, -\beta, -\alpha) = D^{(j)}_{mm'}(\gamma, -\beta, \alpha)$$

$$= D^{(j)*}_{m'm}(-\alpha, \beta, -\gamma) = D^{(j)}_{-m-m'}(-\gamma, \beta, -\alpha)$$

$$= D^{(j)*}_{-m'-m}(\alpha, -\beta, \gamma) \tag{6.3.68}$$

$$= (-1)^{2j+m+m'} D^{(j)}_{-m-m'}(-\gamma, -\beta, -\alpha)$$

$$= (-1)^{2j+m+m'} D^{(j)*}_{-m'-m}(\alpha, \beta, \gamma).$$

Note that $(-1)^{2j+m+m'} = (-1)^{m-m'}$.

With $m = 0$, we obtain from $D^{(j)}_{m'0}(\alpha, \beta, \gamma) = (-1)^{m'} D^{(j)*}_{-m'0}(\alpha, \beta, \gamma)$ the rule $Y^*_{lm}(\beta\alpha) = (-1)^m Y_{l-m}(\beta\alpha)$ (already known from (4.7.29)). Further relations can be derived if one changes any argument by $\pm\pi$.

Finally we consider rotations by 2π about an axis n. This can be achieved by the matrix

$$D^{(j)}(2\pi n) = D^{(j)}(e_3 \to n) D^{(j)}(2\pi e_3) D^{(j)^{-1}}(e_3 \to n).$$

Now the matrix in the middle is (see (6.2.1))

$$D^{(j)}_{m'm}(2\pi, 0, 0) = e^{-2\pi i m'} \delta_{m'm} = (-1)^{2m'} 1 = (-1)^{2j} 1$$

because $2j + 2m' = $ even. Thus $D^{(j)}(2\pi e_3)$ commutes with the other two matrices which together give 1. Hence for any direction n:

$$D^{(j)}(2\pi\, n) = (-1)^{2j} \times 1. \tag{6.3.69}$$

This is the well known result that the half-integer representations are double valued and that consequently the states $|jm\rangle$ multiply by $(-1)^{2j}$ under rotation by 2π.

7

THE JORDAN–SCHWINGER CONSTRUCTION
AND REPRESENTATIONS

In this chapter we shall obtain all the representations of the rotation group and its Lie algebra[1] ($su(2)$) by another very useful method, namely by means of the Jordan–Schwinger construction (Jordan (1935), Schwinger (1952)). The key idea of this approach is the Jordan mapping of the Lie algebra generators into operators made of bosonic creation and annihilation operators acting on a Hilbert space. For the reader's convenience, we start from a short exposition of the main properties of the harmonic oscillator algebra and its representations.

7.1 Bosonic operators

A harmonic oscillator (see, e.g., Landau and Lifschitz (1981)) is an object that is subject to a quadratic potential energy, which produces a restoring force against any displacement from equilibrium that is proportional to the displacement. The Hamiltonian for such a system whose motion is confined to one dimension is

$$H = \frac{p^2}{2m} + \frac{m\omega^2}{2} q^2 \qquad (7.1.1)$$

where q is the displacement of the oscillator (particle) for some fixed origin (equilibrium), p is the momentum, m is the mass of the particle and ω is the (classical) circular frequency of the oscillations. The harmonic oscillator is of basic importance in physics because it provides a model for many kinds of vibrating system, including, e.g., the electromagnetic field (it can be viewed as a collection of infinitely many harmonic oscillators) and because of the following fact. For many more general physical systems, the potential energy $V(q)$ has a minimum at some point q_0 in space. Expanding the potential energy in a series of powers of the distances from that point one can write

$$V(q) = V(q_0) + \frac{1}{2}\left(\frac{\partial^2 V}{\partial q^2}\right)\bigg|_{q_0} (q - q_0)^2 + \cdots \qquad (7.1.2)$$

where the equilibrium point is determined by the condition

$$\left(\frac{\partial V}{\partial q}\right)\bigg|_{q_0} = 0.$$

[1] We denote the group by $SU(2)$, its Lie algebra by $su(2)$.

If a particle of mass m performs small oscillations around the equilibrium position, higher terms in the series (7.1.2) are small and hence up to the inessential constant $V(q_0)$ one has

$$V(q) \sim C(q - q_0)^2$$

which is the harmonic oscillator potential.

Quantization is defined by taking p and q to be self-adjoint operators that satisfy on a dense domain of a Hilbert space the Heisenberg commutation relation

$$[q, p] = i.$$

Let us now introduce the bosonic operators, defined by

$$a^\dagger = \sqrt{\frac{m\omega}{2}} q - i\frac{1}{\sqrt{2m\omega}} p \quad \text{(creation operator)} \qquad (7.1.3)$$

$$a = \sqrt{\frac{m\omega}{2}} q + i\frac{1}{\sqrt{2m\omega}} p \quad \text{(annihilation operator)} \qquad (7.1.4)$$

($a^\dagger$ is adjoint to a). As a consequence of the definition, the bosonic operators obey the commutation relation

$$[a, a^\dagger] = 1 \qquad (7.1.5)$$

and the Hamiltonian H written in terms of bosonic operators takes the form

$$H = \frac{\omega}{2}(a^\dagger a + a a^\dagger) = \omega \left(a^\dagger a + \frac{1}{2} \right). \qquad (7.1.6)$$

Let $|\psi_0\rangle$ be a normalized vector of a Hilbert space $\mathcal{H}$, where the oscillator operators act, such that it satisfies the relation

$$a|\psi_0\rangle = 0. \qquad (7.1.7)$$

The Schrödinger realization of the operators p and q, i.e. $p = -i\,d/dq$ and q is the multiplication operator, converts (7.1.7) into a first-order linear differential equation whose solution is

$$\psi_0(x) = \pi^{-\frac{1}{4}} \exp\left(-\frac{x^2}{2} \right)$$

$$x \equiv \sqrt{m\omega} q.$$

This solution is normalized with respect to the $L^2(-\infty, \infty)$ Hilbert space scalar product

$$\langle \phi | \psi \rangle = \int_{-\infty}^{\infty} \phi^*(x)\psi(x)\,dx$$

so that

$$\langle \phi_0 | \phi_0 \rangle = \int_{-\infty}^{\infty} |\phi_0(x)|^2 \, \mathrm{d}x = 1.$$

The complete set of normalized eigenvectors, $\{|\psi_n\rangle\}$, can be constructed from $|\psi_0\rangle$ by using the definition

$$|\psi_n\rangle = (n!)^{-\frac{1}{2}} (a^\dagger)^n |\psi_0\rangle. \tag{7.1.8}$$

These vectors are normalized eigenstates of the dimensionless number operator N

$$N = (\omega)^{-1} H - \tfrac{1}{2} = a^\dagger a \tag{7.1.9}$$

with the eigenvalues n

$$N|\psi_n\rangle = n|\psi_n\rangle. \tag{7.1.10}$$

The operator $a^\dagger$ acting on $|\psi_n\rangle$ raises the eigenvalues n by one unit (i.e. increases the energy by creation of an 'excitation'), hence, it is called a 'creation operator', and the operator a lowers the eigenvalue n ('annihilation operator').

The spectrum of the number operator N consists of all non-negative integers $\mathbb{Z}_+$; each eigenvalue is nondegenerate, and the eigenvectors $|\psi_n\rangle$ form a complete orthonormal basis for the separable Hilbert space $\mathcal{H} = L^2(-\infty, \infty)$ over the real line.

The obvious generalization of this algebra consists of the consideration of K kinematically independent bosons $a_i^\dagger$, a_i $(i = 1, \ldots, K)$, with the commutation relations

$$[a_i, a_j^\dagger] = \delta_{ij} \qquad i, j = 1, 2, \ldots, K$$

$$[a_i, a_j] = [a_i^\dagger, a_j^\dagger] = 0 \tag{7.1.11}$$

and

$$H = \omega \left(\sum_{i=1}^{K} a_i^\dagger a_i + \frac{K}{2} \right) \tag{7.1.12}$$

so that the eigenstates of H contain n_1 excitations of the type 1, n_2 excitations of the type 2, etc; i.e. they are of the form

$$(a_1^\dagger)^{n_1} (a_2^\dagger)^{n_2} \cdots (a_K^\dagger)^{n_K} |\psi_0\rangle. \tag{7.1.13}$$

To explain the name 'bosonic operators' for $a_i^\dagger$ and a_j we reinterpret the states (7.1.13) as follows. Consider the quantum mechanics of K identical bosons. Let i be an index counting the set of quantum numbers characterizing the states of a single boson (these quantum numbers may be discrete or continuous with appropriate ranges); in other words the complete set of one-boson states. According to a general postulate of quantum theory for indistinguishable particles, the distinct states of the system of K identical bosons are only those characterized by the number of boson n_i in an arbitrary state. Thus the state (7.1.13) can be interpreted as n_1 bosons in the state ψ_1, n_2 bosons in the state ψ_2, etc. Hence $a_k^\dagger$ creates a boson in the state k, a_k annihilates one.

Let $\mathcal{H}^{(K)}$ be the space of states of a system of bosons with the basis vectors $|n_1, n_2, \ldots, n_K\rangle$ labelled by the occupation numbers. In this space the actions of $a_i^\dagger$ and a_j are expressed as follows:

$$a_i|n_1, n_2, \ldots, n_K\rangle = \sqrt{n_i}|n_1, \ldots, n_{i-1}, n_i - 1, n_{i+1}, \ldots, n_K\rangle$$
$$a_i^\dagger|n_1, n_2, \ldots, n_K\rangle = \sqrt{n_i + 1}|n_1, \ldots, n_{i-1}, n_i + 1, n_{i+1}, \ldots, n_K\rangle.$$

Then by direct computations, one can indeed obtain from these equations the commutation relations (7.1.12).

There is another interesting and important realization of the canonical commutation relations (7.1.5).

Let $\mathcal{H}_{BF}$ be a Hilbert space of analytical functions of complex variables z^*, z with the scalar product

$$\langle f|g\rangle = \int (f(z))^* g(z) \, e^{-z^* z} \, dz^* \, dz.$$

Then the map

$$a \to \frac{\partial}{\partial z} \qquad a^\dagger \to z$$

gives the representation of the commutation relation (Bargmann–Fock representation). This was introduced in a complete form by Bargmann (1961). The complete set of eigenfunctions of the number operator N in this realization has the form

$$f_n(z) = \frac{z^n}{\sqrt{n!}} \qquad n \in \mathbb{Z}_+. \tag{7.1.14}$$

The Bargmann–Fock representation is closely related to the so-called coherent states $|\zeta\rangle$

$$|\zeta\rangle = \sum_n f_n(\zeta)|\psi_n\rangle = e^{\zeta a^\dagger}|\psi_0\rangle \tag{7.1.15}$$

which have many remarkable properties (both mathematical and physical ones) (see Klauder and Skagerstam (1985)). Generalization of this representation to the case of a multi-oscillator (7.1.11) is straightforward.

7.2 Realization of $su(2)$ Lie algebra and the rotation matrix in terms of bosonic operators

Now we turn to the construction of $su(2)$ Lie algebra operators and their representations in terms of the bosonic operators.

First of all, we notice that the physical quantities such as energy (7.1.6), (7.1.12), angular momentum $M_{ij} = q_i p_j - q_j p_i$ etc, are bilinear expressions in $a_i^\dagger$ and a_j of the form

$$c_{ij} a_i^\dagger a_j \qquad c_{ij} \in \mathbb{C}.$$

This suggests constructing all basis elements of a Lie algebra L in terms of such bilinear combinations. The cornerstone of this construction is Jordan's observation (Jordan (1935)) that the mapping $\mathcal{L}$ of $K \times K$ matrices X_{ij} into bosonic operators (7.1.11) given by

$$\mathcal{L}: X_{ij} \longrightarrow \mathcal{L}_X = \sum_{i,j=1}^{K} X_{ij} a_i^\dagger a_j \qquad (7.2.1)$$

preserves the operation of commutation of matrices.

The proof of this statement follows directly from the bosonic commutation relations (7.1.11). Let us denote by $\bar{a}$ the vector with the components $(a_1^\dagger, \ldots, a_K^\dagger)$ and by a the vector with the components $(a_1, \ldots, a_K)$, so that the map (7.2.1) can be written in the form

$$\mathcal{L}: X \longrightarrow \mathcal{L}_X = \bar{a}^{\mathrm{T}} X a. \qquad (7.2.2)$$

Then for the commutator $[\mathcal{L}_X, \mathcal{L}_Y]$ one easily finds

$$[\mathcal{L}_X, \mathcal{L}_Y] = [\bar{a}^{\mathrm{T}} X a, \bar{a}^{\mathrm{T}} Y a] = \bar{a}^{\mathrm{T}} [X, Y] a = \mathcal{L}_{[X,Y]}.$$

Thus the Jordan map has the property

$$[\mathcal{L}_X, \mathcal{L}_Y] = \mathcal{L}_{[X,Y]}. \qquad (7.2.3)$$

It is linear over $\mathbb{C}$

$$\lambda \mathcal{L}_X + \mu \mathcal{L}_Y = \mathcal{L}_{\lambda X + \mu Y} \qquad \lambda, \mu \in \mathbb{C}$$

and the unit matrix has the map

$$\mathcal{L}_1 = \bar{a}^{\mathrm{T}} a = \sum_{i=1}^{K} a_i^\dagger a_i.$$

Now consider the map for the spin-$\frac{1}{2}$ representation ($K = 2$):

$$\mathcal{L}: \tfrac{1}{2}\sigma_i \longrightarrow J_i = \mathcal{L}_{\frac{1}{2}\sigma_i} = \tfrac{1}{2}(\bar{a}^{\mathrm{T}} \sigma_i a). \qquad (7.2.4)$$

The property of the Jordan map guarantees that the operators J_i satisfy the commutation relations of the $su(2)$ algebra

$$[J_i, J_j] = \mathrm{i} \epsilon_{ijk} J_k. \qquad (7.2.5)$$

In the explicit form the operators $J_\pm$ and J_3 have the following Jordan–Schwinger realization:

$$J_+ = a_1^\dagger a_2 \qquad J_- = a_2^\dagger a_1$$

$$J_3 = \tfrac{1}{2}(a_1^\dagger a_1 - a_2^\dagger a_2) \equiv \tfrac{1}{2}(N_1 - N_2). \tag{7.2.6}$$

The invariant square of the angular momentum J^2 is expressed in terms of the bosonic operators as follows:

$$J^2 = \frac{1}{2} \sum_{i,j} a_i^\dagger a_j a_j^\dagger a_i - \frac{1}{4} N^2 \tag{7.2.7}$$

where

$$N \equiv \sum_i a_i^\dagger a_i = N_1 + N_2.$$

From the above considered representations of the bosonic algebra it follows that $|\psi_{n_1,n_2}\rangle$ are the eigenvectors for J^2 with the eigenvalues

$$J^2 |\psi_{n_1,n_2}\rangle = \tfrac{1}{2}n(\tfrac{1}{2}n + 1) \tag{7.2.8}$$

where

$$n = n_1 + n_2.$$

This implies the equality for the angular momentum quantum number j:

$$j = \tfrac{1}{2}n = 0, \tfrac{1}{2}, 1, \dots .$$

Thus, together with (7.2.6) we have

$$j = \tfrac{1}{2}(n_1 + n_2) \qquad m = \tfrac{1}{2}(n_1 - n_2).$$

The $su(2)$ operators form a subalgebra of the bosonic operator algebra. So the whole representation Hilbert space of the latter is not irreducible with respect to the former. Indeed, consider the subspace $\mathcal{H}^{(2j)}$ of the vectors of the form

$$P(a_1^\dagger, a_2^\dagger)|0\rangle$$

where $P(a_1^\dagger, a_2^\dagger)$ is a homogeneous polynomial in $a_1^\dagger$ and $a_2^\dagger$ of the degree $2j$ and $|0\rangle$ is the shorthand notation for the vacuum vector $|\psi_{0,0}\rangle$, i.e. $a_1|0\rangle = a_2|0\rangle = 0$. The space $\mathcal{H}^{(2j)}$ is invariant with respect to the operators J_i. Making use of the equalities

$$J_\pm|0\rangle = J_3|0\rangle = 0$$

and

$$A P(a_1^\dagger, a_2^\dagger)|0\rangle = [A, P(a_1^\dagger, a_2^\dagger)]\,|0\rangle$$

for any bosonic operator such that $A|0\rangle = 0$, one can find that the normalized eigenvectors for angular momentum are expressed in terms of the bosonic operators as follows (Schwinger (1952)):

$$|jm\rangle = P_{jm}(a_1^\dagger, a_2^\dagger)|0\rangle \qquad (7.2.9)$$

where

$$P_{jm}(a_1^\dagger, a_2^\dagger) = \frac{(a_1^\dagger)^{j+m}(a_2^\dagger)^{j-m}}{\sqrt{(j+m)!(j-m)!}}$$

and that the operators (7.2.6) have the standard action on these vectors:

$$J_3|jm\rangle = m|jm\rangle$$

$$J_\pm|jm\rangle = \sqrt{(j \mp m)(j \pm m + 1)}|j, m \pm 1\rangle. \qquad (7.2.10)$$

These results show that the Jordan–Schwinger construction gives all the representations of the $su(2)$ Lie algebra in the simplest way.

The finite transformations of the space $\mathcal{H}^{(2j)}$ generated by J_i can be obtained by exponentiation of the matrix $-i\eta \cdot \sigma/2$. To a given η there corresponds the unitary unimodular matrix

$$U(\eta) = \exp(-i\eta \cdot \sigma/2)$$

which itself corresponds to a unitary bosonic operator due to the Jordan map

$$\exp(-i\eta \cdot \sigma/2) \rightarrow \exp(\mathcal{L}_{-i\eta\cdot\sigma/2}) = \exp(-i\eta \cdot \mathcal{L}_{\sigma/2}) = \exp(-i\eta \cdot J).$$
$$(7.2.11)$$

Matrix elements of this operator in $\mathcal{H}^{(2j)}$ give the representations of the rotation group

$$\langle jm'|\exp(-i\eta \cdot \mathcal{L}_{\sigma/2})|jm\rangle = D_{m'm}^{(j)}(\eta).$$

Thus the exponentiated Jordan–Schwinger construction yields all irreducible representations of the rotation group. Moreover, the explicit construction (7.2.9) allows an alternative derivation of the rotation matrices.

Denote, for brevity, $V_U = \exp(\mathcal{L}_{-i\eta\cdot\sigma/2})$. Acting by this operator on $|jm\rangle$ one has

$$V_U|jm\rangle = (V_U P_{jm}(a_1^\dagger, a_2^\dagger)V_{U^{-1}})V_U|0\rangle$$

$$= P_{jm}(a_1'^\dagger, a_2'^\dagger)|0\rangle \qquad (7.2.12)$$

where

$$a_i'^\dagger = V_U a_i^\dagger V_{U^{-1}}$$

can be expressed via matrix elements of U

$$U = \begin{pmatrix} u_{11} & u_{12} \\ u_{21} & u_{22} \end{pmatrix}$$

as components of a vector

$$a_1'^\dagger = u_{11}a_1^\dagger + u_{21}a_2^\dagger$$

$$a_2'^\dagger = u_{12}a_1^\dagger + u_{22}a_2^\dagger. \tag{7.2.13}$$

Equation (7.2.12) can be rewritten in the form

$$V_U|jm\rangle = \sum_{m'} D_{m'm}^{(j)}(U) P_{jm'}|0\rangle \tag{7.2.14}$$

by means of expansion of the expression (7.2.12) in the original monomial basis using (7.2.13):

$$\frac{(u_{11}a_1^\dagger + u_{21}a_2^\dagger)^{j+m}(u_{12}a_1^\dagger + u_{22}a_2^\dagger)^{j-m}}{\sqrt{(j+m)!(j-m)!}}|0\rangle$$

$$= \sqrt{(j+m)!(j-m)!} \sum_{st} \frac{(u_{11})^{j+m-s}(u_{21})^s(u_{12})^{j-m-t}(u_{22})^t}{(j+m-s)!s!(j-m-t)!t!} \tag{7.2.15}$$

$$\times (a_1^\dagger)^{2j-s-t}(a_2^\dagger)^{s+t}|0\rangle.$$

Comparing this with (7.2.14), one finds

$$D_{m'm}^{(j)}(U) = \sqrt{(j+m)!(j-m)!(j+m')!(j-m')!}$$

$$\times \sum_s \frac{(u_{11})^{j+m-s}(u_{21})^s(u_{12})^{m'-m+s}(u_{22})^{j-m'-s}}{(j+m-s)!s!(m'-m+s)!(j-m'-s)!} \tag{7.2.16}$$

$$\times (a_1^\dagger)^{2j-s-t}(a_2^\dagger)^{s+t}|0\rangle.$$

This gives an explicit expression for rotation matrices in an arbitrary representation.

The Jordan–Schwinger construction can be applied to other Lie algebras. Indeed, consider a set $a_i^\dagger, a_i$ ($i = 1, \ldots, n$) of bosonic operators in a Hilbert space $\mathcal{H}$. Define

$$A_{ij} = a_i^\dagger a_j.$$

Then using the commutation relations (7.1.11) one obtains

$$[A_{ij}, A_{kl}] = \delta_{jk}A_{il} - \delta_{il}A_{jk}. \tag{7.2.17}$$

This means that the set A_{ij} ($i, j = 1, \ldots, n$) forms the set of generators of the Lie algebra $gl(n, \mathbb{C})$. Because any Lie algebra is a subalgebra of $gl(n, \mathbb{C})$

(Ado's theorem), any other complex or real Lie algebra is generated by a subset of A_{ij} $(i, j = 1, \ldots, n)$. In particular, the operators

$$M_{kk} = a_k^\dagger a_k \qquad\qquad k = 1, \ldots, n$$

$$M_{kl} = a_k^\dagger a_l + a_l^\dagger a_k \qquad k < l \le n \qquad\qquad (7.2.18)$$

$$\widetilde{M}_{kl} = i(a_k^\dagger a_l - a_l^\dagger a_k) \quad k < l \le n$$

generate the Lie algebra $u(n)$ and the operators

$$X_{kl} = i(a_k^\dagger a_l - a_l^\dagger a_k) \qquad\qquad (7.2.19)$$

generate the Lie algebra $so(n)$.

Non-compact Lie algebras such as $u(p, q)$, $so(p, q)$ and $sp(p, q)$, including the important cases of De Sitter and anti-De Sitter Lie algebras and the Lie algebra $su(2, 2)$ of the conformal group can be also constructed via the Jordan–Schwinger realization in terms of bosonic oscillators, while the Lie superalgebras can be built out of both bosonic and fermionic operators (cf., e.g., Chaichian and Demichev (1996)). For example, generators of $u(p, q)$ Lie algebras, which are often used in particle physics, are made up of two sets of bosonic operators $a_i^\dagger, a_i$ $(i = 1, \ldots, p)$ and $b_\alpha^\dagger, b_\alpha$ $(\alpha = p + 1, \ldots, p + q)$, both obeying the commutation relations (7.1.11) (see, e.g., Barut and Raczka (1977)). To construct the generators, one defines the array of operators $A = (A_{MN})$, $M, N = 1, \ldots, p + q$

$$A = \begin{pmatrix} -a_i^\dagger a_j + r\delta_{ij} & a_i^\dagger b_\beta^\dagger \\ -b_\alpha a_j & b_\alpha b_\beta^\dagger + r\delta_{\alpha\beta} \end{pmatrix}$$

(r is any real number) in terms of which the generators M_{kl}, $N_{k\beta}$ of $u(p, q)$ read as

$$M_{kk} = A_{kk} \qquad\qquad k = 1, \ldots, p + q$$

$$M_{kl} = A_{kl} + A_{lk} \qquad k \le l$$

$$\widetilde{M}_{kl} = i(A_{kl} - A_{lk}) \qquad k \le l$$

$$M_{\alpha\beta} = A_{\alpha\beta} + A_{\beta\alpha} \qquad \alpha < \beta$$

$$\widetilde{M}_{\alpha\beta} = i(A_{\alpha\beta} - A_{\beta\alpha}) \qquad \alpha < \beta$$

$$N_{k\beta} = A_{k\beta} - A_{\beta k}$$

$$\widetilde{N}_{k\beta} = i(A_{k\beta} + A_{\beta k}).$$

Using the known representations for bosonic algebras, one can then construct representations of all such Lie algebras.

7.3 A short note about the new field of quantum groups

To conclude this section, it is worth mentioning the very important fact that the Jordan–Schwinger construction can be generalized to the case of quantum deformed groups and enveloping algebras. The theory of these objects has attracted great attention during the last few years and seems to find many important applications in different areas of physics (for an introduction and review of quantum group theory, see, e.g., Chaichian and Demichev (1996)).

To give an idea of the deformed Jordan–Schwinger construction, we recall that the defining q-deformed commutation relations for the operators H, X^+, X^- of the quantum Lie algebra $su_q(2)$ have the form

$$[H, X^\pm] = \pm 2X^\pm$$

$$[X^+, X^-] = \frac{q^H - q^{-H}}{q - q^{-1}} = \frac{\sinh(\chi H)}{\sinh(\chi)} = [H]_q.$$

(7.3.1)

Here $q \equiv e^\chi$, and $[x]_q$ denotes the so-called q-square bracket

$$[x]_q = \frac{q^x - q^{-x}}{q - q^{-1}}$$

(7.3.2)

with the property

$$[x]_q \xrightarrow[q \to 1]{} x$$

(7.3.3)

so that in the $q \to 1$ limit the commutation relations (7.3.1) coincide with the usual relations for $su(2)$ Lie algebra.

With the help of q-square brackets many formulas of q-group theory can be written in a form similar to the classical non-deformed case. For example, the Casimir operator of $su(2)$ Lie algebra

$$C_2 = J_\mp J_\pm + \left(\frac{1}{2}(J_0 \pm 1)\right)^2$$

becomes after the deformation

$$C_2^q = X^\mp X^\pm + \left[\frac{1}{2}(H \pm 1)\right]_q^2.$$

(7.3.4)

If q is not a root of unity, representations of $su_q(2)$ have the same dimensions and structure as those in the classical case

$$H|j, m\rangle = 2m|j, m\rangle$$

$$X^\pm|j, m\rangle = \sqrt{[j \mp m]_q[j \pm m + 1]_q}|j, m \pm 1\rangle$$

(7.3.5)

(j is an integer or half-integer number, $m = -j, -j + 1, \ldots, j$). Using (7.3.3) it is easy to see that the q-deformed expressions have the correct classical limit $q \to 1$.

To define the deformed Jordan–Schwinger construction, the mapping (7.2.6) is replaced by

$$X^+ = a_1^\dagger a_2 \qquad X^- = a_2^\dagger a_1$$
$$H = \tfrac{1}{2}(N_1 - N_2) \tag{7.3.6}$$

where $a_i^\dagger$ and a_i ($i = 1, 2$) are q-oscillators in the Fock representations with the commutation relations

$$a_i a_i^\dagger - q a_i^\dagger a_i = q^{-N_i}$$
$$a_i a_i^\dagger - q^{-1} a_i^\dagger a_i = q^{N_i}$$
$$\left[N_i, a_i^\dagger \right] = a_i^\dagger \qquad [N_i, a_i] = -a_i \tag{7.3.7}$$
$$[a_i, a_j] = \left[a_i^\dagger, a_j^\dagger \right] = \left[a_i^\dagger, a_j \right] = 0 \qquad i \neq j.$$

Introduce the eigenstates which are analogous to undeformed angular momentum eigenstates

$$|j, m\rangle = \left([j + m]_q! [j - m]_q! \right)^{-1/2} (a_1^\dagger)^{j+m} (a_2)^{j-m} |0\rangle$$
$$j = 0, \tfrac{1}{2}, 1, \ldots \qquad m = -j, -j + 1, \ldots, j$$

with $[n]_q! \equiv [1]_q [2]_q \cdots [n]_q$ and $[0]_q! \equiv 1$. Then the map (7.3.6) gives the representations (7.3.5).

The Jordan–Schwinger construction for $su(2)_q$ confirms the important result: for all $q \in \mathbb{R}_+$, the unitary irreducible representations of $su(2)_q$ are in one-to-one correspondence with those of $su(2)$ and have the same dimensions.

For further details on quantum group theory and especially on the q-deformed Jordan–Schwinger construction, see, for instance, Chaichian and Demichev (1996).

8

IRREDUCIBLE TENSORS AND TENSOR OPERATORS

8.1 Introduction

What does 'irreducible' mean? (It means, first of all, in our context, irreducible with respect to rotations; we shall not mention this any more.) We have already encountered this word earlier in connection with irreducible subspaces $\mathcal{H}_j$. An irreducible subspace was, as we saw, an invariant subspace $\mathcal{H}_j$, whose basis vectors $|jm\rangle$ with $m = -j, \ldots, +j$ transform among themselves and which does not contain in itself another invariant subspace $h \subset \mathcal{H}_j$. We have seen that the linear transformations $D_{m'm}^{(j)}$ are then irreducible also, i.e. they cannot be split up into boxes along the diagonal.

An irreducible tensor is a tensor whose components transform linearly among themselves under rotations such that, if the whole rotation group is considered, *all* components of the tensor enter the linear combination (which does not exclude that for a certain rotation some coefficients in the linear combination vanish). Therefore, irreducible tensors will be defined with respect to a basis in a j-dimensional space which transforms just like $\mathcal{H}_j$. Tensors with respect to a Cartesian basis are not well suited for this: for instance, a vector $v = (v_x, v_y, v_z)$ is an irreducible tensor of rank 1. Cartesian tensors of any rank can be generated by taking the direct product of two or more such tensors of lower rank:

$$T_{ijkl} = a_i S_{jk} v_l \quad \text{etc.}$$

These tensors are not, however, irreducible with respect to rotations, because they contain parts which transform with a lower rank. For instance, the tensor of rank 2

$$T_{ij} = v_i w_j$$

transforms under rotation (if $v' = Mv$) as follows:

$$(T')_{ij} = v_i' w_j' = \sum_{kl} M_{ik} M_{jl} v_k w_l = \sum_{kl} M_{ik} M_{jl} T_{kl}$$

i.e. T transforms with the direct product of M by itself

$$T' = [M \otimes M] T$$

and we know, since M is equivalent to $D^{(1)}$, that $M \otimes M$ can be reduced to irreducible matrices of dimension $2j + 1$ with $j = 0, 1, 2$, i.e. T can be reduced to

- a tensor of rank 0, the trace

$$T^{(0)} = \sum T_{ii}$$

- and a tensor of rank 1, the vector product

$$T_{ij}^{(1)} = \tfrac{1}{2}(v_i w_j - v_j w_i)$$

- and a tensor of rank 2, the symmetric part minus the trace

$$T_{ij}^{(2)} = \tfrac{1}{2}(v_i w_j + v_j w_i) - \delta_{ij} \sum v_k w_k.$$

Because this behaviour of tensors in the Cartesian basis x, y, z is rather annoying, one considers tensors which transform with the irreducible representations $D^{(j)}$ and which by this definition are automatically irreducible tensors of rank j with $2j+1$ components $T(jm)$; $m = -j, \ldots, +j$. A particular kind of such tensors are e.g. the wave functions $\psi_{jm}(x)$. For them we know how we can, by means of the CGCs, build up a state $|jm\rangle$ from states of lower j_1 and j_2 if only $|j_1 - j_2| \le j \le j_1 + j_2$. If then $|j_1 m_1\rangle$ and $|j_2 m_2\rangle$ are combined into

$$|jm\rangle = \sum_{m_1 m_2} |j_1 m_1 j_2 m_2\rangle \langle j_1 m_1 j_2 m_2 | jm\rangle$$

we know that $|jm\rangle$ again transforms according to an irreducible representation, namely with $D^{(j)}$. Thus $\psi_{jm} = \langle x | jm \rangle$ will be an irreducible tensor of rank j, formed by composition of two tensors $\psi_{j_1 m_1}$ and $\psi_{j_2 m_2}$ of lower rank. Hence it is possible to construct tensors of any rank by composition—just as in the case of Cartesian tensors, although in a little more complicated way, but the great advantage is that the tensors so constructed are automatically irreducible if they are combined from two irreducible ones. Thus this kind of tensor is especially suited for all problems in which rotations come up. Of course, the direct product of two irreducible tensors will be reducible; thus there exist not only the irreducible ones. The whole class of tensors which transform according to $D^{(j)}$ or to direct products $D^{(j)} \otimes D^{(j')}$ are called 'spherical tensors'.

> Therefore, the irreducible tensors are quantities which combine and transform like (standard) angular momentum states; if such a tensor has $2j + 1$ components and transforms like $|jm\rangle$ then it is called an irreducible (spherical) tensor of rank j. $\qquad$ (8.1.1)

We shall give only a brief resumé of the spherical tensors; details and applications may be found in books such as those of Fano and Racah (1959), de-Shalit and Talmi (1963), Rose (1957), Edmonds (1957) and Varshalovich *et al* (1988) and, of course, in many nuclear physics papers.

Remark. We also suppose here that we work with standard states $|\gamma jm\rangle$. In such a representation J is diagonal in γ.

8.2 Definition and properties

We define the irreducible tensor $T(kq, x)$ of rank k by its transformation property under the rotation group (active rotation)[1]

$$T'(kq, x) = \sum_{q'} T(kq', M_a^{-1}x)D_{q'q}^{(k)}. \tag{8.2.1}$$

This holds if T is a function of x; we shall, however, from now on suppress x. The above definition is general and includes the possibility that T is an operator. If we wish to speak not of the components $T(kq)$ but of the whole set, we shall write $T(k;)$. The correspondence between $T(k;)$ and $T(kq)$ is the same as between a vector v and its components. Denoting rank and component by k and q (instead of j and m) is current usage; some authors remain, however, with j and m.

From the definition it follows immediately that

if $T(k;)$ is to represent a physical quantity, k must be an integer; otherwise it would be multiplied by -1 under a 2π rotation and thus be double valued (see (6.3.69))[2]. (8.2.2)

Most interesting in quantum theory are irreducible tensors, whose individual components are quantum mechanical operators in Hilbert space. In this case each individual component transforms under an active rotation in two ways:

● as an operator
$$T'(kq) = U_a T(k, q)U_a^{-1} \tag{8.2.3}$$

● but also as a component of an irreducible tensor, according to (8.2.1).

Together these lead to the transformation law of irreducible tensor operators:

$$U_a T(k, q)U_a^{-1} = \sum_{q'} T(kq')D_{q'q}^{(k)}. \tag{8.2.4}$$

Considering active, infinitesimal rotations about the main coordinate axes $e_\alpha; \alpha = x, y, z$, one finds with (3.4.2)

$$U_a(\eta e_\alpha) = 1 - i\eta J_a$$

$$D_{q'q}^{(k)}(\eta e_\alpha) = \langle kq'|1 - i\eta J_a|kq\rangle.$$

---

[1] For the appearance of M_a^{-1} in the argument of T on the r.h.s. of (8.2.1) see the discussion in section 4.1 for wave functions and vector wave functions, of which the above equation (8.2.1) is a straightforward generalization.

[2] The case of half-integer k is, however, also discussed in the literature (see, e.g., Varshalovich *et al* (1988)).

Inserting this in (8.2.4) immediately gives the result

$$
\begin{aligned}
&[J_a, T(kq)] = \sum_{q'} T(kq')\langle kq'|J_a|kq\rangle \\
&[J_\pm, T(kq)] = \sqrt{k(k+1) - q(q \pm 1)}\,T(kq \pm 1) \\
&[J_z, T(kq)] = q\,T(kq).
\end{aligned}
\tag{8.2.5}
$$

Equations (8.2.5) and (8.2.1) are fully equivalent.

All components of an irreducible tensor have the same parity; a tensor $T(k;\,)$ combined from two tensors of lower rank $T_1(k_1;\,)$ and $T_2(k_2;\,)$ has the parity $P = P_1 P_2$.

While the Hermitian conjugate of a single component of an irreducible tensor operator is defined as usual, one must be careful when defining the Hermitian conjugate of a tensor operator as a whole. The notation is important; we write ($\dagger$, Hermitian conjugate; $*$, complex conjugate) $[T(kq)]^\dagger$ for the Hermitian conjugate of the q-component of $T(k;\,)$; $T^\dagger(kq)$ for the q-component of the Hermitian conjugate tensor $T^\dagger(k;\,)$.

Writing down the Hermitian conjugate of (8.2.4) we obtain

$$
U_a[T(k,q)]^\dagger U_a^{-1} = \sum_{q'} [T(kq')]^\dagger D_{q'q}^{(k)*} = \sum_{q'} [T(kq')]^\dagger (-1)^{q'-q} D_{-q'-q}^{(k)} \tag{8.2.6}
$$

where (6.3.68) has been used. Replacing q and q' by $-q$ and $-q'$ we find (q, q' are integers)

$$
U_a(-1)^q[T(k,-q)]^\dagger U_a^{-1} = \sum_{q'} (-1)^{q'} [T(k,-q')]^\dagger D_{q'q}^{(k)} \tag{8.2.7}
$$

which shows that $(-1)^q[T(k,-q)]^\dagger$ are again components of an irreducible tensor operator. We thus define the q component of $T^\dagger$ as

$$
T^\dagger(kq) = (-1)^q[T(k,-q)]^\dagger. \tag{8.2.8}
$$

With this definition $T^\dagger(k;\,)$ is again an irreducible tensor operator. A Hermitian irreducible tensor operator obeys[3]

$$
\left.
\begin{aligned}
T^\dagger(kq) &= (-1)^q[T(k,-q)]^\dagger = T(kq) \\
[T(kq)]^\dagger &= (-1)^q T(k,-q)
\end{aligned}
\right\}
\begin{array}{l} \text{Hermitian irreducible} \\ \text{tensor operator.} \end{array}
\tag{8.2.9}
$$

Note that spherical harmonics share this property (see (4.7.29)); as they are functions and not operators, $*$ replaces $\dagger$. They also have the same transformation rule under rotations, therefore spherical harmonics Y_{lm} are a special realization of Hermitian irreducible tensors $T(lm)$.

---

[3] If the tensor components are not operators but functions, $\dagger$ is replaced by $*$.

8.3 Tensor product; irreducible combination of irreducible tensors; scalar product

The direct product of two irreducible tensors T and S is defined by

$$(T(k;)\otimes S(k;))_{kqk'q'} = T(kq)S(k'q') \tag{8.3.1}$$

in exact analogy to the direct product of angular momentum states (see section 5.3)

$$|kq\rangle \otimes |k'q'\rangle = |kqk'q'\rangle. \tag{8.3.2}$$

Consequently the direct product $T(k;)\otimes S(k';)$ transforms as the direct product $D^{(k)}\otimes D^{(k')}$, that is, as a reducible representation. Just as with the direct product of states we obtain an irreducible tensor V of rank K by means of

$$\{T(k;)S(k';)\}_{(KQ)} = V(KQ) = \sum_{qq'} T(kq)S(k'q')\langle kqk'q'|KQ\rangle \tag{8.3.3}$$

where the analogy to the composition of states is obvious:

$$|(kk')KQ\rangle = \sum_{qq'} |kqk'q'\rangle\langle kqk'q'|KQ\rangle.$$

The presence of CGCs in (8.3.3) implies that $q+q' = Q$ and $|k-k'| \le K \le k+k'$ and this analogy automatically guarantees that $V(KQ)$ is an irreducible tensor of rank K, because the irreducible tensors are defined by their transformation law being the same as that of angular momentum states. Thus we do not need to prove explicitly that $V(KQ)$ transforms with $D^{(K)}$.

In this way we can construct irreducible tensors of any rank. In particular we can construct one of rank 0, i.e. an invariant:

$$V(00) = \sum_{qq'} T(kq)S(k'q')\langle kqk'q'|00\rangle$$

where the k in T and k' in S must be the same if the CGC is to be $\ne 0$; furthermore $q' = -q$. With (5.3.47)

$$\langle kqk, -q|00\rangle = \frac{(-1)^{k-q}}{\sqrt{2k+1}} \tag{8.3.4}$$

we obtain[4]

$$V(00) = \frac{(-1)^k}{\sqrt{2k+1}} \sum (-1)^q T(kq)S(k,-q). \tag{8.3.5}$$

---

[4] If we were pedantic, $V(00)$ should be written $V(00)_k$ (as in the CGC $\langle kqk-q|(kk)00\rangle$, where we suppressed (kk) also).

This invariant $V(00)$ is uniquely defined only if k is an integer. This is seen most easily in the case where T and S are not operators but classical fields or numbers. Then $TS = ST$ is not required; but

$$\sum(-1)^q S(kq)T(k, -q) = \sum(-1)^q T(kq)S(k, -q)$$
$$= \sum(-1)^{-2q}(-1)^q T(kq)S(k, -q).$$

However, since q shares with k the property of being integer or half-integer, we have $(-1)^{-2q} = (-1)^{2k}$ and hence

$$V(00)_{ST} = (-1)^{2k} V(00)_{TS}. \tag{8.3.6}$$

Then $V(00)_{ST} = V(00)_{TS}$ implies $k =$ integer.

Assuming now k to be an integer, we define the scalar product by the requirement that for ordinary vectors the result is the usual scalar product.

To this end we remark two things:

- the spherical harmonics Y_{lm} transform as $T(lm)$

- the spherical harmonics with $l = 1$ can be written

$$\begin{pmatrix} Y_{11} \\ Y_{10} \\ Y_{1-1} \end{pmatrix} = \sqrt{3/(4\pi r^2)} \begin{pmatrix} -\dfrac{1}{\sqrt{2}}(x + iy) \\ z \\ \dfrac{1}{\sqrt{2}}(x - iy) \end{pmatrix}$$

(see table 4.2 and (4.7.32)).

That is, if we wish the vector x to be written as an irreducible tensor of rank 1, we have to give it the 'spherical components'

$$\begin{pmatrix} x_1 \\ x_0 \\ x_{-1} \end{pmatrix} = \begin{pmatrix} -\dfrac{1}{\sqrt{2}}(x + iy) \\ z \\ \dfrac{1}{\sqrt{2}}(x - iy) \end{pmatrix} = x(1, q). \tag{8.3.7}$$

Accordingly, any vector v can be written as an irreducible spherical tensor, namely in spherical components

$$v = V(1, q) = \begin{pmatrix} v_1 \\ v_0 \\ v_{-1} \end{pmatrix} = \begin{pmatrix} -\dfrac{1}{\sqrt{2}}(v_x + iv_y) \\ v_z \\ \dfrac{1}{\sqrt{2}}(v_x - iv_y) \end{pmatrix}$$

and vice versa   (8.3.8)

$$v = \begin{pmatrix} v_x \\ v_y \\ v_z \end{pmatrix} = \begin{pmatrix} \dfrac{1}{\sqrt{2}}(v_{-1} - v_1) \\ \dfrac{i}{\sqrt{2}}(v_{-1} + v_1) \\ v_0 \end{pmatrix} .$$

One then finds

$$v \cdot w = \sum_{i=x,y,z} v_i w_i = \sum_{q=-1,0,1} (-1)^q v_q w_{-q} = \sum_q V(1q)W(1, -q). \quad (8.3.9)$$

Consequently we define the scalar product for any two irreducible tensors by means of (8.3.5)

$$T(k;)S(k;) = \sqrt{2k + 1}(-1)^k V(00) = \sum_q (-1)^q T(kq)S(k, -q). \quad (8.3.10)$$

Remembering (8.3.6) we can state that for *classical* irreducible tensor fields of rank k

$$ST = -TS \quad \text{if } k = \text{half integer} \quad \text{'anticommuting'}$$
$$ST = TS \quad \text{if } k = \text{integer} \quad \text{'commuting'}.$$

8.4 Invariants and covariant equations

A few words on invariants may be added. We said that $T(k;)S(k;)$ is an invariant. Indeed, it transforms like a state $|00\rangle$, that is, it remains unaffected. Invariant means here form invariant. Namely, if

$$TS = \sum T(kq)S(k, -q)(-1)^q$$

in one frame of reference, then it becomes

$$\sum T'(kq)S'(k, -q)(-1)^q$$

in another one and in fact these two sums are equal, even if T and S are fields

$$\sum (-1)^q T(kq; \boldsymbol{x})S(k, -q; \boldsymbol{y}) = \sum (-1)^q T'(kq; \boldsymbol{x}')S'(k, -q; \boldsymbol{y}') \quad (8.4.1)$$

which, although obvious by the method of construction, can be checked by an explicit calculation using (8.2.1) in the form

$$T'(kq; x') = \sum_{q'} T(kq'; x) D^{(k)}_{q'q}.$$

It will turn out that the D cancel altogether. However in fact no such proof is necessary because we *know* that TS transforms with $D^{(0)} = 1$.

Often the following statements are useful.

> If a question is of such a nature that its answer will always be the same, no matter how the coordinate axes are oriented, then it must be possible to answer this question with the help of those invariants which one can build with the available irreducible tensors. One may then find the answer in that particular frame of reference where it becomes easiest. One looks at how the invariants appear in this system, expresses the answer by them and has thereby found the general invariant formulation. Furthermore: if an equation, given in a particular frame of reference, can be written in a manifestly covariant form (namely, both sides of the equation transform in the same way) which in the mentioned particular frame of reference reduces to the equation given there, then this covariant formulation is the unique generalization of the equation given. $\qquad$ (8.4.2)

As an illustration, we derive once more an earlier result: the addition theorem of spherical harmonics.

Is $P_l(\cos \vartheta)$ an invariant?

- NO! if by ϑ is meant the polar angle of r. In this case

$$P_l(\cos \vartheta) = \sqrt{\frac{4\pi}{2l+1}} Y_{l0}(\vartheta, 0)$$

transforms with $D^{(l)}$ into a combination of Y_{lm}.

- YES! if by ϑ the angle between two vectors, v and w, is meant. In this case it must be possible to express $P_l(\cos \vartheta)$ in a manifestly invariant form.

We consider the second case. Let n_1 and n_2 be two unit vectors with directions (ϑ_1, φ_1) and (ϑ_2, φ_2) respectively. Then $\cos \vartheta = n_1 \cdot n_2$. We now take a coordinate system where $\vartheta'_1 = 0$, $\vartheta'_2 = \vartheta$, $\varphi'_2 = 0$. In these coordinates we have

$$P_l(\cos \vartheta) = \sqrt{\frac{4\pi}{2l+1}} Y_{l0}(\vartheta'_2, \varphi'_2).$$

Can we write this in a manifestly invariant form? Yes, because the spherical harmonic of the other angles $\vartheta'_1 = 0$ and φ' can be written (see (6.3.30))

$$Y_{lm}(\vartheta'_1 = 0, \varphi'_1) = \delta_{m0} \sqrt{\frac{2l+1}{4\pi}}.$$

Hence, still in this particular coordinate frame:

$$P_l(\cos\vartheta) = \sqrt{\frac{4\pi}{2l+1}} Y_{l0}(\vartheta_2', \varphi_2') = \sqrt{\frac{4\pi}{2l+1}} \sum_m \delta_{m0} Y_{lm}(\vartheta_2', \varphi_2')$$

$$= \frac{4\pi}{2l+1} \sum_m Y_{lm}(\vartheta_1', \varphi_1') Y_{lm}(\vartheta_2', \varphi_2').$$

Comparing this with the invariant scalar product (8.4.1) we see that this sum would be invariant if with one of the Y we were to change m into $-m$ and furthermore attach a factor $(-1)^m$ to each term of the sum. Since the sum contains only one term, namely the one with $m = 0$, we are free to make these changes without doing anything to the sum, except that this sum becomes now the invariant scalar product of the two irreducible tensors $Y_{lm}(\vartheta_1', \varphi_1')$ and $Y_{lm}(\vartheta_2', \varphi_2')$. Thus we have the expression (valid in any coordinate system)

$$P_l(\cos\vartheta) = \frac{4\pi}{2l+1} \sum_m (-1)^m Y_{lm}(\vartheta_1, \varphi_1) Y_{l,-m}(\vartheta_2, \varphi_2)$$

$$= \frac{4\pi}{2l+1} \sum_m Y_{lm}(\vartheta_1, \varphi_1) Y_{lm}^*(\vartheta_2, \varphi_2)$$

(8.4.3)

where (4.7.29) was used.

A particular application is the expansion of a plane wave into angular momentum states. Since $\mathbf{k} \cdot \mathbf{r} = kr\cos\vartheta$ is invariant, $P_l(\cos\vartheta)$ in the formula

$$e^{i\mathbf{k}\cdot\mathbf{r}} = \sum_{l=0}^{\infty} i^l(2l+1)j_l(kr)P_l(\cos\vartheta)$$

(8.4.4)

is taken in the invariant sense. Hence $P_l(\cos\vartheta)$ may be expressed in the invariant form (8.4.3)

$$e^{i\mathbf{k}\cdot\mathbf{r}} = 4\pi \sum_{l=0}^{\infty} i^l j_l(kr) \sum_m Y_{lm}(\vartheta_k, \varphi_k) Y_{lm}^*(\vartheta_r, \varphi_r).$$

(8.4.5)

8.5 Spinor and vector spherical harmonics

What we are going to address here could have been discussed previously when the addition of angular momenta was considered.

As we found in the discussion of the physical significance of $\mathbf{J}$ for a vector particle (see section 4.1), its wave function consists of three components which transform like a vector. Of course, if the components of

$$\psi(\mathbf{x}) = \begin{pmatrix} \psi_x(\mathbf{x}) \\ \psi_y(\mathbf{x}) \\ \psi_z(\mathbf{x}) \end{pmatrix}$$

(8.5.1)

transform among each other in a prescribed form, then not just any three functions ψ_x, ψ_y, ψ_z will do: there will be a definite relation between them; e.g. at any given point P, we can achieve that there (in these coordinates, x' say) $\psi'_x(x') \neq 0$; $\psi'_y = \psi'_z = 0$: we simply put the x'-axis in direction ψ.

The same is true if we consider a two-component wave function, whose components transform according to $D^{(\frac{1}{2})}$.

We shall now construct functions (not states) corresponding to spin $\frac{1}{2}$ and 1. Since the composition law is the same as for states—just a sum with CGCs—we could already have done it earlier. However, now we can do it somewhat more generally, since we consider the 'wave functions' no longer as representatives of states, but simply as a set of functions with such and such transformation properties, that is: as irreducible spherical tensors. We can then leave open the possibility of giving them operator properties. In this way the quantities we are going to construct may serve in three different ways (and have each time a completely different physical significance—only the transformation properties remain the same):

- as wave functions of spin-$\frac{1}{2}$ or spin-1 states;
- as irreducible tensors $T_{(l\frac{1}{2})}(JM)$ or $T_{(l1)}(JM)$;
- as irreducible tensor operators.

A quantity which depends on the coordinates ϑ, φ and transforms like a state

$$|(ls)JM\rangle = \sum |lms\mu\rangle\langle lms\mu|JM\rangle \qquad (8.5.2)$$

will necessarily have the form

$$T_{(ls)}(JM; \vartheta, \varphi) = \sum_{m,\mu} Y_{lm}(\vartheta, \varphi)\chi_{s\mu}\langle lms\mu|JM\rangle$$

where s is to be an *intrinsic* and l an *orbital* angular momentum.

We have only to specify the $\chi_{s\mu}$ such that they obey

$$\begin{aligned} S^2\chi_{s\mu} &= s(s+1)\chi_{s\mu} \\ S_z\chi_{s\mu} &= \mu\chi_{s\mu}. \end{aligned} \qquad (8.5.3)$$

For the case $s = \frac{1}{2}$ we know the spinors:

$$\chi_{\frac{1}{2},\frac{1}{2}} \equiv \chi_+ = \begin{pmatrix} 1 \\ 0 \end{pmatrix} \qquad \chi_{\frac{1}{2},-\frac{1}{2}} \equiv \chi_- = \begin{pmatrix} 0 \\ 1 \end{pmatrix}$$

(see (4.8.9) and (4.8.10)). Hence we define the *spinor spherical harmonics*

$$\Phi_{(l\frac{1}{2})JM}(\vartheta, \varphi) = i^l \sum_{m\mu} Y_{lm}(\vartheta, \varphi)\chi_{\frac{1}{2}\mu}\langle lm\tfrac{1}{2}\mu|JM\rangle \qquad (8.5.4)$$

where i^l is a conventional factor which has been introduced to give Φ a simple behaviour under time reversal. Since $m + \mu = M$, we can write very explicitly

$$\Phi_{(l\frac{1}{2})JM}(\vartheta, \varphi) = i^l \begin{pmatrix} \langle l, M - \frac{1}{2}, \frac{1}{2}, \frac{1}{2} | JM \rangle\, Y_{l,M-\frac{1}{2}}(\vartheta, \varphi) \\ \langle l, M + \frac{1}{2}, \frac{1}{2}, \frac{1}{2} | JM \rangle\, Y_{l,M+\frac{1}{2}}(\vartheta, \varphi) \end{pmatrix}. \tag{8.5.5}$$

For the vector spherical harmonics we still have to determine the eigenstates $\chi_{1\mu}$ of S. We know S from (4.1.8)

$$S_x = \begin{pmatrix} 0 & 0 & 0 \\ 0 & 0 & -i \\ 0 & i & 0 \end{pmatrix} \quad S_y = \begin{pmatrix} 0 & 0 & i \\ 0 & 0 & 0 \\ -i & 0 & 0 \end{pmatrix} \quad S_z = \begin{pmatrix} 0 & -i & 0 \\ i & 0 & 0 \\ 0 & 0 & 0 \end{pmatrix}. \tag{8.5.6}$$

The most obvious thing to do is to write

$$\begin{pmatrix} 0 & -i & 0 \\ i & 0 & 0 \\ 0 & 0 & 0 \end{pmatrix} \begin{pmatrix} a \\ b \\ c \end{pmatrix}_{(\mu)} = \mu \begin{pmatrix} a \\ b \\ c \end{pmatrix}_{(\mu)} \tag{8.5.7}$$

and to solve these equations for a, b and c; but as we are now in the chapter on irreducible tensors, let us do so with a corresponding technique, as follows.

A vector v (in the abstract sense), if fixed in space, is an invariant quantity under rotations of the coordinates. Rather, it is its representation by means of components (v_x, v_y, v_z) which is not invariant. That v is an invariant must be— according to the statement (8.4.2)— expressible in a manifestly invariant form. Here it is:

$$v = \sum_{i=x,y,z} v_i e_i = \sum_{i=x,y,z} v_i' e_i'. \tag{8.5.8}$$

Nothing prevents us from writing this 'scalar product' in the spherical basis, namely as the invariant (see (8.3.10))

$$v = \sum_{\mu=1,0,-1} (-1)^\mu v_\mu e_{-\mu} = \sum_{\mu=1,0,-1} (-1)^\mu v_\mu' e_{-\mu}'. \tag{8.5.9}$$

But now we know that the three spherical unit vectors e_1, e_0 and e_{-1} must transform among themselves exactly as states $|1\mu\rangle$ would do; and since each of them has three components they must be eigenvectors of S^2 and S_z with eigenvalues $\mu = 1, 0, -1$ respectively. The e_μ are easily found:

$$v \cdot w = \sum_{i=x,y,z} v_i w_i = \sum_{\mu=1,0,-1} (-1)^\mu v_\mu w_{-\mu}$$

is invariant if v_μ and w_μ are both constructed according to the rule (see (4.7.32) and (8.3.8))

$$w_1 = -\frac{1}{\sqrt{2}}(w_x + iw_y)$$

$$w_0 = w_z$$

$$w_{-1} = \frac{1}{\sqrt{2}}(w_x - iw_y).$$

If we require that $v = \sum_{\mu=1,0,-1}(-1)^{\mu} v_{\mu} e_{-\mu}$ be invariant, then the e_{μ} must be constructed just the same way, namely

$$e_1 = -\frac{1}{\sqrt{2}}(e_x + ie_y) = -\frac{1}{\sqrt{2}}\begin{pmatrix} 1 \\ i \\ 0 \end{pmatrix}$$

$$e_0 = e_z = \begin{pmatrix} 0 \\ 0 \\ 1 \end{pmatrix} \qquad\qquad (8.5.10)$$

$$e_{-1} = \frac{1}{\sqrt{2}}(e_x - ie_y) = \frac{1}{\sqrt{2}}\begin{pmatrix} 1 \\ -i \\ 0 \end{pmatrix}.$$

One easily checks that these e_{μ} indeed fulfil (see (8.5.6))

$$\begin{aligned} S^2 e_{\mu} &= 2e_{\mu} \\ S_z e_{\mu} &= \mu e_{\mu} \qquad \mu = 1, 0, -1. \end{aligned} \qquad (8.5.11)$$

They are thus the eigenstates $\chi_{1\mu}$ needed to construct the vector spherical harmonics. Even if this derivation was, admittedly, longer than the simple determination of the eigenvectors of S_z, it has, however, served as an illustration of the concepts of 'invariance' and 'irreducible spherical tensor'. That is, we have just learned that a vector v, taken in the abstract sense, may be considered as an invariant or spherical tensor of rank 0; quite differently, the three unit vectors $e_x\, e_y\, e_z$ are not invariant, since they are by definition fixed to the coordinate axes and rotate with these. Consequently the three vectors e_{μ} are components of an irreducible spherical tensor $T(1\mu)$. Quite generally

> the components $T(kq)$ of an irreducible spherical tensor may still be scalars, vectors, tensors with respect to x, y, z. $\qquad (8.5.12)$

This ends our little digression and we go back to our problem: we define the *vector spherical harmonics* with $\chi_{1\mu} \equiv e_{\mu}$,

$$Y_{(l1)JM}(\vartheta, \varphi) = \sum_{m,\mu} Y_{lm}(\vartheta, \varphi) e_{\mu} \langle lm1\mu | JM \rangle. \qquad (8.5.13)$$

Their use, in particular for the description of the Maxwell field, is well known. They form a complete orthogonal system (on the unit sphere) for vector functions. This follows from the simple fact that the wave functions $\langle x|(l1)JM \rangle$ are a complete set.

The spinor spherical harmonics are used for spin-$\frac{1}{2}$ fields or wave functions. Some care is necessary if relativistic particles are considered. We know from Dirac's relativistic hydrogen atom that L is not conserved and that the eigenfunctions of J^2 and J_z contain $Y_{j+\frac{1}{2},m}$ and $Y_{j-\frac{1}{2},m}$ in one single spinor. This shows already complications we might encounter in a relativistic treatment.

Thus, relativistic spin-$\frac{1}{2}$ particles *cannot* be described by the spinor spherical harmonics $\Phi_{(l\frac{1}{2})JM}$—unless they have zero mass, when again a two-component wave function is sufficient. We shall come back to these questions later. For the moment we only remark that all those CGCs which can occur in spinor and vector spherical harmonics can be obtained from table 5.2 and (5.3.54)–(5.3.56). For the convenience of the reader we give them here explicitly.

Table of $\langle j_1 m_1 \frac{1}{2} m_2 | j m \rangle$

m_2	$\frac{1}{2}$	$-\frac{1}{2}$	
j			
$j_1 + \frac{1}{2}$	$\sqrt{\dfrac{j_1 + m + \frac{1}{2}}{2j_1 + 1}}$	$\sqrt{\dfrac{j_1 - m + \frac{1}{2}}{2j_1 + 1}}$	(8.5.14)
$j_1 - \frac{1}{2}$	$-\sqrt{\dfrac{j_1 - m + \frac{1}{2}}{2j_1 + 1}}$	$\sqrt{\dfrac{j_1 + m + \frac{1}{2}}{2j_1 + 1}}$	

Table of $\langle j_1 m_1 1 m_2 | j m \rangle$

m_2	1	0	-1	
j				
$j_1 + 1$	$\sqrt{\dfrac{(j_1+m)(j_1+m+1)}{(2j_1+1)(2j_1+2)}}$	$\sqrt{\dfrac{(j_1-m+1)(j_1+m+1)}{(2j_1+1)(j_1+1)}}$	$\sqrt{\dfrac{(j_1-m)(j_1-m+1)}{(2j_1+1)(2j_1+2)}}$	(8.5.15)
j_1	$-\sqrt{\dfrac{(j_1+m)(j_1-m+1)}{2j_1(j_1+1)}}$	$\sqrt{\dfrac{m^2}{j_1(j_1+1)}}$	$\sqrt{\dfrac{(j_1-m)(j_1+m+1)}{2j_1(j_1+1)}}$	
$j_1 - 1$	$\sqrt{\dfrac{(j_1-m+1)(j_1-m)}{2j_1(2j_1+1)}}$	$-\sqrt{\dfrac{(j_1-m)(j_1+m)}{j_1(2j_1+1)}}$	$\sqrt{\dfrac{(j_1+m+1)(j_1+m)}{2j_1(2j_1+1)}}$	

8.6 Angular momenta as spherical tensor operators

If we define (notice the difference compared with (4.6.3)!) the spherical components of J as

$$J_{+1} = -\frac{1}{\sqrt{2}}(J_x + iJ_y) = -\frac{1}{\sqrt{2}}J_+$$
$$J_0 = J_z \tag{8.6.1}$$
$$J_{-1} = \frac{1}{\sqrt{2}}(J_x - iJ_y) = \frac{1}{\sqrt{2}}J_-$$

then these $J_\mu (\mu = 1, 0, -1)$ transform under rotations with $D^{(1)}$ and form an irreducible tensor operator of rank 1. Indeed one can check explicitly that they obey the commutation relations (8.2.5) with $T(kq) = T(1q) = J_q$.

If one rewrites the commutation relations (8.2.5) by expressing the $J_\pm$ by J_μ ($\mu = 1, 0, -1$) then one obtains

$$[J_{\pm 1}, T(kq)] = \mp\sqrt{\frac{1}{2}(k \mp q)(k \pm q + 1)} T(k, q \pm 1)$$

$$[J_0, T(kq)] = q T(kq).$$

(8.6.2)

From the table (8.5.15) of $\langle j_1 m_1 1 m_2 | jm \rangle$ we read off that the roots can be expressed by CGCs; we then write (using the symmetry relations (5.3.52))

$$[J_\mu, T(kq)] = -\sqrt{k(k + 1)}\langle 1\mu kq | kq + \mu \rangle T(k, q + \mu).$$

(8.6.3)

Similarly, for the matrix elements of J_μ one obtains, rewriting $\sqrt{j(j + 1) - m(m \pm 1)}$ by means of (8.5.15), and using the symmetry relations (5.3.52)

$$\langle j'm' | J_\mu | jm \rangle = -\delta_{jj'}\sqrt{j(j + 1)}\langle j'm' | 1\mu jm \rangle.$$

(8.6.4)

8.7 The Wigner–Eckart theorem

This famous theorem asserts that in the matrix element of any arbitrary irreducible tensor operator between *standard* angular momentum states

$$\langle \gamma' j'm' | T(kq) | \gamma jm \rangle$$

the dependence on m', q and m is entirely contained in a CGC which multiplies a 'reduced matrix element', which is independent of m', q and m. This has two consequences:

- the factorization of the matrix element into a CGC and a reduced matrix element makes it sometimes unnecessary to calculate the latter;
- if the matrix element must be fully calculated, then one can do it for the m', q, m-values which make it easiest. In this way one obtains the reduced matrix element and the full one is given by multiplying it by the appropriate CGC.

The proof of this theorem is extremely simple and needs hardly any calculations; it follows simply from considering carefully the transformation properties.

The matrix element in question is formed by taking the scalar product (in Hilbert space) of the standard bra state $\langle \gamma' j'm' |$ with the ket vector $T(kq)|\gamma jm\rangle$. This ket vector transforms exactly like a state $|kqjm\rangle$, namely with $D^{(k)} \otimes D^{(j)}$. Hence it can be written as a linear combination of *standard* states $|\psi; j''m''\rangle$ which transform with $D^{(j'')}$ where $|k - j| \leq j'' \leq k + j$; this linear combination is necessarily

$$T(kq)|\gamma jm\rangle = \sum_{j''} |\psi; j''m''\rangle\langle kqjm | j''m''\rangle$$

$$|\psi; j''m''\rangle = \sum_{qm} T(kq)|\gamma jm\rangle\langle kqjm | j''m''\rangle.$$

(8.7.1)

Therefore

$$\langle \gamma' j' m' | T(kq) | \gamma jm \rangle = \langle \gamma' j' m' | \psi; j'm' \rangle \langle kqjm | j'm' \rangle. \tag{8.7.2}$$

All that remains is to show that $\langle \gamma' j' m' | \psi; j'm' \rangle$ is independent of m'. Looking at the second line of (8.7.1) we again stress its significance: it says that $|\psi; j''m''\rangle$ transforms with the irreducible representation $D^{(j'')}$, that is, like a *standard state*. Consequently ψ (whatever ψ means) is not changed by acting with any of the J_i on $|\psi; jm\rangle$. Thus also

$$J_{\pm} |\psi; j'm'\rangle = \sqrt{j'(j'+1) - m'(m' \pm 1)} |\psi; j'm' \pm 1\rangle$$

and, since $J_- = J_+^{\dagger}$

$$\langle \gamma' j' m' | \psi; j'm' \rangle = \frac{\langle \gamma' j' m' | J_- J_+ | \psi; j'm' \rangle}{j'(j'+1) - m'(m'+1)} = \langle \gamma' j', m'+1 | \psi; j', m'+1 \rangle$$

which shows the independence of m'. Note, however, that the set ψ will in general depend on the nature of T as well as of the values of k, γ and j, hence $\psi \equiv \psi_T(\gamma kj)$.

The matrix element $\langle \gamma' j' m' | \psi; j'm' \rangle = \langle \gamma' j' | \psi j' \rangle$ is essentially the 'reduced matrix element'. Unfortunately there is a variety of factors attached to it by various authors. We shall employ here the notation of Racah, which reads

$$\langle \gamma' j' | \psi_T(\gamma kj); j' \rangle \equiv \frac{(-1)^{j'+k-j}}{\sqrt{2j'+1}} \langle \gamma' j' \| T(k;) \| \gamma j \rangle. \tag{8.7.3}$$

We give here only a few of the different notations.

Our reduced matrix element is the same as that of Edmonds (1957), of Racah (1942), of Messiah (1970) and of de-Shalit and Talmi (1963). It differs from that of Fano and Racah (1959) by $(-1)^{2k}$ (which in all practical cases is no difference) and from that of Rose (1957) by

$$\langle j' \| T(k;) \| j \rangle_{\text{ours}} = \sqrt{2j'+1}(-1)^{2k} \langle j' \| T(k;) \| j \rangle_{\text{Rose}}. \tag{8.7.4}$$

Except for Racah, the authors' names listed here refer to the books quoted in table 5.1. Further notations are compiled in table 5.1 of Edmonds (1957). An extensive collection of formulae is given in Varshalovich *et al* (1988). Apart from a possibly different definition of $\langle \gamma' j' \| T(k;) \| \gamma j \rangle$ it is clear that by means of the symmetry relations of the CGCs and/or $3j$-symbols (see (5.3.52) and (5.3.54)) (8.7.3) and the following Wigner–Eckart theorem can be written in various other equivalent forms.

Combining (8.7.2) and (8.7.3), we obtain the Wigner–Eckart theorem (henceforth W–E theorem)

$$\langle \gamma' j' m' | T(kq) | \gamma jm \rangle = \frac{(-1)^{j'+k-j}}{\sqrt{2j'+1}} \langle j'm' | kqjm \rangle \langle \gamma' j' \| T(k;) \| \gamma j \rangle \tag{8.7.5}$$

where $\langle \gamma' j' \| T(k;) \| \gamma j \rangle$ is invariant under rotations, i.e. independent of m', q and m. Thus

- if $\langle \gamma' j'm'|T(kq)|\gamma jm\rangle$ is to be discussed with respect to its dependence on $m'qm$, then the reduced matrix element can be considered just as an unknown constant common factor—only the CGCs need be considered;
- if $\langle \gamma' j'm'|T(kq)|\gamma jm\rangle$ must be be calculated, one chooses to calculate the (m', q, m)-independent reduced matrix element

$$\langle \gamma' j'\|T(k;)\|\gamma j\rangle = (-1)^{j'+k-j}\sqrt{2j'+1}\frac{\langle \gamma' j'm'|T(kq)|\gamma jm\rangle}{\langle j'm'|kqjm\rangle}$$

for those values m', q and m where it is easiest. The full matrix element for general m', q and m then follows from the Wigner–Eckart theorem.

In particular it follows that for any two irreducible tensors of equal rank, say $T(k;)$ and $S(k;)$,

$$\langle \gamma' j'm'|T(kq)|\gamma jm\rangle = \frac{\langle \gamma' j'\|T(k;)\|\gamma j\rangle}{\langle \gamma' j'\|S(k;)\|\gamma j\rangle}\langle \gamma' j'm'|S(kq)|\gamma jm\rangle. \tag{8.7.6}$$

8.8   Examples of applications of the Wigner–Eckart theorem

8.8.1   The trace of $T(kq)$

The trace of an irreducible tensor is defined by

$$\mathrm{Tr}(T(kq)) = \sum_{\gamma jm}\langle \gamma jm|T(kq)|\gamma jm\rangle. \tag{8.8.1}$$

The W–E theorem gives

$$\mathrm{Tr}(T(kq)) = (-1)^k\sum_{\gamma jm}\frac{1}{\sqrt{2j+1}}\langle jm|kqjm\rangle\langle \gamma j\|T(k;)\|\gamma j\rangle \qquad q = 0.$$

With (5.3.47) we put

$$\frac{1}{\sqrt{2j+1}} = (-1)^{j-m}\langle jmj-m|00\rangle$$

and with (5.3.51)

$$\langle k0jm|jm\rangle = (-1)^{k-j+m}\sqrt{\frac{2j+1}{2k+1}}\langle jmj-m|k0\rangle.$$

Then with the orthogonality relation (5.3.14)

$$\sum_m\frac{1}{\sqrt{2j+1}}\langle k0jm|jm\rangle = (-1)^k\sqrt{\frac{2j+1}{2k+1}}\sum_m\langle k0|jmj-m\rangle\langle jmj-m|00\rangle$$

$$= \delta_{k0}\sqrt{2j+1}.$$

Hence

$$\mathrm{Tr}(T(kq)) = \delta_{k0}\sum_{\gamma j}\sqrt{2j+1}\langle \gamma j\|T(k;)\|\gamma j\rangle \tag{8.8.2}$$

that is, only tensors of rank 0 have a non-vanishing trace.

8.8.2   Tensors of rank 0 (scalars, invariants)

The W–E theorem gives

$$\langle \gamma' j'm'|T(00)|\gamma jm\rangle = \frac{\langle j'm'|00jm\rangle}{\sqrt{2j'+1}}\langle \gamma' j'\|T(0)\|\gamma j\rangle$$

$$= \delta_{jj'}\delta_{mm'}\frac{\langle \gamma' j\|T(0)\|\gamma j\rangle}{\sqrt{2j+1}}. \qquad (8.8.3)$$

8.8.3   The angular momentum operators

We defined the spherical components J_μ by (8.6.1):

$$J_1 = -\frac{1}{\sqrt{2}}(J_x + iJ_y) = -\frac{1}{\sqrt{2}}J_+$$

$$J_0 = J_z \qquad (8.8.4)$$

$$J_{-1} = \frac{1}{\sqrt{2}}(J_x - iJ_y) = \frac{1}{\sqrt{2}}J_-$$

and found (see (8.6.4)) that

$$\langle \gamma' j'm'|J_\mu|\gamma jm\rangle = -\delta_{jj'}\delta_{\gamma\gamma'}\sqrt{j(j+1)}\langle j'm'|1\mu jm\rangle.$$

Since J_μ is an irreducible tensor of rank 1, the W–E theorem says

$$\langle \gamma' j'm'|J_\mu|\gamma jm\rangle = \frac{(-1)^{j'+1-j}}{\sqrt{2j'+1}}\langle j'm'|1\mu jm\rangle\langle \gamma' j'\|J\|\gamma j\rangle$$

where, of course, $m' = m + \mu$. Comparison of these two expressions gives

$$\langle \gamma' j'\|J\|\gamma j\rangle = \delta_{\gamma\gamma'}\delta_{jj'} \cdot \sqrt{j(j+1)(2j+1)}. \qquad (8.8.5)$$

8.9   Projection theorem for irreducible tensor operators of rank 1

In this section we shall write $T(1;) \equiv T$ and $J(1;) \equiv J$, i.e. we use the vector notation but we mean the vectors written in spherical coordinates.

Since J is an irreducible tensor of rank 1, any other irreducible tensor of rank 1 obeys equation (8.7.6) in the form

$$\langle \gamma' j'm'|T|\gamma jm\rangle = \frac{\langle \gamma' j'\|T\|\gamma j\rangle}{\langle \gamma' j'\|J\|\gamma j\rangle}\langle \gamma' j'm'|J|\gamma jm\rangle.$$

This equation loses meaning, however, if $\gamma' \neq \gamma$ and/or $j' \neq j$ because J is diagonal in γ and j, and if not $\{\gamma' = \gamma$ and $j' = j\}$, then the equation reads

$\langle \gamma' j' m' | T | \gamma j m \rangle = \frac{0}{0}$. Therefore we restrict ourselves to $\gamma' = \gamma$ and $j' = j$. Then

$$\langle \gamma j m' | T | \gamma j m \rangle = C \langle \gamma j m' | J | \gamma j m \rangle$$

$$C = \frac{\langle \gamma j \| T \| \gamma j \rangle}{\langle \gamma j \| J \| \gamma j \rangle}. \tag{8.9.1}$$

C is simply a number. The above equation says that the two vectors $\langle \gamma j m' | T | \gamma j m \rangle$ and $\langle \gamma j m' | J | \gamma j m \rangle$ are parallel. Therefore, if we decompose T into a component parallel to J, namely $J(T \cdot J)$, and one $T_\perp$ perpendicular to it:

$$\langle \gamma j m' | T | \gamma j m \rangle = \frac{\langle \gamma j m' | J(T \cdot J) | \gamma j m \rangle}{f(\gamma j m' m)} + \underbrace{\langle \gamma j m' | T_\perp | \gamma j m \rangle}_{=0}.$$

Then the matrix elements of $T_\perp$ will vanish. It remains to determine the constant $f(\gamma j m m')$ which, if ordinary vectors were considered, would be $|J|^2$; thus we expect that $f(\gamma j m' m) = j(j + 1)$. We proceed as follows: since J is diagonal in j and $J \cdot J$ as a tensor of rank 0 is diagonal in j and m (see (8.8.3)), the first matrix element can be split into two (where, because of the just mentioned diagonalities of J and $T \cdot J$, no sum over intermediate states occurs):

$$\langle \gamma j m' | T | \gamma j m \rangle = \frac{\langle \gamma j m' | J(T \cdot J) | \gamma j m \rangle}{f(\gamma j m' m)}$$

$$= \langle \gamma j m' | J | \gamma j m \rangle \frac{\langle \gamma j m' | T \cdot J | \gamma j m \rangle}{f(\gamma j m' m)}. \tag{8.9.2}$$

Comparison with (8.9.1) gives

$$\langle \gamma j m | T \cdot J | \gamma j m \rangle = f(\gamma j m' m) \, C$$

which shows already that f does not depend on m'; but, furthermore, we can again split this matrix element and then we use once more (8.9.1):

$$f(\gamma j m) \, C = \sum_{m'} \langle \gamma j m | T | \gamma j m' \rangle \langle \gamma j m' | J | \gamma j m \rangle$$

(since J is diagonal in j and γ, sum only over m')

$$= C \sum_{m'} \langle \gamma j m | J | \gamma j m' \rangle \langle \gamma j m' | J | \gamma j m \rangle$$

$$= C \langle \gamma j m | J^2 | \gamma j m \rangle = C j (j + 1).$$

Hence $f(\gamma j m) = j(j + 1)$. Thus we obtain from (8.9.2), (8.8.3) and (8.9.1) the projection theorem

$$\langle \gamma j m' | T | \gamma j m \rangle = \frac{\langle \gamma j m' | J (T \cdot J) | \gamma j m \rangle}{j(j+1)}$$

$$= \langle \gamma j m' | J | \gamma j m \rangle \frac{\langle \gamma j m | T \cdot J | \gamma j m \rangle}{j(j+1)}$$

$$= \langle \gamma j m' | J | \gamma j m \rangle \frac{\langle \gamma j \| T \cdot J \| \gamma j \rangle}{j(j+1)\sqrt{2j+1}} \qquad (8.9.3)$$

$$= \langle \gamma j m' | J | \gamma j m \rangle \frac{\langle \gamma j \| T \| \gamma j \rangle}{\langle \gamma j \| J \| \gamma j \rangle}$$

from which follows

$$\langle \gamma j \| T \cdot J \| \gamma j \rangle = j(j+1)\sqrt{2j+1} \frac{\langle \gamma j \| T \| \gamma j \rangle}{\langle \gamma j \| J \| \gamma j \rangle}. \qquad (8.9.4)$$

Putting $T = J$ gives

$$\langle \gamma j \| J^2 \| \gamma j \rangle = j(j+1)\sqrt{2j+1}. \qquad (8.9.5)$$

Example: Landé's formula

The magnetic moment is given by

$$\mu = \mu_0[g_L L + g_S S] \equiv \alpha L + \beta S. \qquad (8.9.6)$$

What is the expectation value of μ in a state $|(ls)jm\rangle$? μ is an irreducible tensor of rank 1, hence (8.9.3) applies:

$$\langle (ls)jm | \mu | (ls)jm \rangle = \langle (ls)jm | \mathbf{J} | (ls)jm \rangle \frac{\langle (ls)jm | J \cdot \mu | (ls)jm \rangle}{j(j+1)}. \qquad (8.9.7)$$

Now, for states $|(ls)jm\rangle$:

$$J \cdot \mu = (L + S) \cdot (\alpha L + \beta S) = \alpha l(l+1) + \beta s(s+1) + (\alpha + \beta) L \cdot S$$
$$J^2 = (L + S)^2 = l(l+1) + s(s+1) + 2L \cdot S = j(j+1).$$

Thus

$$L \cdot S = \tfrac{1}{2}[j(j+1) - l(l+1) - s(s+1)].$$

That gives

$$J \cdot \mu = \frac{\alpha + \beta}{2} j(j+1) + \frac{\alpha - \beta}{2}[l(l+1) - s(s+1)].$$

Putting this into (8.9.7) yields with

$$\langle (ls)jm | J_q | (ls)jm \rangle = m \, \delta_{qz}$$

and α and β replaced by $\mu_0 g_L$ and $\mu_0 g_S$

$$\langle (ls)jm|\mu_q|(ls)jm \rangle = \delta_{qz} m \frac{\mu_0}{2} \left[g_L + g_S + (g_L - g_S) \frac{l(l+1) - s(s+1)]}{j(j+1)} \right]$$

(8.9.8)

which is the Landé formula. That $\langle \mu_x \rangle = \langle \mu_y \rangle = 0$ is physically obvious.

9

PECULIARITIES OF TWO-DIMENSIONAL ROTATIONS: ANYONS, FRACTIONAL SPIN AND STATISTICS

9.1 Introduction

In previous chapters we have only considered the three-dimensional rotation group, its generators and their physical realization in the Hilbert space—the angular momentum. In this chapter we shall give a brief introduction to rotations in two-dimensional space and find that rotations in two-dimensional space have some special properties. In particular, spin and statistics have peculiarities not possible in three- or higher-dimensional space—spin is not quantized in integer or half-integer units of $\hbar$ and hence the particles are not necessarily bosons or fermions but may obey any statistics and were thus called anyons (stemming from 'any' and 'on') by Wilczek (1982, 1990). In fact, the concept of fractional statistics had been already put forward by Leinaas and Myrheim in the late 1970s. They found that the origin of the fractional statistics lies in the peculiar topological properties of the configuration space of many identical particles. This space is doubly connected in three or more dimensions but is multiply connected in two dimensions (Leinaas and Myrheim (1977)). Later Goldin, Menikoff and Sharp reached a similar conclusion using a completely different method based on the rigorous study of the unitary representations of current algebra and diffeomorphism groups (Goldin *et al* (1980, 1981)).

Since the real physical space is three dimensional and the observed particles can only be bosons or fermions, anyons may be thought of as purely mathematical objects. In fact, this need not be the case. In recent years the advances in the understanding of the fractional quantum Hall effect and high-temperature superconductivity have led to interest in this kind of quasi-particle with fractional statistics. It has been found that anyons need not be only purely mathematical objects and may play an important role in physical systems which can effectively or approximately be regarded as two dimensional so that the localized excitations should be quasi-particles obeying the laws of two-dimensional physics. Therefore, these quasi-particles should be just anyons and could be observed. One remarkable example is the famous fractional quantum Hall effect in condensed matter systems, where the collective excitations can be identified as localized quasi-particles with fractional charge, fractional spin and fractional statistics and can be naturally regarded as anyons.

This chapter is arranged as follows. In section 9.2 we shall consider the transformation behaviour of wave functions under two-dimensional rotations to explain the reason why there is a possibility of arbitrary spin. In section 9.3 we shall give an explicit example to show that a physical system in two-dimensional space can indeed have fractional spin. There are still a lot of other interesting topics about anyon physics such as the Chern–Simons gauge theory description of anyon dynamics, the many-anyon problem, the statistical mechanics of anyons, and the relation between anyons and braid groups and between anyons and two-dimensional conformal field theory. These topics are outside the scope of this book; the interested reader can consult the advanced reviews written by Lerda (1992), Forte (1992) and Iengo and Lechner (1992), for instance, and the literature cited therein.

9.2   Properties of rotations in two-dimensional space and fractional statistics

The rotations in two-dimensional space form an Abelian group $SO(2)$ which has only one generator. This is the essential reason for the spin to be arbitrary. As we know, spin is the quantum number labelling the irreducible representations of the (universal) covering group of the rotational group. For $SO(2)$ the (universal) covering group is $U(1)^1$, which is isomorphic to the real line, so that spin may be any real number. This is the essential difference between two-dimensional space and three- or higher-dimensional spaces. In three-dimensional space the rotation group is $SO(3)$ and its (universal) covering group is $SU(2)$ and the commutation relations of Lie algebra elements imply that the quantum number characterizing the representation is discrete, i.e. quantized. However, the two-dimensional rotation group is Abelian. Therefore there are no commutation relations and thus no restrictions on the eigenvalues of the rotation generator. Correspondingly, owing to the relation between spin and statistics, the arbitrariness of the spin leads to the idea that statistics may be arbitrary in two-dimensional space. In the following we shall give a detailed exposition of this statement.

In classical physics, statistics refers to the distribution of identical particles with respect to the physical variables (such as position and velocity) describing them. At the quantum level, the state of the physical system is described by a wave function. Consequently, the notion of statistics is usually related to the sign acquired by the wave function describing a system consisting of several

---

[1] In section 4.9 it was found that the elements of the universal covering group can be regarded as elements of the group together with an indication of along which kind of path from 1 to g the element g is reached. Two paths continuously deformable into each other are regarded as equivalent. Since $SO(2)$ is topologically a circle, the element $g = \exp(i\phi)$ can be reached by any path from $t = 0$ to $t = \phi$ which winds around the circle $n \in Z$ times. Paths with different values of n cannot be deformed to each other continuously. Obviously, the path labelled by n can be regarded as a path from $t = 0$ to $t = \phi + n2\pi$, since t and $t + n2\pi$ correspond to same point of $SO(2)$. In the case of the universal covering group of $SO(2)$, the paths $(0 \to t = \phi + n2\pi)$, $n \in Z$ correspond to different group elements, so that the universal covering group is the additive group of real numbers.

identical particles when any two particles are interchanged. This sign determines the distribution of the particles in a given quantum state. As we know, in three-dimensional space only two kinds of statistics are possible. If the wave function obtains a plus sign (i.e. does not change) when we interchange two identical particles, we say that it describes a bosonic system and the corresponding particle is called a boson. Any number of such particles can exist in a given quantum state; this is referred to as the Bose–Einstein statistics. If the wave function obtains a minus sign, we say that it describes a fermionic system and the corresponding particle is called a fermion. The particle distribution obeys the Fermi–Dirac statistics; this leads to the famous Pauli exclusion principle stating that a quantum state can contain at most one fermion with given quantum numbers. In order to make this statement more precise, we now give a more concrete definition of the statistics.

Let $\Psi(1, 2)$ denote the two-particle wave function describing two identical particles with definite quantum numbers such as energy, angular momentum, spin, etc. The arguments 1 and 2 label the two identical particles. When we move particle 2 around particle 1 by an azimuthal angle φ, the new wave function will differ by a unitary transformation from the old wave function (figure 9.1):

$$\Psi(1, 2) \longrightarrow \Psi'(1, 2) = U(\varphi)\Psi(1, 2)U^{-1}(\varphi) = e^{i\nu\varphi}\Psi(1, 2). \qquad (9.2.1)$$

Figure 9.1. In the operation of particle 2 moving around particle 1 by an angle φ, the wave function Ψ is transformed to Ψ'.

One can see that the new wave function acquires a phase factor which depends on a parameter ν. We can refer to this parameter ν as the statistical parameter. The meaning of ν can be made clear by the single-valuedness requirement of the wave function under different rotation operations. We consider two special exchanges of two particles:

(I)   moving particle 2 around particle 1 by an angle $\varphi = \pi$ and then performing a translation to reach the original spatial configuration (see figure 9.2);

(II) moving particle 2 around particle 1 by an angle $\varphi = -\pi$ and then performing the same translation to restore the initial spatial configuration (see figure 9.3).

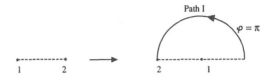

Figure 9.2. The exchange of two particles realized by moving particle 2 counterclockwise around particle 1 by an angle $\varphi = \pi$ (operation I).

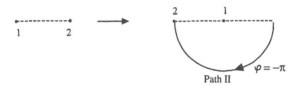

Figure 9.3. The exchange of two particles realized by moving particle 2 clockwise around particle 1 by an angle $\varphi = -\pi$ (operation II).

Owing to the translational invariance of the system, we consider only the effects induced by rotations. In the first case, the new wave function is

$$\Psi'(1, 2) = e^{i\pi\nu}\,\Psi(1, 2). \tag{9.2.2}$$

In the second case, the new wave function is

$$\Psi''(1, 2) = e^{-i\pi\nu}\,\Psi(1, 2). \tag{9.2.3}$$

This simple example clearly illustrates the difference between three- (or higher-) dimensional and two-dimensional space. In three- (or higher-) dimensional space, there is no intrinsic difference between the operations I and II, since we can always deform the operation I into operation II in a continuous way (this is due to the fact that the *fundamental group*[2] of three- or higher-dimensional Euclidian space with one point excluded is trivial). For example, one typical manipulation is to first lift the path I into the third dimension, then to fold it down onto the plane (and, if needed, to deform the path) to fall on the path II (figure 9.4). Therefore, the transformed wave functions should be identical:

$$\Psi''(1, 2) = \Psi'(1, 2) \tag{9.2.4}$$

---

[2] The fundamental group (or the first homotopy group) of a space E consists of homotopy equivalence classes of closed oriented paths starting from a given point x of E, such that two paths are regarded as homotopy equivalent if they can be continuously deformed into each other. The product $\gamma_1 \circ \gamma_2$ of two closed paths γ_1 and γ_2 corresponds to a path such that one first traverses γ_2 and then γ_1. The inverse of a path γ corresponds to the path with opposite direction of traversal. The unit path corresponds to a path continuously deformable to the point x. For the plane with a hole, the fundamental group consists of integers $n \in Z$: n tells how many times the path winds around the hole.

i.e.

$$e^{i\pi\nu} = e^{-i\pi\nu}. \tag{9.2.5}$$

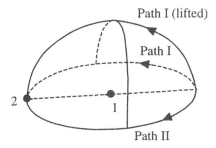

Figure 9.4. In three- or higher-dimensional spaces, the continuous deformation of path I into path II is possible.

This means that in three- or higher-dimensional spaces only $\nu = 0, 1$ (modulo 2) is possible. Therefore, the statistics cannot be arbitrary in three- or higher-dimensional spaces. This implies that under the exchange of two particles, the many-body wave function acquires either a plus sign which corresponds to $\nu = 0$ (the Bose–Einstein statistics) or a minus sign which corresponds to $\nu = 1$ (the Fermi–Dirac statistics). Other values for ν are strictly excluded.

In two-dimensional space the situation is, however, completely different. In two-dimensional space there is no way to continuously deform the path I into path II (the fundamental group of two-dimensional space is not trivial when there is a hole in it), so I and II are two topologically (and physically) distinct operations. Therefore, (9.2.4) and hence (9.2.5) should not necessarily hold any longer and the statistical parameter ν can be arbitrary.

The example above also reflects another important peculiarity of two-dimensional space: the wave function with definite initial and final configuration is not enough to describe the physical system—it is necessary to specify how the identical particles are interchanged. Geometrically, this means that the trajectories of the particles around each other should be specified. Physically, this implies that the directions of the time evolution and space reflection should be indicated (in fact, this means the breaking of parity and time reversal invariances in two-dimensional systems; this is the reason why one can choose the Chern–Simons gauge theory, which has the *same* broken symmetries, to describe the dynamics of anyons in field theory). A second consequence is that in two-dimensional space the exchange operations of particles do not form the permutation group as in three- or higher-dimensional spaces but rather a braid group. Roughly speaking, a braid group is a kind of configuration- and path-dependent permutation group in two-dimensional space. This concept is outside the scope of the present book (it is discussed, e.g., by Lerda (1992) and Forte (1992)).

9.3 Particle–flux system: example of anyon

As found in the previous section, the transformation behaviour of the wave function allows the existence of particles with arbitrary spin and fractional statistics in two-dimensional space. In this section, we shall present a physical model to see how the anyons can be realized in two-dimensional space.

We consider a physical system which consists of an electron interacting with an infinitely long magnetic solenoid (flux tube). Later we can see that this particle–flux system is a typical example for an anyon. In fact, an anyon can be thought of as a 'particle' composed of this charged particle *and* the flux tube. If we ignore the motion along the solenoid, the dynamics is restricted to a plane and should be subject to the rules of two-dimensional physics. In this way we can show how the fractional statistics appears explicitly.

To be specific, we choose the plane in which the electrons move to be the (x, y)-plane and the magnetic field B generated by an infinitely long thin solenoid passing through the origin along the z-direction, i.e.

$$B = \Phi \delta^{(2)}(r) = \Phi \delta(x)\delta(y)e_z \tag{9.3.1}$$

where e_z is the unit vector in the z-direction, $r = (x, y)$ and Φ is the magnetic flux of the solenoid:

$$\Phi = \int dS \cdot B. \tag{9.3.2}$$

Let A denote the corresponding vector potential

$$B = \nabla \times A. \tag{9.3.3}$$

Owing to the gauge transformation freedom

$$A \longrightarrow A + \nabla \Lambda \tag{9.3.4}$$

and (9.3.1) and (9.3.3), one can choose

$$A(x, y) = \frac{\Phi}{2\pi} \left(-\frac{y}{x^2 + y^2}e_x + \frac{x}{x^2 + y^2}e_y \right). \tag{9.3.5}$$

Thus, when restricted to the (x, y)-plane, this particle–flux system possesses explicit $SO(2)$ rotational symmetry, since this is essentially a two-dimensional central force problem. Now, switching on the vector potential A (or equivalently the magnetic field B) adiabatically starting at some time, say $t_0 = 0$, we choose

$$\Phi(0) = 0 \tag{9.3.6}$$

i.e. at the initial time A vanishes everywhere. Then A is slowly turned on until the time t:

$$\Phi(t) = \Phi. \tag{9.3.7}$$

According to Faraday's law, an electric field is generated accordingly:

$$\nabla \times E = -\frac{\partial B}{\partial t} = -\frac{\partial}{\partial t}(\nabla \times A). \tag{9.3.8}$$

Thus there is an electric force acting on the charged particle. As a consequence, the *kinetic* angular momentum, $J_k \equiv r \times m\dot{r}$, will grow slowly according to the equation

$$\frac{dJ_k}{dt} = r \times \frac{e}{c}E = -\frac{e}{2\pi c}\frac{d\Phi}{dt} \qquad \frac{d}{dt}\left(J_k + \frac{e\Phi}{2\pi c}\right) = 0. \tag{9.3.9}$$

Note that in this chapter for clarity we shall write the Planck constant $\hbar$ and the speed of light c explicitly. If we define the *canonical* orbital angular momentum J as

$$J = J_k + \frac{e\Phi}{2\pi c} \tag{9.3.10}$$

we see from (9.3.9) that J is conserved. Therefore, J must have the conventional spectrum, i.e. its eigenvalues are always integers in units of $\hbar$; this follows from the requirement of single-valuedness of the wave function. This is so despite the fact that the algebra of the two-dimensional rotation group is Abelian and an arbitrary constant could be added to the angular momentum operator and thus an arbitrary eigenvalue could be obtained.

Let us give a further interpretation for the difference between J and J_k. One can see that their difference is completely due to the adiabatic introduction of the magnetic flux Φ in the solenoid. As we know, with or without the presence of this flux, only the eigenstates of J can be used to describe the system since J is conserved and only J can be the usual quantum mechanical operator:

$$J \equiv J_z = -i\hbar\frac{\partial}{\partial\varphi} \tag{9.3.11}$$

where φ is the polar angle in the plane. Since the system possesses two-dimensional rotational invariance, the angle-dependent part of the wave function must be proportional to $e^{im\varphi}$ and from the requirement that the wave function is single valued, m must be integer. Therefore, the eigenvalues of J are the same as those of J_z in three-dimensional space:

$$J = \hbar m \qquad m \in Z \tag{9.3.12}$$

where Z denotes the set of integers. Of course, when $\Phi = 0$, the kinetic angular momentum is identical to the canonical angular momentum, so that it also has integer eigenvalues. However, when $\Phi \neq 0$, we see from (9.3.10) that the kinetic angular momentum operator is

$$J_k = J - \frac{e\Phi}{2\pi c} = -i\hbar\frac{\partial}{\partial\varphi} - \frac{e\Phi}{2\pi c} \tag{9.3.13}$$

so that its eigenvalues are

$$J_k = \hbar \left(m - \frac{e\Phi}{2\pi\hbar c} \right) \qquad m \in \mathbb{Z} \tag{9.3.14}$$

i.e. the eigenvalues of J_k are integers shifted by the quantity $-e(\Phi/2\pi c)$ which in general is *not* an integer.

An equivalent way of understanding this phenomenon is to assume that the vector potential takes the special form of (9.3.5) from the onset. Since the potential outside the solenoid is constant, we may set it to zero by a gauge transformation

$$A' = A - \nabla\Omega$$

$$\Omega = \frac{\varphi}{2\pi} \Phi. \tag{9.3.15}$$

Under this gauge transformation the wave function is transformed as

$$\psi' = e^{-i(e\Omega/\hbar c)} \psi. \tag{9.3.16}$$

Note that this gauge transformation is in general singular, i.e. the exponential phase factor is not single valued[3]. The single-valuedness of the wave function means that the original wave function ψ satisfies the periodic boundary condition

$$\psi(r, \varphi + 2\pi) = \psi(r, \varphi) \tag{9.3.17}$$

(here r denotes the distance from the origin in the plane), whereas the gauge-transformed wave function ψ' satisfies

$$\psi'(r, \varphi + 2\pi) = e^{-i(e\Phi/\hbar c)} \psi'(r, \varphi) \tag{9.3.18}$$

since $\psi'(r, \varphi) = e^{-ie(\varphi\Phi/2\pi\hbar c)} \psi(r, \varphi)$. Therefore, the spectrum is shifted into that of J_k according to (9.3.14).

Of course, one may think that nothing new has happened. In quantum physics it is the canonical angular momentum J that generates the rotations of the wave functions so the kinetic angular momentum J_k is not an observable (Jackiw and Redlich (1983)). It is a general fact that the kinetic angular momentum is not equal to the canonical angular momentum when external fields are present. However, as pointed out by Goldhaber and Mackenzie (1988), one can think of (9.3.14) from another viewpoint. The integer canonical angular momentum is divided into two pieces: a piece which is localized near the electron–flux system and is in general fractional, and a piece which is located at the spatial infinity and is also fractional. Furthermore, they argued that this diffused angular

---

[3] Only in the special case when the magnetic flux is quantized, i.e. $\Phi = 2\pi n\hbar c/e$ with integer n, is the gauge transformation in (9.3.15) continuous and thus not singular, and in this case both ψ' and ψ describe the same physical system.

momentum plays no role in describing the local physics phenomena and thus the piece localized on the electron system is actually identical to the kinetic angular momentum. Their argument goes as follows.

From (9.3.2), (9.3.3) and (9.3.10), we can see that

$$J = J_k + \frac{e}{c} r \times A = J_k - \frac{1}{c} \int d^3 r' r' \cdot E(r', t) B(r', t)$$

$$+ \frac{1}{c} \int d^3 r' \nabla' \cdot \left(E(r', t) r' \times A(r', t) \right). \quad (9.3.19)$$

Here the notation $T_{ij} \leftrightarrow EC$ for the tensor of type $T_{ij} = E_i C_j$ has been used with $C = r' \times A(r', t)$ and $i, j = 1, 2, 3$. $E(r, t)$ is the electric field created by the moving charge, which satisfies the Gauss law

$$\nabla \cdot E(r, t) = e \delta^{(3)}(r - r(t)) \quad (9.3.20)$$

where $r(t)$ is the particle position at the time t. From (9.3.1), we can see that the second term on the r.h.s. of (9.3.19) vanishes identically; thus, for this electron–flux system, we have

$$J = J_k + \frac{1}{c} \int d^3 r' \nabla' \cdot \left(E(r', t) r' \times A(r', t) \right) \quad (9.3.21)$$

which explicitly shows that J and J_k differ by just a surface term. We cannot, however, neglect it—its value is $e\Phi/(2\pi c)$! Recalling that the magnetic flux through the solenoid is slowly switched on from a zero initial value to the final value Φ, we can see the role played by such a surface term. At the initial time $\Phi = 0$, $J = J_k$, both of their eigenvalues are integers. When Φ is switched on slowly, J remains constant as a conserved quantity and thus its eigenvalues are integers. After the flux Φ reaches its final value Φ, a piece of J (i.e. the surface term in (9.3.21)) is radiated away by the vector potential. Therefore for the physical phenomena on a finite length scale, after a finite time, this 'dissipated' piece of the angular momentum has nothing to do with the electron–flux system and only J_k is left on it. Thus, despite the fact that the total canonical angular momentum remains integer, the angular momentum of the electron–flux system retains just a part of it. From (9.3.14), it is in general fractional. This is the reason why we can say that this electron–flux system possesses fractional spin.

Owing to the above analysis, we can regard the eigenvalue of the kinetic angular momentum as the spin of the electron–flux system (Wilczek (1982, 1990)). More precisely, it is defined as

$$s = \frac{J_k(m = 0)}{\hbar} = -\frac{e\Phi}{2\pi\hbar c}. \quad (9.3.22)$$

In general s is neither integer nor half-integer. If the magnetic flux is quantized so that the gauge transformation transforming the vector potential A to zero in the region outside the solenoid is non-singular, then s is integer.

Above we have given one example of a physical system with fractional spin. Now let us consider its statistics. One can regard it as an anyon only if it has fractional statistics. In what follows, we can see that this is indeed the case.

As stated in section 9.2, to establish the statistical properties of a quasi-particle, one must consider the wave function of a system consisting of at least two identical quasi-particles and discuss its behaviour under the exchange of two quasi-particles. Without loss of generality, we can consider the system consisting of only two such quasi-particles (figure 9.5) and denote its wave function by $\Psi(1, 2)$. Furthermore, we assume that the magnetic flux and the electron are tightly bound on each quasi-particle and charge–charge and flux–flux interactions can be ignored. Suppose that we slowly (adiabatically) move one electron–flux around the second one by a full loop, say, in our case, electron 1 around the flux 2 on a closed loop Γ (see figure 9.5). Then, owing to the Aharonov–Bohm effect (Aharonov and Bohm (1959)), the new wave function will acquire a phase factor

$$\exp\left(-i\frac{e}{2\pi\hbar c}\oint_{\Gamma} dr \cdot A\right). \tag{9.3.23}$$

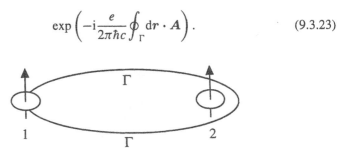

Figure 9.5. Two particle–flux systems.

With the aid of Stokes' theorem, we can write this phase factor in terms of the magnetic flux, namely

$$\exp\left(-i\frac{e}{2\pi\hbar c}\oint_{\Gamma} dr \cdot A\right) = \exp\left(-i\frac{e}{2\pi\hbar c}\int dx\,dy\,B\right)$$
$$= \exp\left(-i\frac{e\Phi}{\hbar c}\right). \tag{9.3.24}$$

Remember that in two-dimensional space, the exchange of two particles means in fact the moving of the particles around each other, in this case two electron–flux systems. When these two electrons are rotated around each other, there are actually two contributions to the phase: one due to the motion of the first electron around the second flux and one due to the motion of the second electron around the first. The total phase factor acquired by the wave function $\Psi(1, 2)$ under a full 2π rotation is then

$$\exp\left(-i\frac{2e\Phi}{\hbar c}\right). \tag{9.3.25}$$

Thus we have

$$\Psi(2, 1) = \exp\left(-i\frac{2e\Phi}{\hbar c}\right)\Psi(1, 2). \qquad (9.3.26)$$

Comparing (9.3.26) with (9.2.1), we obtain in the case of $\varphi = 2\pi$ that the statistical parameter of the electron–flux system is

$$\nu = -\frac{e\Phi}{\pi\hbar c}. \qquad (9.3.27)$$

Moreover, the spin s and the statistical parameter ν are related in the conventional way:

$$\nu = 2s. \qquad (9.3.28)$$

So the statistics is like the spin s—in general neither integer nor half-integer. Therefore, the particle–flux system is an anyon and the standard spin-statistics connection is satisfied. Of course, in some special cases a quasi-particle may behave effectively as a boson or fermion. For example, a particle–flux system composed of a bosonic particle (not electron) with a flux $\Phi = 1/(2e)$ behaves effectively as a fermion.

9.4 Possible role of anyons in physics

In the above sections, we have analysed the peculiarities of rotations in two-dimensional space and have shown the possibility for the existence of quasi-particles with fractional spin and statistics. Then we have given a non-relativistic quantum mechanical model of an anyon. To conclude this chapter, we briefly discuss the possible role of anyons in physics.

Up to now, the only known physical objects which can be described as anyons are quasi-particle and quasi-hole excitations of planar systems of electrons exhibiting the fractional quantum Hall effect (fractional QHE) (Prange and Girvin (1990)). The QHE is observed in two-dimensional systems of electrons at very low temperatures and in very strong magnetic fields orthogonal to the plane where the particles move. In the Hall conductor, the electrons are usually trapped in a thin layer at the interface between two different semiconductors or between a semiconductor and an insulator. The low temperature and the strong magnetic field freeze the motion along the direction perpendicular to the layer, so that this is a typical two-dimensional problem. As we know, the external magnetic field organizes the energy spectrum of the electrons into Landau levels (see for instance Landau and Lifschitz (1981)) and forces the particles to fill such levels from bottom to top. A quantity which plays a key role in the QHE is the filling factor, which is defined as the number of electrons divided by the number of the Landau levels available. The Hall conductance which characterizes the QHE is in fact this filling factor in units of

$e^2/(2\pi\hbar)$. When the filling factor is an integer, there is an integer number of Landau levels completely filled and the Hall conductance is quantized in integer units of $e^2/(2\pi\hbar)$. In this case the Hall effect is called the integer QHE: this phenomenon is essentially a direct manifestation of the Landau quantization for non-interacting electrons in a magnetic field and is now completely understood theoretically.

In the case of the fractional QHE, only a fraction of the Landau levels is filled and the situation becomes much more complicated since the electrons condense into a new type of collective ground state driven by the Coulomb repulsion. This is a strongly correlated two-dimensional electron system. Usually one has no way to solve this kind of system. It is, however, found in some special cases (when the inverse of the filling factor is an odd integer) that the ground state is described very exactly by the Laughlin wave function (Laughlin (1983)). This wave function describes strongly correlated properties of ordinary electrons. It is not simply the product of single electron wave functions, but a complicated superposition of such products. It is just the quasi-particle or quasi-hole excitations over this ground state that turn out to have fractional charge and fractional statistics and can thus be regarded as anyons. This can be seen from the way they are excited. The situation is similar to the particle–flux system described in section 9.3: for the ground state described by the Laughlin wave function, one introduces an infinitesimally thin flux tube near one electron, then turns on the flux adiabatically from zero to the final value $\pm\Phi$, in such a way that the system remains an (instantaneous) eigenstate of the changing Hamiltonian. Due to the Faraday law, the variation of the flux from zero to the value $\pm\Phi$ will generate a (circular) electric field around the electron. The particles will then flow inwards or outwards (depending on the sign of the flux), and a net positive or negative charge will accumulate around this electron. However, since the change of the flux by Φ can be compensated by a gauge transformation (see (9.3.15)), the final state can be considered as an excited state of the original Hamiltonian. Like the phonon in solid state physics, this kind of excitation is a quasi-particle or a quasi-hole. From the discussion in section 9.3, we know that this excited state possesses fractional statistics. Indeed, this kind of excitation can give a good explanation of the fractional filling of electrons into Landau energy levels. The exploration of this aspect is still in progress.

10

A BRIEF GLANCE AT RELATIVISTIC PROBLEMS

10.1 Introduction

So far we have dealt with non-relativistic situations. This is not sufficient, because in high-energy and elementary particle physics the particles have in general relativistic velocities. In that case one cannot restrict one's consideration to the components J_x, J_y and J_z of angular momentum, since these components form—at least—part of a four-vector, if not of a tensor. In particular the spin of a particle with relativistic velocity is not well defined: the spin component of such a particle, with respect to a given direction, is no longer simply m ($-s \leq m \leq s$); in fact it depends on the velocity of the particle. For a single particle one can, of course, always transform to its rest system, where the spin component in a given direction is well defined, but that rarely helps, because in general we have to do with more than one particle; only one of these particles can be at rest. This then leads to difficulties in applying conservation laws, for instance for J_z. An example: consider a decay $a \to b + c$; let the spins be s_a, s_b and s_c. Then assume that a was with spin in the z-direction ($m_a = s_a$) and that this particle was at rest. Now it decays and the state is characterized by a momentum $p_b = -p_c = p$ and a total angular momentum $j = s_a, m = m_a$. This m has to be shared between particles b and c, but is it correct to say $m = m_b + m_c$? What are m_b and m_c, when both particles are in relativistic motion with respect to the rest system of the decaying particle a, that is, the system where m has a clear significance? There is in this case no Lorentz system where all three operators J_z, $J_{z.b}$ and $J_{z.c}$ are well defined; at least it will be wrong simply to couple the states of particles b and c by CGCs to the state of the particle a.

There are several possible approaches to this problem and we are here going to discuss the simplest one, which is known under the name of the 'helicity formalism'. We shall give here an introduction which will help the reader to understand what is behind it and to prepare him to read without difficulty the paper by the inventors Jacob and Wick (1959).

A much more general approach would be to treat the homogeneous Lorentz group in the same way as we did the three-dimensional rotation group. This means: one considers the Lorentz group, finds its generators and selects among them a complete set of commuting observables (analogues to J^2 and J_z). Then one writes down the representation of the Lorentz group by means of the eigenstates of the complete set of commuting observables, discusses the direct product of two such representations and finds the CGC of the Lorentz group, namely the matrix elements of the transformation between coupled and uncoupled states, which at the same time transform the direct product of two representations into its reduced form (see chapter 6). In this completely covariant description the conservation laws can be properly taken into account.

The present chapter is much less explicit than the other parts of the book. It should only serve as an appetizer; the reader will find enough literature in the form of reviews and books which, after this short glance, might hopefully look less technical to him than without having read this chapter. For some references on the subject one can look, for instance, at Schweber (1961), Gelfand *et al* (1963), Naimark (1964) and Vilenkin (1968). For relativistic kinematics see, for instance, Hagedorn (1963).

Our treatment of relativistic problems will be similar to what we did in the case of rotations: we shall consider the infinitesimal Lorentz transformations and corresponding generators, find their commutation relations and select a complete set of commuting observables, one of which will be the helicity. We shall see that the helicity presents several advantages and almost eliminates the above-mentioned difficulties inherent in the relativistic theory of angular momentum.

Our notation is:

four-vectors:
$$a \equiv (a^0 \, a^1 \, a^2 \, a^3) \equiv (a^0, \mathbf{a})$$
where $\mathbf{a}$ is an ordinary three-vector.

invariant product:
$$a_\mu a^\mu = a_\mu g^{\mu\nu} a_\nu = a^{0^2} - \mathbf{a}^2; \text{ hence}$$

metric tensor:
$$g^{\mu\nu} = \begin{pmatrix} 1 & & & 0 \\ & -1 & & \\ & & -1 & \\ 0 & & & -1 \end{pmatrix} = g_{\mu\nu}$$

summation convention: Greek indices $0\ldots3$
Latin indices $1\ldots3$
sum over double indices of which one is an upper and the other a lower one:
$$a_\mu b^\mu = a_0 b^0 + a_1 b^1 + a_2 b^2 + a_3 b^3$$
$$= a_0 b_0 - a_1 b_1 - a_2 b_2 - a_3 b_3$$

raising and lowering indices:
$$a_\mu = g_{\mu\nu} a^\nu \qquad a^\mu = g^{\mu\nu} a_\nu$$
$$g^\mu_\nu = g^{\mu\rho} g_{\rho\nu} = \delta^\mu_\nu.$$

(10.1.1)

10.2 The generators of the inhomogeneous Lorentz group (Poincaré group)

In general one considers angular momentum for relativistic particles in view of applications to scattering. The experimental situation is then in almost all cases that one or more particles are to be characterized by their momentum, their total spin and their polarization. In analysing angular distributions the total angular momentum has to be considered too, but, since the linear momentum is so important, we have to discuss it as one family of operators which may be used in a complete set of commuting observables.

10.2.1 Translations; four-momentum

Physics is supposed to be invariant under translations in space and time. Let $a = (a^0, a)$ be a constant four-vector, which describes the translation $x^\mu \to x'^\mu = x^\mu + a^\mu$; then, if the whole system under consideration (if necessary, with inclusion or after removal of part of its environment) undergoes such a translation, the states of the translated system are again possible states of the untranslated one; in fact, if T_a designates an active translation then there is unitary transformation U_a (the subscript refers to 'active')

$$|\psi\rangle \to U(T_a)|\psi\rangle$$

which for infinitesimal a becomes

$$U_a(a) = 1 + i p_\mu a^\mu. \tag{10.2.1}$$

The physical significance of the generators p^μ is found by considering the Schrödinger function

$$\psi(x, t) \equiv \langle x|\psi\rangle \tag{10.2.2}$$

of the transformed state, $U_a(a)|\psi\rangle$. The physical system and consequently its Schrödinger function has been bodily transformed by the translation a from x^μ to $x^\mu + a^\mu$ and from $x^\mu - a^\mu$ to x^μ. Hence

$$\psi'(x) \equiv \langle x|U_a(a)|\psi\rangle$$

and

$$\psi'(x) = \psi(x - a) = \langle x|U_a(a)|\psi\rangle. \tag{10.2.3}$$

For infinitesimal a^μ one finds

$$\psi(x - a) = \psi(x) - a^\mu \frac{\partial \psi}{\partial x^\mu} = \langle x|1 + i p_\mu a^\mu|\psi\rangle$$

hence

$$-a^\mu \frac{\partial \psi}{\partial x^\mu} = i a^\mu \langle x|p_\mu|\psi\rangle.$$

Thus

$$p^\mu \leftrightarrow i \frac{\partial}{\partial x^\mu} \quad \text{or} \quad (p^0, p) \leftrightarrow \left(i \frac{\partial}{\partial t}, -i\nabla \right) \tag{10.2.4}$$

which are the familiar formulae for energy and three-momentum.

As all translations and therefore the p^μ commute, it is clear that we may choose all four p^μ together in our complete set of commuting observables. However, $p_\mu p^\mu = m^2$ also commutes with all p^μ and can be used. Thus another useful set will be m, p; and, since m is invariant, this is a very convenient choice, because in many cases we need not mention m explicitly.

10.2.2 The homogeneous Lorentz group; angular momentum

The space components of angular momentum were found from considering the infinitesimal rotations about x, y and z. In a four-dimensional space $(t, \boldsymbol{x})$ it is no longer sufficient to consider the rotations about the four coordinate axes, because there are more than four independent rotations. They are most conveniently characterized by the plane they map onto itself. Indeed, in a three-dimensional space there is only one plane orthogonal to a given vector (e.g. the xy-plane to the z-axis) and therefore a rotation can be characterized by an axis as well as by a plane. In four dimensions there is a whole three-dimensional space orthogonal to any given axis, thus 'I rotate the system by an angle η about the nth axis (e.g. about the time axis)' is not a complete statement because it can still be any rotation in three-space, whereas, e.g., 'I rotate the system by an angle η in the $x^0 x^1$-plane' is a complete statement: it means a pure Lorentz transformation in the x-direction, without any rotation in three-space.

Of course, we could have introduced the characterization of rotations by a plane already in three-space but we did not need to. Now, however, we must do it.

10.2.2.1 Introducing a new notation adapted to space-time

Consider first the rotations in three-space. Instead of saying that we rotate by η about the z-axis, we shall say that we rotate by η in the xy-plane etc. Hence we replace η by η_{ik} and J by J_{ik} such that

$$\eta_{12} = \eta_3 \qquad J_{12} = J_3 \text{ and cycl. perm.} \tag{10.2.5}$$

In fact this is the most adequate notation, because J is not truly a vector, but a pseudovector; namely its orbital part is $\boldsymbol{L} = \boldsymbol{r} \times \boldsymbol{p}$, i.e. $L_{12} = L_3 = x_1 p_2 - x_2 p_1$. It follows then that J_{ik} is skew symmetric: $J_{ik} = -J_{ki}$. On the other hand, an infinitesimal rotation can be written in three-space as

$$M(R) = 1 - \mathrm{i}\boldsymbol{\eta} \cdot \boldsymbol{J} \tag{10.2.6}$$

and, if P is the space inversion ('parity'), then

$$M(P) = -1 \tag{10.2.7}$$

so that it follows that rotations and the parity operation commute:

$$PRP \rightarrow M(P)M(R)M(P) = (-1)M(R)(-1) = M(R) \rightarrow R. \tag{10.2.8}$$

Therefore, $\boldsymbol{\eta} \cdot \boldsymbol{J}$ must commute with P and this implies (since $\boldsymbol{J}$, as a pseudovector, commutes with P) that also $\boldsymbol{\eta}$ does. Hence also $\boldsymbol{\eta}$ is a pseudovector and should be adequately written as a skew-symmetric tensor $\eta_{ik} = -\eta_{ki}$; indeed, giving an axis $\boldsymbol{\eta} = \eta\boldsymbol{n}$ we always had to add the prescription that a positive angle η forms a right-handed screw with the direction of n. This screwedness of n will be contained in it automatically if we define $n_3 = e_1 \times e_2$ etc, where now the (polar) vectors e_1 and e_2 denote that plane which is mapped onto itself under the rotation characterized by $e_1 \times e_2 = n_3$.

In this notation we have

$$\eta \cdot J = \sum_i \eta_i J_i = \frac{1}{2} \sum_{kl} \eta_{lk} J_{lk} = -\frac{1}{2} \sum_{lk} \eta_{lk} J_{kl}.$$

According to our summation convention (10.1.1) we omit the summation sign and write

$$\eta \cdot J = -\tfrac{1}{2} \eta_{lk} J^{kl}$$

$$U_a(\eta) = e^{\frac{i}{2} \eta_{lk} J^{kl}} \equiv e^{-i\eta \cdot J} \tag{10.2.9}$$

where $J_{kl} = J^{kl}$ has been used (this is an identity if only three-space is considered and it follows from the form of $g_{\mu\nu}$ also for space-time).

It should be clear that each J^{kl} is a Hermitian operator which may be written in a matrix representation as

$$\langle jm | J^{kl} | jm' \rangle = J^{kl}_{(j)mm'}.$$

Let us translate the commutation relation

$$[J_1, J_2] = iJ_3 \quad \text{and cycl. perm.}$$

into the new notation. It becomes

$$[J_{23}, J_{31}] = -iJ_{21}.$$

Calling the subscripts $i\,k\,l\,m$, we see that obviously the numbers 1, 2 and 3 must all appear, and one of them twice. Thus, if we exhibit the two equal subscripts by underlining them, then

$$i\underline{kl}m \rightarrow -i\delta_{kl} J_{im}$$
$$i\underline{klm} \rightarrow -i\delta_{im} J_{kl}$$
$$i\underline{kl}m \rightarrow +i\delta_{il} J_{km}$$
$$i\underline{klm} \rightarrow +i\delta_{km} J_{il}.$$

The sign is determined by the number of exchanges of two subscripts which are necessary to make the two inner ones equal; each exchange implies a factor of -1 because $J_{ik} = -J_{ki}$. Since the above four possibilities are exclusive, only one can happen and we may simply add up all the expressions on the right-hand side:

$$[J_{ik}, J_{lm}] = -i (\delta_{kl} J_{im} + \delta_{im} J_{kl} - \delta_{il} J_{km} - \delta_{km} J_{il}) \tag{10.2.10}$$

which is the desired new form of the commutation relations.

In this notation an infinitesimal rotation of a three-vector v about $\eta = (\eta_{23}, \eta_{31}, \eta_{12})$ will be written

$$M_a(\eta) = 1 - i\eta \cdot M = 1 - i\eta n \cdot M = 1 + \tfrac{1}{2}\eta n_{ik} M^{ki} \qquad (10.2.11)$$

where the components of the vector $M \equiv (M^{23}, M^{31}, M^{12})$ are 3×3 matrices, which can be read off from (3.3.13)

$$M^{23} = \begin{pmatrix} 0 & 0 & 0 \\ 0 & 0 & -i \\ 0 & i & 0 \end{pmatrix} \qquad M^{31} = \begin{pmatrix} 0 & 0 & i \\ 0 & 0 & 0 \\ -i & 0 & 0 \end{pmatrix}$$

$$M^{23} = \begin{pmatrix} 0 & -i & 0 \\ i & 0 & 0 \\ 0 & 0 & 0 \end{pmatrix}. \qquad (10.2.12)$$

They are identical to the spin-1 matrices S (4.1.8). We observe that in the ik-position of M^{ik} appears $-i$ and in the ki-position $+i$, all other matrix elements being zero.

10.2.2.2 Extension from space to space-time

We now extend the definition to the full inhomogeneous Lorentz group, by writing for infinitesimal η

$$M_a(\eta) = 1 + \tfrac{i}{2}\eta_{\mu\nu} M^{\nu\mu}. \qquad (10.2.13)$$

These $M_a(\eta)$ are now 4×4 matrices. Writing

$$(M_a(\eta))^\rho{}_\lambda = \delta^\rho{}_\lambda + \tfrac{i}{2}\eta_{\mu\nu}(M^{\nu\mu})^\rho{}_\lambda \equiv \delta^\rho{}_\lambda + \varepsilon^\rho{}_\lambda \qquad (10.2.14)$$

it follows from the invariance of $x_\mu y^\mu$ that

$$(x_\mu y^\mu)' = (M(\eta))_\mu{}^\nu (M(\eta))^\mu{}_\sigma x_\nu y^\sigma = x_\nu y^\nu$$

which implies for the matrices $M(\eta)$:

$$M_\mu{}^\nu M^\mu{}_\sigma = \delta^\nu{}_\sigma. \qquad (10.2.15)$$

Hence, with $M_\mu{}^\nu = \delta_\mu{}^\nu + \varepsilon_\mu{}^\nu$, neglecting higher orders, $(\delta_\mu{}^\nu + \varepsilon_\mu{}^\nu)(\delta^\mu{}_\sigma + \varepsilon^\mu{}_\sigma) = \delta^\nu{}_\sigma + \varepsilon^\nu{}_\sigma + \varepsilon^\nu{}_\sigma = \delta^\nu{}_\sigma$ or $\varepsilon^{\sigma\nu} = -\varepsilon^{\nu\sigma}$. That is, as long as the $(M^{\mu\nu})^{\rho\sigma}$ are still 4×4 matrices, they must be skew symmetric in ρ and σ. It has to be kept in mind, however, that the matrix which acts on the four-vector x in the sense of ordinary matrix multiplication is not $(M(\eta))^{\rho\sigma}$ but $(M(\eta))^\rho{}_\sigma$; namely

$$x'^\rho = (M(\eta))^\rho{}_\sigma x^\sigma$$

is 'ordinary matrix multiplication'—remember (10.1.1). Thus, with the notation

$$(M)^\rho{}_\sigma =$$

ρ \ σ	0	1	2	3
0				
1		$M^\rho{}_\sigma$		
2				
3				

we have to write the *generators of space rotations* (see (10.2.12))

$$(M^{23})^\rho{}_\sigma = \begin{pmatrix} 0 & 0 & 0 & 0 \\ 0 & 0 & 0 & 0 \\ 0 & 0 & 0 & -i \\ 0 & 0 & i & 0 \end{pmatrix} \qquad (M^{31})^\rho{}_\sigma = \begin{pmatrix} 0 & 0 & 0 & 0 \\ 0 & 0 & 0 & i \\ 0 & 0 & 0 & 0 \\ 0 & -i & 0 & 0 \end{pmatrix}$$

$$(M^{12})^\rho{}_\sigma = \begin{pmatrix} 0 & 0 & 0 & 0 \\ 0 & 0 & -i & 0 \\ 0 & i & 0 & 0 \\ 0 & 0 & 0 & 0 \end{pmatrix}$$

(10.2.16)

in order to retain our formulae. If we compare any two matrices $A^{\rho\sigma}$ and $A^\rho{}_\sigma = A^{\rho\mu} g_{\mu\sigma}$ then, because $g_{00} = -g_{11} = -g_{22} = -g_{33} = 1$ (all others zero), the two matrices differ in the shaded region by the sign:

$$A^{\rho\sigma} = \quad \begin{array}{c|cccc} & 0 & 1 & 2 & 3 \\ \hline 0 & & & & \\ 1 & & & + & \\ 2 & & & & \\ 3 & & & & \end{array} \quad , \qquad A^\rho{}_\sigma = \quad \begin{array}{c|cccc} & 0 & 1 & 2 & 3 \\ \hline 0 & & & & \\ 1 & & & - & \\ 2 & & & & \\ 3 & & & & \end{array}$$

(10.2.17)

Thus

$$(M^{23})^{\rho\sigma} = \begin{pmatrix} 0 & 0 & 0 & 0 \\ 0 & 0 & 0 & 0 \\ 0 & 0 & 0 & i \\ 0 & 0 & -i & 0 \end{pmatrix} \text{ etc.}$$

This can be written quite generally

$$(M^{\mu\nu})^{\rho\sigma} = i\delta^{\mu\rho}\delta^{\nu\sigma} - i\delta^{\nu\rho}\delta^{\mu\sigma}$$

and leads to

$$(M^{01})^{\rho\sigma} = \begin{pmatrix} 0 & i & 0 & 0 \\ -i & 0 & 0 & 0 \\ 0 & 0 & 0 & 0 \\ 0 & 0 & 0 & 0 \end{pmatrix} \qquad (M^{02})^{\rho\sigma} = \begin{pmatrix} 0 & 0 & i & 0 \\ 0 & 0 & 0 & 0 \\ -i & 0 & 0 & 0 \\ 0 & 0 & 0 & 0 \end{pmatrix}$$

$$(M^{03})^{\rho\sigma} = \begin{pmatrix} 0 & 0 & 0 & i \\ 0 & 0 & 0 & 0 \\ 0 & 0 & 0 & 0 \\ -i & 0 & 0 & 0 \end{pmatrix}.$$

Going back now to $(M^{0k})^\rho{}_\sigma$ implies the mentioned changes in sign (10.2.17); we thus obtain the *generators of pure Lorentz transformations*

$$(M^{01})^\rho{}_\sigma = \begin{pmatrix} 0 & -i & 0 & 0 \\ -i & 0 & 0 & 0 \\ 0 & 0 & 0 & 0 \\ 0 & 0 & 0 & 0 \end{pmatrix} \qquad (M^{02})^\rho{}_\sigma = \begin{pmatrix} 0 & 0 & -i & 0 \\ 0 & 0 & 0 & 0 \\ -i & 0 & 0 & 0 \\ 0 & 0 & 0 & 0 \end{pmatrix}$$

$$(M^{03})^\rho{}_\sigma = \begin{pmatrix} 0 & 0 & 0 & -i \\ 0 & 0 & 0 & 0 \\ 0 & 0 & 0 & 0 \\ -i & 0 & 0 & 0 \end{pmatrix}.$$

$$(10.2.18)$$

We shall call the set $(M^{\mu\nu})^\rho{}_\sigma$ (rather than $(M^{\mu\nu})^{\rho\sigma}$), i.e. (10.2.16) and (10.2.18); the generators of the homogeneous Lorentz group (space rotations and pure Lorentz transformations).

The most general homogeneous Lorentz transformation (including ordinary rotations) in the active interpretation will be written

$$M_a(\eta) = e^{\frac{1}{2}\eta_{\mu\nu}M^{\nu\mu}} = e^{-\frac{1}{2}\eta_{\mu\nu}M^{\nu\mu}}$$

$$x'^\rho = (M_a(\eta))^\rho{}_\sigma x^\sigma$$

$$(10.2.19)$$

where the correspondence to the old three-space formulae is $\eta_{12} = \eta_3$; $M_{12} = M_3$ etc. It is noteworthy that by adopting the technique of writing covariant (x_μ) and contravariant (x^μ) components and the summation convention $x_\mu x^\mu = x^0 x^0 - \boldsymbol{x} \cdot \boldsymbol{x}$, we could extend our three-space formulae to space-time by simply including a zeroth component.

10.2.2.3 Physical significance of the new generators

We work out the physical significance of the finite Lorentz transformation

$$x'^\rho = (M(\eta_{10}))^\rho{}_\sigma x^\sigma$$

$$(M(\eta_{10}))^\rho{}_\sigma = (e^{i\eta_{10}M^{01}})^\rho{}_\sigma$$

as follows: we call $\eta_{10} = \eta$, then

$$M(\eta) = e^{i\eta M^{01}} = \sum_{n=0}^{\infty} \frac{(i\eta M^{01})^n}{n!}$$

$$i\eta M^{01} = \eta \begin{pmatrix} 0 & 1 & 0 & 0 \\ 1 & 0 & 0 & 0 \\ 0 & 0 & 0 & 0 \\ 0 & 0 & 0 & 0 \end{pmatrix} \qquad (i\eta M^{01})^2 = \eta^2 \begin{pmatrix} 1 & 0 & 0 & 0 \\ 0 & 1 & 0 & 0 \\ 0 & 0 & 0 & 0 \\ 0 & 0 & 0 & 0 \end{pmatrix}.$$

Hence all odd powers go with the upper, all even (except zero) powers with the lower matrix. Thus

$$M(\eta) = 1 + \begin{pmatrix} 1 & 0 & 0 & 0 \\ 0 & 1 & 0 & 0 \\ 0 & 0 & 0 & 0 \\ 0 & 0 & 0 & 0 \end{pmatrix} \left(-1 + \sum \frac{\eta^{2n}}{2n!} \right)$$

$$+ \begin{pmatrix} 0 & 1 & 0 & 0 \\ 1 & 0 & 0 & 0 \\ 0 & 0 & 0 & 0 \\ 0 & 0 & 0 & 0 \end{pmatrix} \sum \frac{\eta^{2n+1}}{(2n+1)!}$$

or

$$M_a(\eta) = \begin{pmatrix} \cosh \eta & \sinh \eta & 0 & 0 \\ \sinh \eta & \cosh \eta & 0 & 0 \\ 0 & 0 & 1 & 0 \\ 0 & 0 & 0 & 1 \end{pmatrix} = (M(\eta))^\mu{}_\nu$$

$$x'^\mu = M^\mu{}_\nu x^\nu \quad \text{gives} \tag{10.2.20}$$

$$x'^0 = x^0 \cosh \eta + x^1 \sinh \eta$$

$$x'^1 = x^0 \sinh \eta + x^1 \cosh \eta.$$

This is in the active interpretation. It is, however, conceptually simpler to discuss Lorentz transformations in the passive sense. We only have to replace η by $-\eta$; we find

$$x'^0 = x^0 \cosh \eta - x^1 \sinh \eta$$

$$x'^1 = -x^0 \sinh \eta + x^1 \cosh \eta.$$

In this interpretation one and the same space-time point P has coordinates $x^0 = t$; $x^1 = x$; $y = z = 0$ in a frame of reference K and $x'^0 = t'$; $x'^1 = x'$; $y' = z' = 0$ in another one, K'. Let K' move with velocity β in the positive x-direction (figure 10.1).

Assume P is at rest in K'; e.g. let $x'^1 = 0$; then

$$\frac{x^1}{x^0} = \frac{x}{t} = \beta = \tanh \eta$$

is the velocity of P seen from K. This gives with

$$\cosh \eta = \frac{1}{\sqrt{1 - \tanh^2 \eta}} = \frac{1}{\sqrt{1 - \beta^2}} \equiv \gamma$$

$$\tag{10.2.21}$$

$$\sinh \eta = \frac{\tanh \eta}{\sqrt{1 - \tanh^2 \eta}} = \beta \gamma \qquad \text{(passive interpretation)}$$

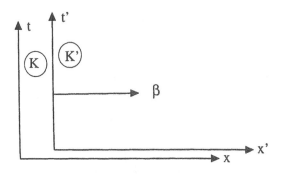

Figure 10.1. Action of Lorentz transformation.

the passive Lorentz transformation

$$P \equiv (x^0 x^1)_K \equiv (x'^0 x'^1)_{K'}$$
$$x'^0 = \gamma x^0 - \beta\gamma x^1 \tag{10.2.22}$$
$$x'^1 = -\beta\gamma x^0 + \gamma x^1.$$

10.3 The angular momentum operators

We now generalize the matrices $M_a(\eta)$ acting on four-vectors to the induced unitary transformations $U_a(\eta)$ acting on states; the correspondence is the same as for the rotation group:

$$M_a(\eta) = e^{\frac{1}{2}\eta_{\mu\nu}M^{\nu\mu}}$$
$$U_a(\eta) = e^{\frac{1}{2}\eta_{\mu\nu}J^{\nu\mu}} \qquad J^{\nu\mu} = -J^{\mu\nu}. \tag{10.3.1}$$

The $J^{\nu\mu}$ are then the generalized angular momentum operators.

In order to define states, we need a complete set of commuting observables.

As relativistic problems are most frequently encountered in high-energy scattering and elementary particle physics, the plane wave states $|p^\mu \ldots\rangle$ are particularly important. We shall therefore derive the commutation relations of the $J^{\nu\mu}$ among each other and with the p^μ. We will then be able to find how we can achieve the best compromise between non-commutativity and usefulness of the various operators.

10.3.1 Commutation relations of the $J^{\mu\nu}$ with each other

The $J^{\mu\nu}$, being the generators of the induced unitary transformation, must follow the same commutation relations as do the 4×4 matrices $M^{\mu\nu}$. We could work them out by straightforward calculation. It is easier, however, to generalize the commutation relation for the J_{ik} to those of the $J^{\mu\nu}$ with the help of (10.2.10).

We only need to remember that $g_{00} = +1$ and $g_{ik} = -\delta_{ik}$. We then obtain immediately

$$[J^{\mu\nu}, J^{\rho\sigma}] = i(g^{\nu\rho}J^{\mu\sigma} + g^{\mu\sigma}J^{\nu\rho} - g^{\mu\rho}J^{\nu\sigma} - g^{\nu\sigma}J^{\mu\rho}). \qquad (10.3.2)$$

It is left to the reader to check that these are indeed the commutation relations for $M^{\mu\nu}$ too.

10.3.2 Commutation relations of the $J^{\mu\nu}$ with p^{ρ}

In a similar way we obtain the commutator between $J^{\mu\nu}$ and p^{ρ}: we first write it down for $L^{jk} = L_{jk}$ and p^l. Namely, with $L^{jk} = x^j p^k - x^k p^j$ we have

$$[L^{jk}, p^l] = [x^j p^k, p^l] - [x^k p^j, p^l]$$

$$= i\delta^{jl}p^k - i\delta^{kl}p^j.$$

This gives, when generalized to $jkl \to \mu\nu\rho$ and $L \to J$ (remember $\delta^{ik} = -g^{ik}$)

$$[J^{\mu\nu}, p^{\rho}] = i(g^{\nu\rho}p^{\mu} - g^{\mu\rho}p^{\nu}). \qquad (10.3.3)$$

10.4 A complete set of commuting observables

For the description of a free particle (one-particle state) we wish to have its momentum p and its spin component in a given direction, altogether four quantum numbers. We have not much doubt that the corresponding operators commute. The set is, however, not complete, because in relativistic quantum mechanics there are four components p^{μ} with $p_{\mu}p^{\mu} = m^2$. Furthermore we expect the total spin s also to be a quantum number available together with the p^{μ}.

The programme is therefore that we wish to retain the p^{μ} for characterizing the one-particle states and we hope to add to them total spin and the spin component in a given direction (instead of p^{μ} we can take m, p). There will be, as usual, other quantum numbers labelling our one-particle states, besides those just mentioned; for instance, isospin, nucleon number, strangeness etc, which will be written in one symbol, γ, as before.

The problem is now to find some operators for the spin. In three-space, i.e. non-relativistic quantum mechanics, this was easy: we only had to write

$$J = L + S.$$

We could try here $J^{\mu\nu} = L^{\mu\nu} + S^{\mu\nu}$ where $L^{\mu\nu} = x^{\mu}p^{\nu} - x^{\nu}p^{\mu}$. But we are faced with the difficulty of saying what the operator x^{μ} means—and that becomes quite troublesome. We cannot go into these questions here; suffice it to say that relativistic quantum theory is essentially field theory and that there the x^μ obtain the new interpretation of being labels on field operators and no longer operators themselves. However, even if we do not take the step from the one-particle theory to field theory and rather try to remain in 'ordinary'—but relativistic—quantum mechanics (e.g. Dirac's theory), we have the difficulty that $L^{\mu\nu} = x^\mu p^\nu - x^\nu p^\mu$ and $S^{\mu\nu}$ are not separately conserved. These difficulties and the construction of a conserved spin for a Dirac particle are discussed in the article by Hilgerved and Wouthuysen (1963).

What is spin? Spin is that part of the angular momentum which survives the transformation to the rest system, just because there no orbital part is left (let us assume $m \neq 0$; the case of massless particles will be taken up later). Whereas in general $J^{\mu\nu}$ and p^ρ do not commute, the situation is different in the subspace of particles at rest ($p^k = 0$; $p^0 = m$). Call these states $|m, \mathbf{0}, \ldots\rangle$. Then from the commutation relation (10.3.3)

$$[J^{\mu\nu}, p^\rho]|m, \mathbf{0}, \ldots\rangle = \mathrm{i}(g^{\nu\rho}p^\mu - g^{\mu\rho}p^\nu)|m, \mathbf{0}, \ldots\rangle.$$

The commutator vanishes (on this set of states) unless $\mu = 0$ or $\nu = 0$ (only one is possible since $J^{\nu\nu} \equiv 0$). Let $\mu = 0$. Then

$$[J^{0\nu}, p^\rho]|m, \mathbf{0}, \ldots\rangle = (\mathrm{i}g^{\nu\rho}p^0 - \mathrm{i}g^{0\rho}p^\nu)|m, \mathbf{0}, \ldots\rangle$$

where the second term vanishes because $\nu \neq 0$. Again the commutator vanishes (on this set) unless $\nu = \rho$ because of $g^{\nu\rho}$, and, since $\nu \neq 0$, we find with $p^0|m, \mathbf{0}, \ldots\rangle = m|m, \mathbf{0}, \ldots\rangle$

$$[J^{0k}, p^k]|m, \mathbf{0}, \ldots\rangle = -\mathrm{i}m|m, \mathbf{0}, \ldots\rangle.$$

That is, of the whole family of commutors $[J^{\mu\nu}, p^\rho]$ only $[J^{0k}, p^k] = -\mathrm{i}m$ is different from zero for particles at rest.   (10.4.1)

This means that the operators J^{ik} commute with the p^μ for particles at rest; they can be measured without disturbing the particle. Of course, they do not commute among each other. Loosely speaking, we can thus say that the J^{ik} go over to the spin operators S^{ik} and commute with the momenta if we apply them to particles at rest, whereas the J^{0k} do not commute with the p^k and therefore cannot be measured without giving a momentum to the particle.

In what follows we shall formalize these arguments and construct operators representing spin, consider their commutation relations among each other and with p and finally try to find a suitable set of commuting observables to label our states.

10.4.1 The spin four-vector w^μ and the spin tensor $S^{\mu\nu}$

A sensible question is then, 'what is the total spin and its z-component for this particle at rest?'

And now, since this question must have an answer which is independent of the Lorentz frame in which it is asked, there must be covariant formulation available; and this covariant formulation must be expressible by the operators p^ρ and $J^{\mu\nu}$ (see the statement (8.4.2)).

We consider what p^ρ and $J^{\mu\nu}$ look like for a particle at rest:

$$p^\rho \rightarrow \{p^0 = m, 0, 0, 0\}$$

$$J^{\mu\nu} \rightarrow \{J^{0k}, S^{23}, S^{31}, S^{12}\}.$$

Can we construct a quantity from p^ρ and $J^{\mu\nu}$ where the inconvenient J^{0k} drop out, whereas the $(S^{23}, S^{31}, S^{12}) \equiv S$ remain? As this quantity has to reduce to the three-vector S in the rest frame, we are led to seek a four-vector operator. The four-vector $J^{\mu\nu} p_\nu = Z^\mu$ does not work: it is $Z^\mu = (0, J^{k0})$ in the rest system—just the operator we do not want. We must construct it the other way round: the J^{ik} must be multiplied by p^0 and the J^{0k} by $p^i = 0$. Combinations of this kind, $J^{ik} p^0$ and $J^{0k} p^i$, can be achieved by means of the completely antisymmetric tensor of rank 4:

$$\varepsilon_{\mu\nu\rho\sigma} = \begin{cases} +1 & \text{for } \mu\,\nu\,\rho\,\sigma\text{=even permutation of 0123} \\ -1 & \text{for } \mu\,\nu\,\rho\,\sigma\text{=odd permutation of 0123} \\ 0 & \text{otherwise (namely if any indices are equal).} \end{cases} \qquad (10.4.2)$$

It follows that under cyclic permutation (being odd) $\varepsilon_{\mu\nu\rho\sigma} = -\varepsilon_{\nu\rho\sigma\mu}$ and that raising or lowering one index implies a change of sign if it is 1, 2 or 3 and no change if it is 0. We put

$$w^\mu = \frac{1}{2}\varepsilon^\mu_{\ \nu\rho\sigma} J^{\nu\rho} p^\sigma = (\boldsymbol{J} \cdot \boldsymbol{p}, J p^0 + \boldsymbol{J}^0 \times \boldsymbol{p}) \equiv (w^0, \boldsymbol{w}) \qquad (10.4.3)$$

where $\boldsymbol{J} \equiv (J^{23}, J^{31}, J^{12})$; $\boldsymbol{J}^0 \equiv (J^{01}, J^{02}, J^{03})$. The explicit form follows by working out the components of w^μ. Indeed, applying this operator w^μ to a particle at rest gives, as is seen immediately from the explicit form

$$w^\mu \rightarrow m(0, J^{23}\, J^{31}\, J^{12}) \rightarrow m(0, S^1, S^2, S^3) \qquad (10.4.4)$$

since then $p^0 = m$ and $\boldsymbol{p} = 0$.

Applying a Lorentz transformation in direction β to $(w^\mu)_{\text{Rest}}$ (see (10.4.22)) such that the particle is at rest in a frame K' and moves with β in K, yields with $\gamma = (1 - \beta^2)^{-\frac{1}{2}}$

$$w^\mu = \left(m\gamma\beta \cdot S_R, mS_R + \beta\frac{m\gamma^2}{\gamma+1}\beta \cdot S_R \right)$$
$$= \left(\boldsymbol{p} \cdot S_R, mS_R + \frac{\boldsymbol{p}}{E+m}\boldsymbol{p} \cdot S_R \right) \qquad (10.4.5)$$

where S_R is the spin at rest and $p = (E, \boldsymbol{p}) = (m\gamma, m\beta\gamma)$ is understood as an ordinary (not operator) four-vector. Assuming for a moment that we remain in the frame of relativistic one-particle quantum mechanics; then $L^{\mu\nu} = x^\mu p^\nu - x^\nu p^\mu$ has a meaning and we can write $J^{\mu\nu} = L^{\mu\nu} + S^{\mu\nu}$. Then

$$w^\mu = \tfrac{1}{2}\varepsilon^\mu_{\ \nu\rho\sigma} J^{\nu\rho} p^\sigma = \tfrac{1}{2}\varepsilon^\mu_{\ \nu\rho\sigma} S^{\nu\rho} p^\sigma \qquad (10.4.6)$$

since $\varepsilon^\mu_{\ \nu\rho\sigma} L^{\nu\rho} p^\sigma = \varepsilon^\mu_{\ \nu\rho\sigma}(x^\nu p^\rho - x^\rho p^\nu)p^\sigma \equiv 0$ on account of the antisymmetry of ε. The inverse of (10.4.6) exists; it is

$$S^{\mu\nu} = -\frac{1}{m^2}\varepsilon^{\mu\nu}_{\ \ \rho\sigma} w^\rho p^\sigma \quad \text{in components}$$

$$
\begin{aligned}
S^0 \equiv (S^{01}, S^{02}, S^{03}) &= \frac{1}{m^2}[(\boldsymbol{J}^0 \times \boldsymbol{p}) \times \boldsymbol{p} + (\boldsymbol{J} \times \boldsymbol{p})p^0] \\
&= \frac{1}{m^2}[(\boldsymbol{J}^0 \cdot \boldsymbol{p})\boldsymbol{p} - \boldsymbol{J}^0(\boldsymbol{p}^2) + (\boldsymbol{J} \times \boldsymbol{p})p^0]
\end{aligned}
\qquad (10.4.7)
$$

$$S = (S^{23}, S^{31}, S^{12}) = \frac{1}{m^2}[-(\boldsymbol{J} \cdot \boldsymbol{p})\boldsymbol{p} + \boldsymbol{J}(p^0)^2 + (\boldsymbol{J}^0 \times \boldsymbol{p})p^0].$$

That this is the inverse of (10.4.6) is immediately seen in the rest system $S^0 \Rightarrow 0$; $S \Rightarrow J$; and from the covariance of the formula it follows that it is true in any Lorentz frame. That $S^0 \Rightarrow 0$ in the rest system is a typical property of spin.

Inserting w^α from (10.4.6) into (10.4.7) we find

$$S^{\mu\nu} = -\frac{1}{m^2}\varepsilon^{\mu\nu}_{\ \ \alpha\beta}\varepsilon^\alpha_{\ \lambda\rho\sigma} J^{\lambda\rho} p^\sigma p^\beta \qquad (10.4.8)$$

which no longer refers to 'orbital angular momentum' and can be considered as a covariant definition of spin: it 'projects out' the spin $S^{\mu\nu}$ of $J^{\mu\nu}$. It has this property, however, only for one-particle states.

Thus $S^{\mu\nu}$ and w^μ are two operators which can be used to describe spin (for one-particle states only; otherwise 'spin' has no meaning). The antisymmetry of ε ensures another typical property of spin:

$$w_\mu p^\mu = 0 \qquad S^{\mu\nu} p_\nu = 0. \qquad (10.4.9)$$

10.4.2 Commutation relations for w^μ and $S^{\mu\nu}$

We shall prove all commutation relations first for states of particles at rest and then generalize the result covariantly. This procedure is unique (see statement (8.4.2)).

We shall say loosely 'in the rest system the operators become...' when we really mean 'in the subspace $\mathcal{H}_{m,0}$ spanned by states of particles at rest, the operators behave like...'.

(i) The tensor $T^{\mu\lambda} = [w^{\mu}, p^{\lambda}]$ vanishes in the rest system, where $w^{\mu} \to m(0, J^{23}, J^{31}, J^{12})$ and $p^{\lambda} \to (m, 0, 0, 0)$ commute (see (10.4.1)). A tensor $T^{\mu\lambda}$ which vanishes with all components in one Lorentz system vanishes everywhere. Hence

$$[w^{\mu}, p^{\lambda}] \equiv 0. \tag{10.4.10}$$

Consequently any component of w^{μ} or any invariant linear combination $a_{\mu}w^{\mu}$ can be measured together with p^{μ}. In particular, since p^0 is the generator of the time translation, $[w^{\mu}, p^0] = 0$ implies that w^{μ} is conserved. The commutativity of w^{μ} with p^{ρ} is a further indication that w^{μ} describes the spin part of the total angular momentum. Orbital angular momentum is not invariant under translations (and this is expressed by (10.3.3)) but spin is. Indeed, also $S^{\mu\nu}$ commutes with p^{ρ} since this is obviously so in the rest frame:

$$[S^{\mu\nu}, p^{\rho}] = 0. \tag{10.4.11}$$

This observation offers a different derivation of the operator w^{μ}. We could have defined the spin as that part of the total angular momentum which is invariant under translations. This part of the angular momentum must then commute with p^{μ}. If then this part is used to write down the unitary transformation for a Lorentz transformation, this transformation must leave p^{μ} invariant. Let us do this for an infinitesimal $\eta_{\mu\nu}$: $U = 1 + (i/2)\eta_{\mu\nu}J^{\nu\mu}$; $Up^{\rho}U^{\dagger} = p^{\rho}$ gives on the subspace of states $|p^{\mu}, \ldots\rangle$ (see (10.3.3))

$$\eta_{\mu\nu}[J^{\nu\mu}, p^{\rho}] = i\eta_{\mu\nu}(g^{\mu\rho}p^{\nu} - g^{\nu\rho}p^{\mu}) = 2i\eta^{\rho\mu}p_{\mu} = 0.$$

Thus only such $\eta^{\rho\mu}$ are admitted which give zero if contracted with p_{μ}. Such a condition can be fulfilled by means of the ε-tensor, one arbitrary constant four-vector n and the four-vector p itself (here the eigenvalues are meant)

$$\eta_{\mu\nu} = \varepsilon_{\mu\nu\rho\sigma}p^{\rho}n^{\sigma}$$

where n may be taken orthogonal to p, since any parallel component will not contribute. With such an η our unitary transformation becomes, with (10.4.6)

$$U = 1 + \frac{i}{2}\eta_{\mu\nu}J^{\nu\mu} = 1 + \frac{i}{2}\varepsilon_{\mu\nu\rho\sigma}J^{\nu\mu}p^{\rho}n^{\sigma} = 1 + iw_{\sigma}n^{\sigma}. \tag{10.4.12}$$

These transformations $U = \exp(iw_{\sigma}n^{\sigma})$ with arbitrary four-vector n are a subgroup of the Lorentz group. They leave the eigenstates $|p^{\mu}, \ldots\rangle$ invariant if p^{μ} is the one contained in w^{σ}. This subgroup is called 'Wigner's little group of p^{μ}'. It is the analogue of the rotations which leave a given axis in three-space invariant.

(ii) Next consider $[w^\mu, J^{\tau\lambda}]$. In the rest system

$$w^\mu \to (0, mS^1, mS^2, mS^3)$$

$$J^{\tau\lambda} \to J^{0k}, S^1, S^2, S^3$$

$$[w^j, J^{kl}] = i(g^{jk}w^l - g^{jl}w^k).$$

The covariant generalization is

$$[w^\mu, J^{\tau\lambda}] = i(g^{\mu\tau}w^\lambda - g^{\mu\lambda}w^\tau). \tag{10.4.13}$$

(iii) Furthermore, consider $[w^\mu, w^\rho]$. In the rest system we obtain with $w^i \to mS^i$

$$[w^j, w^k] = i\varepsilon^{jkl0}w^l p^0$$

which is obviously true for $j = 1, k = 2$ ($\varepsilon^{1210} = \delta_{l3}$). Lowering the indices l and 0 and generalizing covariantly give (see (10.4.7))

$$[w^\mu, w^\rho] = -i\varepsilon^{\mu\rho}{}_{\lambda\sigma}w^\lambda p^\sigma = -im^2 S^{\mu\rho}. \tag{10.4.14}$$

(iv) The most important commutation relation will be that of the operator $W = w_\mu w^\mu$ with p^ρ and $J^{\rho\sigma}$. The commutator of any operator A with p^ρ and $J^{\rho\sigma}$ is just constructed such that A is correctly Lorentz transformed by means of $UAU^\dagger$. Indeed, we could have found the commutation relations of w with J and J with J by requiring the correct Lorentz transformation, e.g.

$$U(\eta)p^\rho U^\dagger(\eta) = (M(\eta))^\rho{}_\sigma p^\sigma$$

with $U(\eta)$ and $M(\eta)$ given by (10.3.1), and (10.2.16) and (10.2.18).

It is no miracle, therefore, that the commutation relations of w^μ with $J^{\rho\sigma}$ and p^μ with $J^{\rho\sigma}$ are the same: both have to transform like four-vectors.

It is clear then, that invariant combinations such as e.g. $w_\mu w^\mu$ or $w_\mu p^\mu$ or $S_{\mu\nu}S^{\mu\nu}$ will commute with $J^{\rho\sigma}$. As neither w_μ nor $S_{\mu\nu}$ contain 'orbital parts', it follows that these operators and, of course, the invariants made of them commute with p^ρ, whereas the invariant $J_{\mu\nu}J^{\nu\mu}$ will not commute with p^ρ (as seen in the rest system, this is the total angular momentum squared, and that depends on the location of the origin). It is therefore not useful for us.

Since $S^{\mu\nu}$ and w^ρ become the same in the rest system, $(w)_{\mathrm{Rest}} = m(0, S^{23}, S^{31}, S^{12})_{\mathrm{Rest}}$, the invariants $w_\mu w^\mu$ and $S_{\mu\nu}S^{\mu\nu}$ are essentially the same. The invariant $w_\mu p^\mu$ is zero. We thus may take $w_\mu w^\mu$, which we know then commutes with $J^{\rho\sigma}$ and p^λ. Its physical significance follows from

$$-\frac{1}{m^2}w_\mu w^\mu = -\frac{1}{m^2}(w_\mu w^\mu)_{\mathrm{Rest}} = (S^2)_{\mathrm{Rest}} = \text{total spin squared.} \tag{10.4.15}$$

Since

$$[w_\mu w^\mu, p^\rho] = [w_\mu w^\mu, J^{\rho\sigma}] = [w_\mu w^\mu, w^\rho] = 0 \tag{10.4.16}$$

(the last equation holds because w^ρ is constructed from J and p), it follows that the total spin squared is Lorentz invariant, conserved (for one-particle states; only for such states is spin defined!) and can be measured simultaneously with momentum or angular momentum. We shall therefore use it to label our states; since the space components $S_x = S^{23}$; $S_y = S^{31}$; $S_z = S^{12}$ obey the usual angular momentum commutation relations, it follows that if we denote the states by $|m, p, s, \ldots\rangle$

$$w_\mu w^\mu |m, p, s, \ldots\rangle = -m^2 s(s+1)|m, p, s, \ldots\rangle. \qquad (10.4.17)$$

10.4.3 Construction of a complete set of commuting observables; helicity

Our set of commuting operators contains so far p^μ and $w_\mu w^\mu$: momentum and magnitude of spin. In three-space we did not have the momentum, but the magnitude (j) of the total angular momentum. To this one could add the component of the angular momentum in an arbitrary direction. Can we do something similar also here? That a 'component $\eta_{\mu\nu} J^{\nu\mu}$ of the total angular momentum in a given direction' will commute with $w_\mu w^\mu$ is obvious since each $J^{\rho\sigma}$ commutes with it. But will such a component commute with p? In three-space the component $n \cdot J$ commutes with p if n is parallel to p: the angular momentum in the direction of motion can be measured simultaneously with the momentum. (Strictly speaking, the commutator $[p_i, n \cdot J]$ vanishes on the subspace of states $|p, \ldots\rangle$ for n parallel to the (eigenvalue) vector p.) Geometrically, a rotation commutes only with translations along the rotation axis n.

Let us consider the subspace of states $|p^\mu, s, \ldots\rangle$ with p^μ fixed. Which component $\eta_{\mu\nu} J^{\nu\mu}$ of the total angular momentum does commute with the operators p^μ on this subspace and can thus be measured simultaneously with the four-momentum? This component will leave p^μ invariant. The condition is $\eta_{\mu\nu}[J^{\nu\mu}, p^\rho] = 0$ and we have already seen that it leads to $\eta_{\mu\nu} J^{\nu\mu} = 2w_\sigma n^\sigma$, the 'little group of p^μ'. Apart from a factor depending on the normalization of the four-vector n, we can then say that the operator

$$\Lambda(n, p) \equiv -\frac{1}{m} w_\sigma n^\sigma \qquad (10.4.18)$$

with arbitrary n is 'the component of the total angular momentum in direction p and n'. That there are two four-vectors, p and n, to define a component of the angular momentum, is a consequence of the four dimensions: a four-dimensional rotation is defined if a plane is given which is left invariant. The unitary transformation

$$U(n, p) = e^{iw_\mu n^\mu}$$

corresponds to a Lorentz transformation which leaves p and n and therefore the plane spanned by p and n invariant.

This is very nice, but still unsatisfactory from the physicist's point of view. Namely the operator $w_\mu n^\mu$, whose eigenvalues we can measure simultaneously with p, is still too arbitrary because of the four-vector n. We therefore should restrict the arbitrariness of n. Three possibilities offer themselves.

(i)   We define n with respect to a particular Lorentz frame, e.g., by saying that in the rest system we put $n = (0, n)$ where n is a unit vector; in particular we might choose n in the z-direction and have thus related $w_\mu n^\mu$ to the 'spin component in the z-direction'. Such a choice is not very lucky, because this operator is then not invariant (in spite of its form) since one vector (n) is attached to the coordinate frame and the other one (w) to the physical system. From this point of view already in non-relativistic quantum mechanics the choice of one component J_z rather than of an invariant combination can be considered unlucky.

(ii)   We define n as before, i.e. by $n = (0, n)$ in one particular Lorentz system of our particle. We might give n any direction of e.g. perpendicular or parallel to the direction of p (before the transformation to rest) and then require that n be transformed as a four-vector. Then the transformation of the operator $\Lambda(n, p)$ is somewhat unusual:

$$\Lambda'(n, p) = -\frac{1}{m} \left(M(\eta)^\sigma{}_\lambda n^\lambda \right) U(\eta) w_\lambda U^\dagger(\eta) = -\frac{1}{m} n'^\lambda w'_\lambda \tag{10.4.19}$$

$$= \Lambda(n, p) = \text{invariant.}$$

That is, not only is w_λ transformed by the unitary transformation U, but n is transformed by M and that means that Λ is *not* transformed like ordinary operators:

$$\Lambda'(n, p) \neq U(\eta)\Lambda(n, p)U^\dagger(\eta).$$

The invariance of Λ thus achieved guarantees that in whichever Lorentz system $\Lambda(n, p)$ is calculated, it always has the same meaning

$$\Lambda(n, p) = \Lambda(n, p)_{K_0} = (S \cdot n)_{K_0}. \tag{10.4.20}$$

(iii) We define n by attaching it bodily to the physical system. Then $w_\mu n^\mu$ is invariant, since n and w have to be transformed simultaneously. This 'attaching n to the system' has its difficulties, however, because if we attach n to the system, we must do it in a unique way. That is, the system must have some marks on it which can serve to fix the four-vector n; in other words, there must already be a four-vector exhibited by the system in question, or at least by its state. Two such vectors might do: w and p. Now w drops out for two reasons. Firstly, its components do not commute and thus no eigenvalue four-vector w^μ with all components given exists. If nevertheless we try the operator four-vector w^μ and just put $n^\mu \equiv w^\mu$, then $w_\mu w^\mu$ results and this operator is already contained in our set. Secondly, we may try p^μ, of which all components can be measured simultaneously and which thus can be considered as an eigenvalue four-vector. Unfortunately this does not work either, because $w_\mu p^\mu = 0$ in all Lorentz frames (10.4.9).

Of these three possibilities the last one would have been the most attractive, but it does not work. Jacob and Wick (1959), however, forced it to work; they had to pay for it, of course: the price was a certain loss of invariance; a loss, however, which is just acceptable. The point is this: if n cannot be parallel to p—since then it does not contribute to $w_\mu n^\mu$—why cannot it be orthogonal to p and still have a space part n which is parallel to the space part p? Let us then take a 'four-vector' n, which we call l from now on (longitudinal) and require that

$$l_\mu p^\mu = 0 \quad l \text{ parallel to } p.$$

We have then

$$l_\mu p^\mu = l^0 p^0 - l \cdot p = 0 \quad \text{hence} \quad l^0 = \frac{l \cdot p}{p^0}$$

$$l_\mu l^\mu = \left(\frac{l \cdot p}{p_0}\right)^2 - l^2 = |l|^2 \left(\left(\frac{p}{p_0}\right)^2 - 1\right) = |l|^2 (\beta^2 - 1) < 0$$

if β denotes the three-velocity of our particle. In that case

$$p = (m\gamma, m\beta\gamma) \quad \gamma^2 = (1 - \beta^2)^{-1}.$$

Putting $|l|^2 = \gamma^2$, we find

$$l = \left(\beta\gamma, \frac{\beta}{\beta}\gamma\right) = \left(\frac{|p|}{m}, \frac{p}{|p|} \frac{p_0}{m}\right) \quad l_\mu l^\mu = -1. \tag{10.4.21}$$

It should be stressed that this l is not a genuine four-vector, because, if it is given in one particular Lorentz system by the definition (10.4.21), then, if we apply the same Lorentz transformation to p and to l, the Lorentz-transformed l is in general not the same as that l which is constructed, according to (10.4.21), from the Lorentz-transformed p. That is, in general $l(p') \neq l'(p)$.

Thus

- *either* l transforms as a genuine four-vector, then the definition (10.4.21) does not hold in all Lorentz frames (it is indeed not covariant!)
- *or* the definition (10.4.21) is required in all Lorentz frames, then l is not a four-vector, but a quantity which in each new Lorentz frame has to be calculated anew from p, using the definition.

There is a class of Lorentz transformations for which this difficulty does not arise: the spatial rotations, because they rotate p and l in the same way, and the pure Lorentz transformations in the direction of p (or l), because they leave these directions invariant. In particular, we can transform to rest. All these statements may easily be checked by an explicit application of the formula for the general Lorentz transformation: if a is any four-vector and if the frame K' moves with β against the frame K, then

$$a = a' + \beta\gamma\left(\frac{\gamma}{\gamma+1}\beta a' + a'^0\right)$$

$$a^0 = \gamma(a'^0 + \beta \cdot a').$$

(10.4.22)

That by a general Lorentz transformation on p *and* l the relation (10.4.21) will be destroyed is seen by carrying it out in two steps: first to the rest system and then to the required one. In the second step the parallelism of l and p will be lost unless this second Lorentz transformation is in the direction of l.

In the rest system we obtain for the operator $\Lambda(lp)$

$$\Lambda(l, p) \equiv -\frac{1}{m}w_\mu l^\mu$$

$$(\Lambda)_{\text{Rest}} = \left(\frac{1}{m}w \cdot l\right)_{\text{Rest}} = \left(\frac{S \cdot p}{|p|}\right)_{\text{Rest}} = \begin{array}{l}\text{spin component in} \\ \text{the direction of} \\ \text{motion} = \text{helicity}\end{array}$$

(10.4.23)

where $(p/|p|)_{\text{Rest}}$ is understood in the sense of

$$\lim_{|p|\to 0}\frac{p}{|p|} = \left(\frac{p}{|p|}\right)_{\text{Rest}}.$$

According to the above-mentioned peculiarities of l we can then

- *either* define l by (10.4.21) in *one* particular Lorentz frame K_0 and require it to transform like a genuine four-vector (then the operator $\Lambda(l, p) = -(1/m)w_\mu l^\mu$ is invariant and, calculated in whichever Lorentz frame, answers the question, 'what is the component of the spin in direction p in the frame K_0?', but *not* the question, 'what is the component of the spin in the direction of the *actual* momentum p?'; with this definition of l we would be back to the second possibility of choosing n—see (ii) above);

- *or* we require that l should not be Lorentz transformed, but in any frame of reference be constructed from the four-vector p by means of (10.4.21). In that case $\Lambda(l, p)$ is *not* invariant under all Lorentz transformations, but it *always* answers the question, 'what is the component of the spin in the direction of the *actual* momentum p?'. $\Lambda(l, p)$ is, however, invariant under pure Lorentz transformation in the direction of p (unless such a Lorentz transformation 'transforms beyond the rest system' and thereby changes the sign of p; then also $\Lambda(l, p)$ will change sign, as is seen immediately in (10.4.23)) and under space rotations. This class of transformations is sufficient to transform any four-momentum p^μ into any other one belonging to the same mass: let p'^μ be the new four-momentum; first we transform from p^μ to rest, then rotate such that l^μ points in the direction of p'

and finally make a pure Lorentz transformation in direction p' until the prescribed magnitude of p' is attained. Under this whole procedure $\Lambda(l, p)$ is invariant.

With Jacob and Wick (1959) we shall choose the second alternative, where $\Lambda(l, p)$ is not generally Lorentz invariant, but invariant enough to serve our purposes. In that case l is a unique function of p and can be omitted. The eigenvalues of $\Lambda(p)$ are $\lambda = s, s - 1, \ldots, -s$, as follows from its significance in the rest system and its invariance under the mentioned class of Lorentz transformations. This class is not a subgroup of the Lorentz group, as one sees easily: let $L(p)$ be a pure Lorentz transformation in direction p and R any space rotation; then $RL(p)$ belongs to the class, but not $L(p)R$. However $L(Rp)R$ belongs again to it.

To sum up:

we define the helicity operator (not by the misleading 'invariant' $w_\mu l^\mu$, but explicitly) by

$$\Lambda(p) \equiv -\frac{1}{m^2}\left(w^0|p| - w \cdot p\frac{p^0}{|p|}\right) = \frac{S \cdot p}{|p|} = \frac{J \cdot p}{|p|}$$

with eigenvalues $\lambda = -s, -s + 1, \ldots, +s$ on states $\qquad$ (10.4.24)

$|p, s, \lambda, \ldots\rangle \qquad \lambda = $ spin component in the direction of p
$\qquad\qquad\qquad = $ total angular momentum component
$\qquad\qquad\qquad\quad$ in the direction of p.

That $\Lambda(p) = (J \cdot p)/|p|$ follows immediately if one inserts into the first part of (10.4.24) the explicit form of w^μ as it appears in (10.4.3): $w^\mu = (J \cdot p, J p^0 + J^0 \times p)$. $\Lambda(p)$ is invariant under $L(p)$ (pure Lorentz transformations in direction p) (unless the sign of p changes, when $\lambda \to -\lambda$) and under space rotations R, as well as under all Lorentz transformations which can be written as a product of the type

$$L(R_n \cdots R_1 p)R_n \cdots L(R_2 R_1 p)R_2 L(R_1 p)R_1$$

i.e. by Lorentz transformations generated by continued application of space rotations and pure Lorentz transformations in the direction of the last momentum. These Lorentz transformations suffice to transform a given p into any p' with the same mass, but they do not constitute a subgroup of the Lorentz group.

If $\Lambda(p)$ is measured with eigenvalue λ, then, in any Lorentz frame which cannot be reached by a transformation leaving Λ invariant, any one eigenvalue $\lambda' = -s, -s + 1, \ldots, +s$ may be found with the amplitude $\langle p', s, \lambda', \ldots | U(\eta) | p, s, \lambda, \ldots\rangle$ where $U(\eta)$ connects these two Lorentz frames, e.g. if $U(\eta) = \exp(iw_\mu n^\nu)$ (i.e. $U(\eta)$ belongs to the little group of p), p^μ is left invariant but λ not.

Our one-particle states $|m, p, s, \lambda, \gamma\rangle$ are then characterized by the complete set of commuting observables (for $m \neq 0$)

four-momentum p (or m, p)

total spin squared $-\dfrac{1}{m^2} w^\mu w_\mu = s(s+1) = (S^2)_{\text{Rest}}$ (10.4.25)

helicity $\Lambda(p) = \dfrac{S \cdot p}{|p|} = \lambda(= -s, -s+1, \ldots, +s)$

further observables (charge etc) $\Gamma = \gamma$.

The second possibility ((ii) above) for the definition of n, namely such that $\Lambda(np) = -(1/m)n_\mu w^\mu = (S \cdot n)_{K_0}$ is invariant, should not be forgotten. It might suit well for some purposes.

10.4.4   Zero-mass particles

A particular situation arises for particles of mass zero. We shall prove the following statement.

For particles with spin but with rest mass zero the helicity is either $+\lambda_0$ or $-\lambda_0$, where the λ_0 is 'the spin' of the particle. That is, the spin of mass zero particles is always parallel or antiparallel to the direction of p.   (10.4.26)

Why 'the spin' is in quotation marks and how this comes about is seen in the following.

Recall (10.4.5) and look at it as if it were a classical equation

$$(w^0, w) = m(S^0, S) = m\left(\gamma\beta \cdot S_R, S_R + \beta\frac{\gamma^2}{\gamma+1}\beta \cdot S_R\right).$$

If $m \to 0$, then $\beta \to 1$ and $\gamma \to \infty$. What happens, if $\gamma \to \infty$? The direction of S is turned more and more parallel (if $\beta \cdot S_R > 0$) or antiparallel (if $\beta \cdot S_R < 0$) to the direction of β, because for large γ

$$S = S_R + \beta\gamma\lambda_0 \to \beta\gamma\lambda_0.$$

(Here we have put $\beta \cdot S_R = \lambda_0 =$ longitudinal spin component in the rest system.) This shows that 'direction and magnitude of the spin of a moving particle' both depend on the velocity and have a clear meaning only in the rest system. Hence, for $\gamma \to \infty$ one has

$$(w^0, w) \approx \lambda_0(m\gamma, m\beta\gamma) \approx \lambda_0(p^0, p).$$

If we now go with $m \to 0$ and $\gamma \to \infty$ such that $m\gamma = p_0$ remains finite we obtain

$$w^\mu = \lambda_0 p^\mu; \quad w_\mu w^\mu = w_\mu p^\mu = p_\mu p^\mu = 0.$$

Now, since no rest system exists any longer, λ_0 can be called 'the spin component in direction p' only in the sense of an analogy.

After this classical discussion we turn to an exact formulation. For $m = 0$ we have from (10.4.15) $w_\mu w^\mu = -m^2(S^2)_{\text{Rest}} = 0^1$; furthermore for $m = 0$ the identity $w_\mu p^\mu = 0$ holds and also $p_\mu p^\mu = 0$. Therefore in this case w^μ may well have a component in direction p^μ. Let n_μ and m_μ be two unit four-vectors orthogonal to each other and to p^μ: $n_\mu m^\mu = n_\mu p^\mu = m_\mu p^\mu = 0$. Then put

$$w^\mu = an^\mu + bm^\mu + cp^\mu.$$

$w_\mu w^\mu = 0$ yields $a = b = 0$, hence

$$w^\mu = cp^\mu \text{ (for } m = 0). \tag{10.4.27}$$

Since this is a covariant equation, it follows that c must be Lorentz invariant. Its significance follows from (10.4.3), which we repeat here (using the fact that in $w^\mu = \frac{1}{2}\epsilon^\mu{}_{\nu\rho\sigma}J^{\nu\rho}p^\sigma$ we can replace $J^{\nu\rho}$ by $S^{\nu\rho}$):

$$w^\mu = (w^0, \boldsymbol{w}) = (\boldsymbol{S} \cdot \boldsymbol{p}, \boldsymbol{S}p^0 + \boldsymbol{S}^0 \times \boldsymbol{p}) = c(p^0, \boldsymbol{p}) \tag{10.4.28}$$

where the last equation is true for $m = 0$. Since now also $p^0 = |\boldsymbol{p}|$, we read off (compare (10.4.24): $\Lambda(p) = (\boldsymbol{S} \cdot \boldsymbol{p})/|\boldsymbol{p}|$)

(i) $cp^0 = \boldsymbol{S} \cdot \boldsymbol{p}$ or $c = \dfrac{\boldsymbol{S}\boldsymbol{p}}{|\boldsymbol{p}|} \equiv \Lambda(p)$

(ii) $c\boldsymbol{p} = \boldsymbol{S}p^0 + \underbrace{\boldsymbol{S}^0 \times \boldsymbol{p}}_{=0} = \boldsymbol{S}p^0$ (since $\boldsymbol{S}^0 \times \boldsymbol{p}$ is orthogonal to $\boldsymbol{p}$).

From $c = \Lambda(p)$ and the invariance of c it follows that for $m = 0$ the helicity operator $\Lambda(p)$ must be a true invariant under all Lorentz transformations and not only under the restricted class discussed above. This is indeed the case, since for $m \to 0$ our 'four-vector' l^μ becomes a genuine four-vector: for $m \to 0$ (10.4.21) transforms into

$$l^\mu = \frac{1}{m}\left(|\boldsymbol{p}|, \boldsymbol{p}\frac{p_0}{|\boldsymbol{p}|}\right) \to \frac{1}{m}(p^0, \boldsymbol{p}) = \frac{p^\mu}{m}. \tag{10.4.29}$$

Hence for $m \to 0$ our l^μ transforms like p^μ. That its components tend to infinity does not matter: we have defined (equation (10.4.23)) $\Lambda(l, p) = -(1/m)w_\mu l^\mu$

---

[1] This is not quite correct; since a rest system no longer exists, we cannot know whether '$(S^2)_{\text{Rest}}$' might not be ∞; then $w_\mu w^\mu$ could be anything. In fact, $w_\mu w^\mu \neq 0$ leads to 'continuous spin' which seems not to be realized in Nature and will not be discussed here.

and this quantity is now a true invariant (and even finite). The invariance of $\Lambda(p)$ for $m = 0$ (we shall write $\Lambda_0(p)$ for this particular Λ) implies that its eigenvalue λ_0, found in one Lorentz frame, will be the same in all Lorentz frames. Furthermore this is equivalent to

$$[\Lambda_0(p), p^\mu] = [\Lambda_0(p), J^{\mu\nu}] = 0 \qquad (10.4.30)$$

and since p^μ and $J^{\mu\nu}$ (and combinations thereof) are the only observables (apart from those others, Γ, which are supposed to commute anyway with p^μ and $J^{\mu\nu}$) it follows that no measurement can induce any transition from λ^0 to $\lambda'^0 \neq \lambda^0$. In other words: for $m = 0$ only one eigenvalue λ^0 of the helicity exists (and not a spectrum $\lambda = -s, -s + 1, \ldots, +s$ as for $m \neq 0$). The question arises: which of the possible values between $-s$ and $+s$ does the particle choose? In fact this question has no sense, because s is no longer defined. The point is this: as long as $m \neq 0$ we define the total spin by

$$(S^2)_{\text{Rest}} = -\frac{1}{m^2} w_\mu w^\mu = S_{\mu\nu} S^{\nu\mu} = s(s + 1).$$

This is clearly invariant, commutes with all p and J and thus could be written $s(s + 1)$. If $m = 0$ no such invariant exists any longer; the formula gives simply $s(s + 1) = 0/0$. No limiting process works, because $m \neq 0$ and $m = 0$ are qualitatively different: however small m is, a rest system exists—but it does not exist when $m = 0$. Between these two possibilities there is no continuous connection. Thus we cannot define $s(s + 1)$ either by the invariant or by a limiting process. Of course, we could say that we call the quantity $S_x^2 + S_y^2 + S_z^2$ 'the spin squared', but we must then accept that this is not an invariant quantity and does not therefore 'belong to the particle'. We thus see that the quantum number $s =$ spin cannot be invariantly defined and the question of which value λ_0 the particle selects, between $-s$ and $+s$, makes no sense. However, $\Lambda_0(p) = \lambda_0$ is now itself an invariant, which never changes and therefore characterizes the particle rather than the state. We may then, since our old invariant s cannot be defined, simply call λ_0 'the spin' of the mass zero particle.

There is even a simple argument in favour of this definition: from (10.4.28) we found that

$$S p^0 = \Lambda_0(p)p. \qquad (10.4.31)$$

Now for $m = 0$ we can always find a coordinate system in which the eigenvalue four-momentum is $(p, 0, 0, p)$. On such a state, call it $|p, \lambda_0\rangle$, we have then

$$p_0|p, \lambda_0\rangle = p_z|p, \lambda_0\rangle = p|p, \lambda_0\rangle$$

$$p_x|p, \lambda_0\rangle = p_y|p, \lambda_0\rangle = 0.$$

Hence, since $p^0 \neq 0$ commutes with S and $p_z \neq 0$ with $\Lambda_0(p)$

$$S_x|p, \lambda_0\rangle = S_y|p, \lambda_0\rangle = 0$$

$$S_z|p, \lambda_0\rangle = \lambda_0|p, \lambda_0\rangle$$

so that the eigenvalue of the non-vanishing space component of S is just λ_0.

λ_0 can be positive, negative or zero. If it is zero, only one state exists. If it is different from zero, then

- *either* for the particle a parity operation is defined (the particle has a definite parity ± 1), then by applying the parity operation p goes to $-p$ but S goes to S (because it is a pseudovector), and therefore $\lambda_0 \rightarrow -\lambda_0$. Thus for particles with $m = 0$ both helicity states with $\lambda_0 = \pm|\lambda_0|$ exist, if the particle has a definite parity (e.g. the light quantum);

- *or* the particle has no parity, then only one value λ_0 exists and the particle has always a definite helicity (neutrino). In the neutrino case, however, CP is a good quantum number and since charge conjugation changes neither S nor p, it follows that the antineutrino has helicity opposite to that of the neutrino. Note that the possibility of having two states with $\pm\lambda_0$ is not in contradiction to the above statement that λ_0 could not change: the above statement was based on the fact that $\Lambda_0(p)$ commutes with all p^μ and $J^{\mu\nu}$, whereas the change from λ_0 to $-\lambda_0$ was induced by the parity (or CP) operation, which does not belong to the proper Lorentz group and was therefore excluded from that statement.

10.5 The use of helicity states in elementary particle physics

In this last section we select some parts of the paper by Jacob and Wick (1959).

10.5.1 Construction of one-particle helicity states of arbitrary p

Since the helicity $\Lambda(p)$ is invariant under rotations and Lorentz transformation in the direction p, we may construct helicity states starting from a system in which p points in the z-direction. Call such a state (we omit γ)

$$|p, s, \gamma\rangle = \text{state with } p^\mu = (p^0, 0, 0, p). \qquad (10.5.1)$$

If then $p = p(\sin \vartheta \cos \varphi, \sin \vartheta \sin \varphi, \cos \vartheta)$, we obtain the state $|p, \vartheta, \varphi, \lambda\rangle$ with the same λ by an active rotation $R_a(\alpha, \beta, \gamma)$ with $\alpha = \varphi, \beta = \theta, \gamma = -\varphi$ where $\gamma = -\varphi$ is not really necessary but has been chosen to make $R = 1$ if $\vartheta = 0$. Thus

$$|p, \vartheta, \varphi, s, \lambda\rangle = U_a(\varphi, \vartheta, -\varphi)|p, s, \lambda\rangle = e^{-i\varphi J_z} e^{-i\vartheta J_y} e^{i\varphi J_z}|p, s, \lambda\rangle. \qquad (10.5.2)$$

If the mass $m \neq 0$, we can even start all that in the rest system of the particle and generate $|p, s, \lambda\rangle$ by a Lorentz transformation in the z-direction. For states at rest—call them $|0, s, \lambda\rangle$—λ is the eigenvalue of S_z. The relative phases of the states with $\lambda = -s, -s + 1, \ldots, +s$ will then be defined as usual by (see (4.6.12)):

$$(S_x \pm iS_y)|0, s, \lambda\rangle = \sqrt{s(s + 1) - \lambda(\lambda \pm 1)}|0, s, \lambda \pm 1\rangle. \qquad (10.5.3)$$

This does not work for $m = 0$, since no rest system exists and, since $S_x = S_y = 0$, even $S_+ = S_x + iS_y$ and $S_- = S_x - iS_y$ commute on the states $|p, \lambda_0\rangle$. But we can define the relative phase between $|p, \lambda_0\rangle$ and $|p, -\lambda_0\rangle$ by means of the parity operation, which only exists when both values of λ_0 are allowed. Let us first see how we can define $|p, s, -\lambda\rangle$ for a particle with $m \neq 0$ and then for $m = 0$.

For $m \neq 0$ there are three possibilities.

(i) 2λ *times applying* $S_- = S_x - iS_y$.

$$|p, s, -\lambda\rangle = \frac{(s - \lambda)!}{(s + \lambda)!} S_-^{2\lambda} |p, s, \lambda\rangle. \qquad (10.5.4)$$

This does not apply for $m = 0$, since then S_- and s are not defined.

(ii)   *A Lorentz transformation in the z-direction*, which carries $p \to -p$ (passing over the rest system) and thereby $\lambda \to -\lambda$ and a rotation by π about x and y. For such a rotation about x we had the result $R_x|jm\rangle = (-1)^{-j}|j - m\rangle$ (see footnote to (6.3.65)) and, since $R_y = e^{-i(\pi/2)J_z} R_x e^{(\pi/2)J_z}$, we obtain $R_y|jm\rangle = (-1)^{m-j}|j - m\rangle$. Since the correspondence is (remember m is S_z, λ is S_p)

$$|p, s, \lambda\rangle = |p, s, m\rangle$$
$$|-p, s, \lambda\rangle = |-p, s, -m\rangle \qquad (10.5.5)$$

we have $R_y|p, s, \lambda\rangle = (-1)^{s-\lambda}|-p, s, \lambda\rangle$. We prefer R_y to R_x since $s - \lambda$ is always integer; the factor $(-1)^{s-\lambda} = (-1)^{\lambda-s}$ is real. Then the state $|p, s, -\lambda\rangle$ can be defined by

$$|p, s, \lambda\rangle \overset{L}{\to} |-p, s, -\lambda\rangle \overset{R_y}{\to} |p, s, -\lambda\rangle.$$

The whole procedure still works if we let $p \to 0$ afterwards. If we combine the L-transformation (which reverses the sign of p and of λ) and the rotation R_y in one symbol $\bar{R}_y$, then for all p (including $p = 0$)

$$|p, s, -\lambda\rangle = (-1)^{s-\lambda} \bar{R}_y |p, s, \lambda\rangle. \qquad (10.5.6)$$

This procedure does not work for $m = 0$, because no rest system exists and consequently p and λ cannot be reversed by a Lorentz transformation.

(iii) *The parity operation combined with* R_y. Defining the intrinsic parity of the particle by the phase factor η in

$$P|p, s, \lambda\rangle = \eta|-p, s, -\lambda\rangle$$

we obtain

$$Y := PR_y = R_y P$$
$$|p, s, -\lambda\rangle = (-1)^{s-\lambda} \eta Y |p, s, \lambda\rangle. \qquad (10.5.7)$$

This procedure works even for $m = 0$ if the particle is parity invariant and if we put $s = \lambda_0$. Thus we adopt it for all particles. When $m \neq 0$, all three definitions are equivalent.

$$|p, s, -\lambda\rangle = (-1)^{s-\lambda} R_y P\eta |p, s, \lambda\rangle = (-1)^{s-\lambda} \bar{R}_y |p, s, \lambda\rangle$$

$$= \frac{(s-\lambda)!}{(s+\lambda)!} S_-^{2\lambda} |p, s, \lambda\rangle \tag{10.5.8}$$

$$|p, -\lambda_0\rangle = R_y P\eta |p, \lambda_0\rangle \quad \text{for } m = 0.$$

In (ii) above we found that our old result $R_x |jm\rangle = (-1)^{-j} |j, -m\rangle$ implies

$$|-p, s\lambda, \rangle = (-1)^{s-\lambda} R_y |p, s, \lambda\rangle \quad \text{for } m \neq 0$$

$$|-p, \lambda_0\rangle = R_y |p, \lambda_0\rangle \quad \text{for } m = 0 \tag{10.5.9}$$

where both definitions become equal if we define the spin by $s = \lambda_0$ for massless particles. This defines completely all one-particle helicity states for $m \neq 0$ and $m = 0$.

10.5.2 Two-particle helicity states

We define these states conveniently in the centre-of-momentum frame by a direct product of one state $|p_a, s_a, \lambda_a\rangle$ and another one $|p_b, s_b, \lambda_b\rangle$ where $p_a = -p_b$. That is, the state where $p_a = +p$ and $p_b = -p$ may be written

$$|ps_a\lambda_a s_b\lambda_b\rangle = |ps_a\lambda_a\rangle \otimes |-ps_b\lambda_b\rangle.$$

This state may be transformed into a state $|p\vartheta\varphi s_a\lambda_a s_b\lambda_b\rangle$ by rotating it by $U(\varphi, \vartheta, -\varphi) = e^{-i\varphi J_z} e^{-i\vartheta J_y} e^{i\varphi J_z}$ where $J = J_a + J_b$. Noting that (see (10.5.5))

$$J_{z,a} |ps_a\lambda_a\rangle = \lambda_a |ps_a\lambda_a\rangle$$

$$J_{z,b} |-ps_b\lambda_b\rangle = -\lambda_b |-ps_b\lambda_b\rangle$$

we find

$$|p\vartheta\varphi s_a\lambda_a s_b\lambda_b\rangle = e^{i\varphi(\lambda_a-\lambda_b)} e^{-i\varphi J_z} e^{-i\vartheta J_y} |ps_a\lambda_a s_b\lambda_b\rangle \tag{10.5.10}$$

where the last rotation $e^{-i\varphi J_z}$ can no longer be written as a simple numerical factor, since $R_y(\vartheta) |ps_a\lambda_a s_b\lambda_b\rangle$ is not an eigenstate of J_z.

10.5.3 Eigenstates of the total angular momentum

If we apply our projection formula (6.3.52)

$$\frac{2j+1}{8\pi^2} \int d\omega\, D_{mm'}^{(j)*}(\omega) R(\omega) |\psi\rangle = \sum_{\gamma'} |\gamma' jm\rangle\langle\gamma' jm'|\psi\rangle \tag{10.5.11}$$

to a state $|\psi\rangle = |ps_a\lambda_a s_b\lambda_b\rangle$, then we should obtain an eigenstate of total angular momentum $|pjms_a\lambda_a s_b\lambda_b\rangle$, where λ_a and λ_b are conserved in the state, since they are invariant under rotations. Now, $|ps_a\lambda_a s_b\lambda_b\rangle$ is an eigenstate of J_z with $m' = \lambda = \lambda_a - \lambda_b$, hence in (10.5.11) only $m' = \lambda$ will give a non-zero projection. We shall furthermore write a normalization factor $N_j/2\pi$ instead of $(2j+1)/8\pi^2$ and obtain

$$|pjms_a\lambda_a s_b\lambda_b\rangle = \frac{N_j}{2\pi} \int d\alpha \, d\gamma \, \sin\beta \, d\beta \, D^{(j)*}_{m\lambda}(\alpha, \beta, \gamma) R(\alpha, \beta, \gamma) |ps_a\lambda_a s_b\lambda_b\rangle.$$
$$(10.5.12)$$

Now from (10.5.10)

$$R(\alpha, \beta, \gamma)|ps_a\lambda_a s_b\lambda_b\rangle = e^{-i\gamma\lambda} e^{-i\alpha\lambda} |p\alpha\beta s_a\lambda_a s_b\lambda_b\rangle$$

with (6.2.16)

$$D^{(j)*}_{m\lambda}(\alpha, \beta, \gamma) = e^{i(m\alpha+\lambda\gamma)} d^{(j)}_{m\lambda}(\beta)$$

so that $e^{i\lambda\gamma}$ drops out and the γ-integration yields 2π. Hence

$$|pjms_a\lambda_a s_b\lambda_b\rangle = N_j \int d\alpha \, \sin\beta \, d\beta \, D^{(j)*}_{m\lambda}(\alpha, \beta, -\alpha)|p\alpha\beta s_a\lambda_a s_b\lambda_b\rangle. \quad (10.5.13)$$

The normalization factor N_j is determined by requiring that

$$\langle p'j'm's'_a\lambda'_a s'_b\lambda'_b|pjms_a\lambda_a s_b\lambda_b\rangle = \delta(p'-p)\delta_{j'j}\delta_{m'm}$$
$$\times \delta_{s'_a s_a}\delta_{s'_b s_b}\delta_{\lambda'_a\lambda_a}\delta_{\lambda'_b\lambda_b} \qquad (10.5.14)$$

whereas for plane wave states we require

$$\langle p'\beta'\alpha's'_a\lambda'_a s'_b\lambda'_b|p\beta\alpha s_a\lambda_a s_b\lambda_b\rangle = \delta(p'-p)\delta(\alpha'-\alpha)\delta(\cos\beta'-\cos\beta)$$
$$\times \delta_{s'_a s_a}\delta_{s'_b s_b}\delta_{\lambda'_a\lambda_a}\delta_{\lambda'_b\lambda_b}. \qquad (10.5.15)$$

In the following calculation all δ except $\delta_{jj'}\delta_{m'm}$ and $\delta(\alpha'-\alpha)\delta(\cos\beta'-\cos\beta)$ drop out and the quantum numbers are suppressed in the states. Then we have, after integrating over α' and $\sin\beta' \, d\beta'$:

$$\langle j'm'|jm\rangle = N_{j'}N_j \int d\alpha \, \sin\beta \, d\beta \, D^{(j')}_{m'\lambda}(\alpha, \beta, -\alpha) D^{(j)*}_{m\lambda}(\alpha, \beta, -\alpha).$$

Since the factors $e^{i\lambda\alpha} e^{-i\lambda\alpha}$ cancel, we may replace $-\alpha$ by anything, e.g. by γ, and integrate with $d\gamma/2\pi$, without changing the expression. Then the integral becomes (6.3.47)

$$\int d\alpha \, \sin\beta \, d\beta \frac{d\gamma}{2\pi} D^{(j)*}_{m\lambda}(\alpha, \beta, \gamma) D^{(j')}_{m'\lambda}(\alpha, \beta, \gamma) = \frac{4\pi}{2j+1}\delta_{jj'}\delta_{m'm}$$

and we obtain

$$\langle j'm'|jm \rangle = N_j^2 \frac{4\pi}{2j+1} \delta_{jj'} \delta_{m'm} = \delta_{jj'} \delta_{m'm}$$

if we put $N_j = \sqrt{(2j+1)/4\pi}$. Finally, then the eigenstate of total angular momentum J^2 and J_z and helicities λ_a, λ_b is

$$|pjms_a\lambda_a s_b\lambda_b\rangle = \sqrt{\frac{2j+1}{4\pi}} \int \sin\vartheta\, d\vartheta\, d\varphi\, D_{m\lambda}^{(j)*}(\varphi\vartheta - \varphi)|p\vartheta\varphi s_a\lambda_a s_b\lambda_b\rangle$$

$$|p\vartheta\varphi s_a\lambda_a s_b\lambda_b\rangle \equiv |ps_a\lambda_a\rangle \otimes |-ps_b\lambda_b\rangle \qquad p_a = p(\vartheta,\varphi).$$

The normalizations (10.5.14) and (10.5.15) imply

$$\langle p'\vartheta\varphi s_a'\lambda_a' s_b'\lambda_b'|pjms_a\lambda_a s_b'\lambda_b\rangle = \delta_{s_a's_a}\delta_{s_b's_b}\delta_{\lambda_a'\lambda_a}\delta_{\lambda_b'\lambda_b}\delta(p'-p)$$

$$\times \sqrt{\frac{2j+1}{4\pi}} D_{m\lambda}^{(j)*}(\varphi,\vartheta,-\varphi). \qquad (10.5.16)$$

The transformation between these two bases, $|\vartheta,\varphi\rangle$ and $|jm\rangle$, is unitary: one easily verifies (we omit the s_a, s_b, etc)

$$\int \sin\vartheta\, d\vartheta\, d\varphi\, \langle j'm'\lambda_a\lambda_b|\vartheta\varphi\lambda_a\lambda_b\rangle\langle\vartheta\varphi\lambda_a\lambda_b|jm\lambda_a\lambda_b\rangle = \delta_{jj'}\delta_{mm'}$$

$$\sum_{jm}\langle\vartheta'\varphi'\lambda_a\lambda_b|jm\lambda_a\lambda_b\rangle\langle jm\lambda_a\lambda_b|\vartheta\varphi\lambda_a\lambda_b\rangle = \delta(\cos\vartheta'-\cos\vartheta)\delta(\varphi'-\varphi).$$

$$(10.5.17)$$

The corresponding projection operators are

$$\int \sin\vartheta\, d\vartheta\, d\varphi\, |\vartheta\varphi\lambda_a\lambda_b\rangle\langle\vartheta\varphi\lambda_a\lambda_b| = P_{\lambda_a\lambda_b}$$

$$\sum_{jm}|jm\lambda_a\lambda_b\rangle\langle jm\lambda_a\lambda_b| = P_{\lambda_a\lambda_b}. \qquad (10.5.18)$$

10.5.4 The S-matrix; cross-sections

Let a reaction be of the type $a + b \rightarrow c + d$; then the S-matrix element is given by

$$\langle p_c p_d \lambda_c \lambda_d|S|p_a p_b \lambda_a \lambda_b\rangle \qquad (10.5.19)$$

where we suppressed the spins $s_a s_b s_c s_d$ since they remain constant anyway. With the normalization (10.5.14), (10.5.15) this is equivalent to

$$\langle p_c p_d \lambda_c \lambda_d|S|p_a p_b \lambda_a \lambda_b\rangle = (2\pi)^6 \delta(p'^\mu - p^\mu)\frac{\sqrt{vv'}}{pp'} \qquad (10.5.20)$$

$$\times \langle \vartheta'\varphi'\lambda_c\lambda_d|S(p^\mu)|\vartheta\varphi\lambda_a\lambda_b\rangle$$

with

$$v = \frac{p}{\sqrt{p^2 + m_a^2}} + \frac{p}{\sqrt{p^2 + m_b^2}} = v_1 + v_2$$

(note that because of the presence of vv' this normalization is not Lorentz covariant). In the centre-of-momentum frame $p = p_a + p_b = p_c + p_d = 0$.

The cross-section is then related to the transition-matrix element $S = 1 - iT$ by the formula

$$\left(\frac{d\sigma}{d\Omega}\right)_{c.m.} = \left(\frac{2\pi}{p}\right)^2 |\langle \vartheta \varphi \lambda_c \lambda_d | T(E) | 00 \lambda_a \lambda_b \rangle|^2 \qquad (10.5.21)$$

where $E = E_{c.m.} = [(p_a + p_b)^\mu (p_a + p_b)_\mu]^{\frac{1}{2}} = \sqrt{p^2 + m_a^2} + \sqrt{p^2 + m_b^2}$. We now go to the jm-representation, where the rotational invariance of S can be used explicitly. Namely, as a rotationally invariant quantity $S(E)$ is necessarily an irreducible tensor of rank 0, therefore energy conservation and the Wigner–Eckart theorem give (equation (8.8.3))

$$\langle E' j' m' \lambda_c \lambda_d | S | E j m \lambda_a \lambda_b \rangle = \delta(E' - E)\delta_{j'j}\delta_{m'm}\langle \lambda_c \lambda_d | S^j (E) | \lambda_a \lambda_b \rangle \quad (10.5.22)$$

where the remaining matrix element is (up to a different notation) the reduced matrix element.

With the help of the projection operators (10.5.18)

$$\sum_{jm} |jm\lambda_a\lambda_b\rangle\langle jm\lambda_a\lambda_b| = P_{\lambda_a\lambda_b} \qquad (10.5.23)$$

we obtain (omitting $\delta(E' - E)$ and writing already $S(E)$)

$$\langle \vartheta \varphi \lambda_c \lambda_d | S(E) | 00 \lambda_a \lambda_b \rangle = \langle \vartheta \varphi \lambda_c \lambda_d | P_{\lambda_c \lambda_d} S(E) P_{\lambda_a \lambda_b} | 00 \lambda_a \lambda_b \rangle$$

$$= \sum_{jm} \sum_{j'm'} \langle \vartheta \varphi \lambda_c \lambda_d | j'm' \lambda_c \lambda_d \rangle \langle j'm' \lambda_c \lambda_d | S(E) | jm\lambda_a\lambda_b \rangle$$

$$\times \langle jm\lambda_a\lambda_b | \vartheta_0 \varphi_0 \lambda_a \lambda_b \rangle; \ \vartheta_0 = \varphi_0 = 0 \qquad (10.5.24)$$

with (10.5.22)

$$= \sum_{jm} \langle \vartheta \varphi \lambda_c \lambda_d | jm \lambda_c \lambda_d \rangle \langle jm\lambda_a\lambda_b | \vartheta_0 \varphi_0 \lambda_a \lambda_b \rangle \langle \lambda_c \lambda_d | S^j (E) | \lambda_a \lambda_b \rangle.$$

(10.5.16) now gives

$$\langle \vartheta \varphi \lambda_c \lambda_d | jm\lambda_c\lambda_d \rangle = \sqrt{\frac{2j+1}{4\pi}} D_{m\mu}^{(j)*}(\varphi, \vartheta, -\varphi) \qquad \mu = \lambda_c - \lambda_d$$

$$\langle jm\lambda_a\lambda_b | \vartheta_0 \varphi_0 \lambda_a \lambda_b \rangle = \sqrt{\frac{2j+1}{4\pi}} D_{\lambda m}^{(j)}(0, 0, 0) = \sqrt{\frac{2j+1}{4\pi}} \delta_{m\lambda}.$$

If we insert this into (10.5.24) we obtain

$$\langle \vartheta \varphi \lambda_c \lambda_d | S(E) | 00 \lambda_a \lambda_b \rangle = \frac{1}{4\pi} \sum_j \left[(2j+1) \langle \lambda_c \lambda_d | S^j(E) | \lambda_a \lambda_b \rangle D_{\lambda\mu}^{(j)^*}(\varphi, \vartheta, -\varphi) \right]$$

$$\lambda = \lambda_a - \lambda_b \qquad \mu = \lambda_c - \lambda_d.$$

(10.5.25)

If we write the differential cross-section

$$\frac{d\sigma}{d\Omega} = |f(\lambda_c \lambda_d, \lambda_a \lambda_b, \vartheta, \varphi)|^2 \tag{10.5.26}$$

then the scattering amplitude f is given by

$$f(\lambda_c \lambda_d, \lambda_a \lambda_b, \vartheta, \varphi) = \frac{1}{2p} \sum_j (2j+1) \langle \lambda_c \lambda_d | T^j(E) | \lambda_a \lambda_b \rangle D_{\lambda\mu}^{(j)^*}(\varphi, \vartheta, -\varphi)$$

(10.5.27)

where again $S^j(E) = 1 - iT^j(E)$.

As an illustration we compare this with the usual formula. The comparison is made by specializing to the case $\lambda_a = \lambda_b$, $\lambda_c = \lambda_d$, when $\lambda = \mu = 0$ and j must be an integer l. In this case (6.3.33) gives $D_{00}^{(l)}(\varphi, \vartheta, -\varphi) = P_l(\cos \vartheta)$ and we obtain for the scattering amplitude

$$f(\lambda_a \lambda_a, \lambda_c \lambda_c, \vartheta, \varphi) = -\frac{1}{2ip} \sum_l (2l+1) \langle \lambda_c \lambda_c | S^l(E) - 1 | \lambda_a \lambda_a \rangle P_l(\cos \vartheta)$$

(10.5.28)

which is very similar to the well known formula for spinless particles. The usual formula for particles with spin is, however, much more complicated. The helicity formalism simultaneously achieves simpler formulae and greater generality. This is another example of the experience that a fully relativistic treatment is often simpler than the non-relativistic one.

10.5.5 Evaluation of cross-section formulae

For the calculation of cross-sections (10.5.27) has to be squared. Either one first calculates explicitly the amplitude using the formula (6.2.16) for D, and then squares, or one squares directly. In this case (we abbreviate $\langle \lambda_c \lambda_d | T^j(E) | \lambda_a \lambda_b \rangle \equiv \langle j \rangle$)

$$|f|^2 = \frac{1}{4p^2} \sum_{jj'} (2j+1)(2j'+1) \langle j \rangle \langle j' \rangle^* D_{\lambda\mu}^{(j)^*} D_{\lambda\mu}^{(j')}. \tag{10.5.29}$$

The product of two D can be reduced to a sum over D by means of (6.3.68), (6.3.4) and (6.3.33):

(6.3.68): $\quad D_{\lambda\mu}^{(j)*} = D_{-\lambda-\mu}^{(j)}(-1)^{\lambda-\mu}$

(6.3.4): $\quad D_{-\lambda-\mu}^{(j)} D_{\lambda\mu}^{(j')} = \sum_j \langle j - \lambda j'\lambda|J0\rangle\langle J0|j - \mu j'\mu\rangle D_{00}^{(J)} \quad$ (10.5.30)

(6.3.33): $\quad D_{00}^{(J)}(\varphi, \vartheta, -\varphi) = P_J(\cos\vartheta) \quad (J = \text{integer} = l).$

Thus, if all helicities are assumed to be given

$$\frac{d\sigma}{d\Omega}(\lambda_c\lambda_d, \lambda_a\lambda_b, \vartheta) = \frac{1}{4p^2}\sum_{jj'}(2j + 1)(2j' + 1)(-1)^{\lambda-\mu}$$

$$\times\langle\lambda_c\lambda_d|T^j(E)|\lambda_a\lambda_b\rangle\langle\lambda_c\lambda_d|T^{j'}(E)|\lambda_a\lambda_b\rangle^* \qquad (10.5.31)$$

$$\times\sum_l\langle j - \lambda j'\lambda|l0\rangle\langle l0|j - \mu j'\mu\rangle P_l(\cos\vartheta)$$

where $\lambda = \lambda_a - \lambda_b$; $\mu = \lambda_c - \lambda_d$. If not all polarizations are specified, then one carries out in this formula the corresponding averages over λ_a and/or λ_b as well as the sums over λ_c and/or λ_d. If particles with $m = 0$ occur, only two states are possible if parity is conserved; only one state if parity is violated.

10.5.6 Discrete symmetry relations: parity, time reversal, identical particles

Invariance of elementary processes under the inhomogeneous Lorentz group including rotations has been used to build up all this formalism. Thus these invariances cannot give us any further simplifications or relations.

The discrete groups, however, lead to further relations.

10.5.6.1 Parity

For a one-particle state we found with $Y = PR_y = P\,e^{-i\pi J_y}$ (see (10.5.7))

$$P|ps\lambda\rangle = (-1)^{s-\lambda}\eta e^{i\pi J_y}|ps, -\lambda\rangle.$$

For a two-particle state $(J_y = J_{ya} + J_{yb})$ with (10.5.5)

$$P|ps_a\lambda_as_b\lambda_b\rangle = (-1)^{s_a+s_b-\lambda_a+\lambda_b}\eta_a\eta_b\,e^{i\pi J_y}|ps_a, -\lambda_as_b, -\lambda_b\rangle. \qquad (10.5.32)$$

This implies for total angular momentum states (10.5.12)

$$P|pjms_a\lambda_as_b\lambda_b\rangle = (-1)^{s_a+s_b-\lambda_a+\lambda_b}\eta_a\eta_b\sqrt{\frac{2j+1}{4\pi}}$$

$$\times\int\sin\beta\,d\beta\,d\alpha\,d\gamma\,D_{m\lambda}^{(j)*}(\alpha, \beta, \gamma)R(\alpha, \beta, \gamma)R(0 - \pi 0)|ps_a, -\lambda_a, s_b, -\lambda_b\rangle$$

$$(10.5.33)$$

where $PR(\alpha, \beta, \gamma) = R(\alpha, \beta, \gamma)P$ has been used. We can now change the integration variables by defining $R(\alpha, \beta, \gamma)R(0, -\pi, 0) \equiv R(\alpha', \beta', \gamma')$; $\sin \beta \, d\beta \, d\alpha \, d\gamma = \sin \beta' \, d\beta' \, d\alpha' \, d\gamma'$ (that $d\omega = d\omega'$ is not trivial but can be shown). Then we have to relate D to the new variables:

$$D(\alpha, \beta, \gamma)D(0, -\pi, 0) = D(\alpha', \beta', \gamma')$$

or

$$D^{(j)}_{m\lambda}(\alpha, \beta, \gamma) = \sum_\rho D^{(j)}_{m\rho}(\alpha', \beta', \gamma')D^{(j)}_{\rho\lambda}(0, \pi, 0).$$

Now $R_y|jm\rangle = (-1)^{j-m}|j - m\rangle$ (see (10.5.7)) so that

$$D^{(j)}_{\rho\lambda}(0, \pi, 0) = (-1)^{j-\lambda}\delta_{-\rho,\lambda}$$

thus

$$D^{(j)*}_{m\lambda}(\alpha, \beta, \gamma) = (-1)^{j-\lambda}D^{(j)*}_{m,-\lambda}(\alpha', \beta', \gamma').$$

Then (10.5.33) changes into (remember $\lambda = \lambda_a - \lambda_b$ and $j - \lambda$ integer)

$$P|pjms_a\lambda_a s_b\lambda_b\rangle = (-1)^{j-s_a-s_b}\eta_a\eta_b\sqrt{\frac{2j+1}{4\pi}}$$

$$\times \int \sin \beta \, d\beta \, d\alpha \, d\gamma \, D^{(j)*}_{m,-\lambda}(\alpha, \beta, \gamma)R(\alpha, \beta, \gamma)|ps_a - \lambda_a s_b - \lambda_b\rangle.$$

Comparing this with the definition (10.5.16), one sees that

$$P|pjms_a\lambda_a s_b\lambda_b\rangle = (-1)^{j-s_a-s_b}\eta_a\eta_b|pjms_a, -\lambda_a, s_b, -\lambda_b\rangle \qquad (10.5.34)$$

where $j - s_a - s_b$ is always integer. If the S-operator is invariant under parity, then

$$\langle \text{final}|S|\text{initial}\rangle = \langle \text{final}|P^{-1}SP|\text{initial}\rangle$$

$$= \langle \text{final}|PSP^{-1}|\text{initial}\rangle$$

implies

$$\langle -\lambda_c - \lambda_d|S^j| - \lambda_a - \lambda_b\rangle = \eta\langle\lambda_c\lambda_d|S^j|\lambda_a\lambda_b\rangle$$

$$\eta = (-1)^{s_a+s_b-s_c-s_d}\frac{\eta_c\eta_d}{\eta_a\eta_b} = \eta^{-1}. \qquad (10.5.35)$$

This relation approximately halves the number of independent S-matrix elements.

If we insert this into the scattering amplitude (10.5.27), then we would like to have the subscripts $-\lambda - \mu$ with the D. This can be done; the second and fourth line of (6.3.68) yield

$$D^{(j)*}_{\lambda\mu}(\varphi, \vartheta, -\varphi) = (-1)^{\lambda-\mu}D^{(j)*}_{-\lambda-\mu}(-\varphi, \vartheta, \varphi).$$

The factor $(-1)^{\lambda-\mu}$ can be compensated by $\varphi \to \pi - \varphi$ so that $D^{(j)*}_{\lambda\mu}(\varphi, \vartheta, -\varphi) = D^{(j)*}_{-\lambda-\mu}(\pi - \varphi, \vartheta, \varphi - \pi)$. This gives in (10.5.27)

$$f(-\lambda_c, -\lambda_d, -\lambda_a, -\lambda_b, \vartheta, \varphi) = \eta f(\lambda_a, \lambda_b, \lambda_c, \lambda_d, \vartheta, \pi - \varphi) \qquad (10.5.36)$$

if parity is conserved.

10.5.6.2 Time reversal

Time reversal is represented by an antilinear and anti-unitary operator T (do not confuse with the transition operator T). Time reversal obviously changes momentum and angular momentum into their negatives; hence, since λ refers to the direction of p, no change of λ occurs and hence $|ps\lambda\rangle \rightarrow |-ps\lambda\rangle$. The same can be done by a rotation R_y, only both operations will give rise to different phase factors. Combining these into one symbol, ϵ, we have

$$T|ps\lambda\rangle = \epsilon\, e^{-i\pi J_y}|ps\lambda\rangle. \tag{10.5.37}$$

Is ϵ independent of p and λ? Yes, it is. Let us apply T^2 to $|ps\lambda\rangle$. It restores $|ps\lambda\rangle$ up to a phase factor η, which obviously cannot depend on p and λ. Then

$$T^2|ps\lambda\rangle = \eta|ps\lambda\rangle = \epsilon^2\, e^{-2\pi i J_y}|ps\lambda\rangle.$$

Now $|ps\lambda\rangle$ can be expanded into a sum over $|pjms\lambda\rangle$ and each of these obtains a factor $(-1)^{2j}$ upon 2π-rotation. Since these j are either all integers or all half odd integers, it follows that

$$\epsilon^2\, e^{-2\pi i J_y}|ps\lambda\rangle = \pm\epsilon^2|ps\lambda\rangle = \eta|ps\lambda\rangle.$$

Hence $\epsilon^2 = \pm\eta$ independent of p and γ. We may even eliminate ϵ by multiplying all states $|ps\lambda\rangle$ by a phase $\alpha^*(s)$. Then (10.5.37) gives (since T is anti-unitary)

$$T\alpha^*(s)|ps\lambda\rangle = \alpha(s)T|ps\lambda\rangle = \alpha(s)\epsilon\, e^{-i\pi J_y}|ps\lambda\rangle$$

$$= \alpha^2(s)\epsilon\, e^{-i\pi J_y}\alpha^*(s)|ps\lambda\rangle.$$

Choosing $\alpha(s) = 1/\sqrt{\epsilon}$ and calling $\alpha^*|ps\lambda\rangle$ again $|ps\lambda\rangle$, we have no ϵ any longer. For such states $\eta = \pm 1$. Putting $\epsilon = 1$, we have

$$T|ps_a\lambda_a s_b\lambda_b\rangle = e^{-i\pi J_y}|ps_a\lambda_a s_b\lambda_b\rangle. \tag{10.5.38}$$

This gives, applied to the $|pjms_a\lambda_a s_b\lambda_b\rangle$ states (see (10.5.12) and remember that $TD^{(j)*} = D^{(j)}T$, whereas T commutes with the abstract rotation $R(\alpha, \beta, \gamma)$: e.g. $TR(\alpha, 0, 0)|jm\rangle = T\, e^{-i\alpha m}|jm\rangle = e^{i\alpha m}|j, -m\rangle = R(\alpha, 0, 0)T|jm\rangle)$

$$T|pjms_a\lambda_a s_b\lambda_b\rangle = \sqrt{\frac{2j+1}{4\pi}}\int \sin\beta\, d\beta\, d\alpha\, d\gamma \tag{10.5.39}$$
$$\times D_{m\lambda}^{(j)}(\alpha, \beta, \gamma)R(\alpha, \beta, \gamma)R(0, \pi, 0)|ps_a\lambda_a s_b\lambda_b\rangle.$$

We define $R(\alpha', \beta', \gamma') = R(\alpha, \beta, \gamma)R(0, \pi, 0)$ and

$$D_{m\lambda}^{(j)}(\alpha, \beta, \gamma) = \sum_\rho D_{m\rho}^{(j)}(\alpha'\beta'\gamma')D_{\rho\lambda}^{(j)}(0 - \pi 0).$$

From page 271 we have $D_{\rho\lambda}^{(j)}(0, \pi, 0) = (-1)^{j-\lambda}\delta_{-\rho,\lambda}$; and since $D^{(j)}(2\pi) = (-1)^{2j}$, it follows that $D_{\rho\lambda}^{(j)}(0, -\pi, 0) = (-1)^{j+\lambda}\delta_{-\rho,\lambda}$. Therefore

$$T|pjms_a\lambda_a s_b\lambda_b\rangle = \sqrt{\frac{2j+1}{4\pi}}(-1)^{j+\lambda}\int \sin\beta\, d\beta\, d\alpha\, d\gamma$$
$$\times D_{m,-\lambda}^{(j)}(\alpha, \beta, \gamma)R(\alpha, \beta, \gamma)|ps_a\lambda_a s_b\lambda_b\rangle.$$

Now use (6.3.68) $D_{m,-\lambda}^{(j)} = (-1)^{-m-\lambda}D_{-m\lambda}^{(j)*}$ and obtain

$$T|pjms_a\lambda_a s_b\lambda_b\rangle = (-1)^{j-m}|pj, -m, s_a\lambda_a s_b\lambda_b\rangle. \tag{10.5.40}$$

Now, for any anti-unitary operators A and B it holds that $\langle\psi|(AB)|\varphi\rangle = \{((\langle\psi|A)(B|\varphi))\}^*$; furthermore, for the S-matrix $TS^*T = S$; thus with $T = A$, $ST = B$

$$\langle\psi|S|\varphi\rangle = \langle\psi|(TS^*T)|\varphi\rangle = \{((\langle\psi|T)(S^*T|\varphi))\}^*.$$

We then obtain with (10.5.40)

$$\langle jm\lambda_c\lambda_d|S|jm\lambda_a\lambda_b\rangle = \langle j-m\lambda_c\lambda_d|S^*|j-m\lambda_a\lambda_b\rangle^*$$
$$= \langle j-m\lambda_a\lambda_b|S|j-m\lambda_c\lambda_d\rangle$$

and with (10.5.22) on both sides

$$\langle\lambda_c\lambda_d|S^j|\lambda_a\lambda_b\rangle = \langle\lambda_a\lambda_b|S^j|\lambda_c\lambda_d\rangle \tag{10.5.41}$$

if T-invariance holds.

10.5.6.3 Identical particles

We use the following notation: we distinguish the numerical values of the quantum numbers by an additional prime, whereas the subscripts a and b indicate which particle actually has this numerical value. Consider then a state (we omit $s_a = s_b = s$)

$$|p\lambda_a\lambda_b'\rangle \equiv |p_a\lambda_a\rangle \otimes |-p_b\lambda_b'\rangle.$$

By exchanging the particles we obtain

$$P_{ab}|p_a\lambda_a\rangle \otimes |-p_b\lambda_b'\rangle = |p_b\lambda_b\rangle \otimes |-p_a\lambda_a'\rangle. \tag{10.5.42}$$

Now from (10.5.9)

$$|p_b\lambda_b\rangle = (-1)^{-s+\lambda}e^{i\pi J_y^{(b)}}|-p_b\lambda_b\rangle$$
$$|-p_a\lambda_a'\rangle = (-1)^{s-\lambda'}e^{-i\pi J_y^{(a)}}|p_a\lambda_a'\rangle$$
$$= (-1)^{-s-\lambda'}e^{i\pi J_y^{(a)}}|p_a\lambda_a'\rangle$$

where $R(2\pi)|ps\lambda\rangle = (-1)^{2s}|ps\lambda\rangle$ (see (6.3.69)) was used. We then obtain $(2s - \lambda + \lambda'$ is integer!)

$$P_{ab}|p\lambda_a\lambda_b'\rangle = (-1)^{2s-\lambda+\lambda'} e^{i\pi J_y} |p\lambda_a'\lambda_b\rangle. \qquad (10.5.43)$$

If we compare this with (10.5.32) then we see that apart from the additional $\eta_a\eta_b$ and the different state on the r.h.s. the structure is the same. Carrying through the same calculation as after (10.5.32) would then lead to the equivalent of (10.5.34), namely

$$P_{ab}|pjm\lambda_a\lambda_b'\rangle = (-1)^{j-2s}|pjm\lambda_a'\lambda_b\rangle \quad \text{(identical particles)}. \qquad (10.5.44)$$

Since integer spin gives Bose–Einstein and half odd integer Fermi–Dirac statistics, the correct states are given by applying the operator $\frac{1}{\sqrt{2}}(1+(-1)^{2s}P_{ab})$ to the state $|pjm\lambda_a\lambda_b'\rangle$. Then the factor $(-1)^{2s}$ is eliminated and the state with correct symmetry for two identical particles is

$$|pjm\lambda\lambda'\rangle \equiv \frac{1}{\sqrt{2}}\left\{|pjm\lambda_a\lambda_b'\rangle + (-1)^j|pjm\lambda_a'\lambda_b\rangle\right\} \qquad (10.5.45)$$

whatever the spin s of these two particles may be. This implies that in a total angular momentum eigenstate $|pjm\lambda\lambda'\rangle$ of two identical particles only $\lambda \neq \lambda'$ is allowed for odd j.

Example

If a particle of spin j decays into two identical decay products ($\pi^0 \to \gamma\gamma$), then for j odd $\lambda \neq \lambda'$. Assume $j_{\pi^0} = 1$, then $|\lambda_\gamma| = 1$ and then $\lambda \neq \lambda'$ requires, e.g., $\lambda_a = 1, \lambda_b = -1$; then $J_z \to m = +2$ in contradiction to $j_\gamma = 1$.

11

SUPERSYMMETRY IN QUANTUM MECHANICS AND PARTICLE PHYSICS

11.1 What is supersymmetry?

The concept of supersymmetry (Golfand and Likhtman (1971), Volkov and Akulov (1973), Wess and Zumino (1974)), which relates bosonic and fermionic states in quantum mechanics, i.e. combines integer and half-integer spin states (particles) in one multiplet, has played a central role in the development of quantum field theory for two decades (see, e.g., Wess and Bagger (1983), Gates *et al* (1983), West (1987)). Supersymmetric models of unification of the fundamental interactions are the most promising candidates to extend the standard model of strong and electroweak interactions. Gravity was also generalized by incorporating supersymmetry (SUSY) into a theory called supergravity. In this theory, Einstein's general theory of relativity turns out to be a necessary consequence of a local gauged SUSY. Thus, local SUSY theories provide a natural framework for the unification of gravity with the other fundamental interactions of nature.

Another theoretical motivation for studying supersymmetry is offered by string theory (Green *et al* (1987)). The presence of fermionic string states together with bosonic ones imposes a supersymmetric structure on the theory. In effective field theories which approximate string theory in the energy domain below the Planck mass equal to 10^{19} GeV, this structure manifests itself as a supersymmetry among particles. In the string domain supersymmetry is essential in order to define a tachyon-free theory (tachyons are hypothetical particles with unphysical imaginary mass).

Thus supersymmetry is a necessary ingredient in any unification of all basic interactions of nature, i.e. strong, electroweak and gravitational interactions. One more important fact which explains why supersymmetry is so important for the unification is the following: SUSY offers a possible way to avoid the *no-go* theorem of Coleman and Mandula (1967) which was based on the assumption of a Lie algebraic realization of symmetries (supersymmetric or, in more mathematical language, *graded* Lie algebras were unfamiliar to particle theorists at the time of the proof of the no-go theorem). More precisely, this theorem states that any Lie group containing the Poincaré group and an internal symmetry group is the trivial product of both. In other words, internal symmetry transformations always commute with the Poincaré transformations. The hypotheses of this theorem are quite general. They consist of the axioms of relativistic quantum field theory and of the assumption that all symmetries are realized in terms of Lie groups. A way to circumvent it was, however, found by Haag *et al* (1975). These authors simply relaxed one of the hypotheses of the no-go theorem, namely the one which concerns the groups of symmetry. They assumed that the infinitesimal generators of the symmetry obey a *superalgebra* (called also *supersymmetric Lie algebra* or *graded Lie algebra*). A superalgebra is a generalization of the notion of a Lie algebra, where some of the infinitesimal generators are fermionic, which means that some of the commutation rules are replaced by anticommutation rules.

A further motivation for supersymmetry is found in the solution of the *hierarchy problem* of the grand unified theories (see, e.g., Ross (1984)). In these theories, which tend to unify all the particles and forces except gravitational ones, two energy scales must be introduced, typically of the order of 10^3–10^4 GeV, the electroweak scale, and 10^{15}–10^{16} GeV, the grand unification scale. This means that masses of the particles in the model must be fine tuned with a precision of 10^{-12}. Such a tuning is not possible in ordinary quantum field theory, since the presence of the so-called ultraviolet divergences of the quantum mass corrections induces a strong instability of the mass differences. In supersymmetric theories ultraviolet divergences are milder; in particular the quantum mass corrections depend only on the logarithm of the ultraviolet cut-off, instead of its square. The huge mass differences in grand unified theories are then much more stable.

Thus the attractive property is that SUSY relates bosonic and fermionic degrees of freedom and has the virtue of taming ultra-violet divergences. Moreover, some ultraviolet finite supersymmetric models have been known for a long time. Most of these models have the so-called *N-extended supersymmetry*: $N = 2$ or $N = 4$, where N counts the fermionic generators. The basis of the extended N-supersymmetry algebra consists of

- bosonic (*even*) Hermitian generators T_a, $a = 1, \ldots, \dim(G)$, of some Lie group G

- the (*even*) generators P_μ and $M_{\mu\nu}$ of the four-dimensional Poincaré group

- fermionic (*odd*) generators Q_α^i, $\alpha = 1, 2$; $i = 1, \ldots, N$, belonging to a dimension N representation of G, and their conjugates $\bar{Q}_i^{\dot\alpha}$ and

- central charges Z^{ij}, i.e. bosonic (*even*) operators commuting with all the T_a and all the Q_α and $\bar{Q}^{\dot\alpha}$, as well as with the Poincaré generators.

The T_a and Z^{ij} are scalars, whereas Q_α^i and $\bar{Q}_i^{\dot\alpha}$ belong to the two inequivalent fundamental (two-component spinor) representations of the Lorentz group which are conjugate to each other. The dotted and undotted indices of two-component spinors can be raised or lowered with the help of the invariant antisymmetric tensors $\varepsilon_{\alpha\beta} = \varepsilon^{\alpha\beta} = -\varepsilon_{\dot\alpha\dot\beta} = -\varepsilon^{\dot\alpha\dot\beta}$, $\varepsilon_{12} = +1$.

The general superalgebra of N-extended supersymmetry, also called the

N-super-Poincaré algebra, reads as (see (10.3.2) and (10.3.3))

$$[M_{\mu\nu}, M_{\rho\sigma}] = -\mathrm{i}(g_{\mu\rho}M_{\nu\sigma} - g_{\mu\sigma}M_{\nu\rho} + g_{\nu\sigma}M_{\mu\rho} - g_{\nu\rho}M_{\mu\sigma})$$
$$[M_{\mu\nu}, P_{\lambda}] = \mathrm{i}(P_{\mu}g_{\nu\lambda} - P_{\nu}g_{\mu\lambda}) \tag{11.1.1}$$

$$[T_a, T_b] = \mathrm{i}f_{ab}{}^c T_c \tag{11.1.2}$$

$$\left\{Q_{\alpha}^i, Q_{\beta}^j\right\} = \varepsilon_{\alpha\beta}Z^{ij} \tag{11.1.3}$$

$$\left\{Q_{\alpha}^i, \bar{Q}_j^{\dot{\alpha}}\right\} = 2\delta_j^i(\sigma^{\mu})_{\alpha}{}^{\dot{\alpha}}P_{\mu} \tag{11.1.4}$$

$$[Q_{\alpha}^i, M_{\mu\nu}] = \tfrac{1}{2}(\sigma_{\mu\nu})_{\alpha}{}^{\beta}Q_{\beta}^i \tag{11.1.5}$$

$$[\bar{Q}_i^{\dot{\alpha}}, M_{\mu\nu}] = -\tfrac{1}{2}(\bar{\sigma}_{\mu\nu})_{\dot{\beta}}^{\dot{\alpha}}\bar{Q}_i^{\dot{\beta}} \tag{11.1.6}$$

$$[Q_{\alpha}^i, T_a] = (R_a)^i{}_j Q_{\alpha}^j \tag{11.1.7}$$

$$[\bar{Q}_i^{\dot{\alpha}}, T_a] = -\bar{Q}_j^{\dot{\alpha}}(\bar{R}_a)^j{}_i. \tag{11.1.8}$$

The curly brackets $\{\cdot, \cdot\}$ denote an anticommutator:

$$\{Q_1, Q_2\} \equiv Q_1 Q_2 + Q_2 Q_1.$$

Other (anti)commutators are equal to zero. The four-vector σ^{μ} is defined as

$$(\sigma^0, \sigma^1, \sigma^2, \sigma^3) = \left(\begin{pmatrix} 1 & 0 \\ 0 & 1 \end{pmatrix}, \begin{pmatrix} 0 & -1 \\ -1 & 0 \end{pmatrix}, \begin{pmatrix} 0 & \mathrm{i} \\ -\mathrm{i} & 0 \end{pmatrix}, \begin{pmatrix} -1 & 0 \\ 0 & 1 \end{pmatrix}\right).$$
$$\tag{11.1.9}$$

The matrices $\sigma_{\mu\nu} = (\mathrm{i}/4)\left[\sigma_{\mu}, \sigma_{\nu}\right]$ and their complex conjugates $\bar{\sigma}^{\mu\nu}$ provide two inequivalent two-dimensional representations for the Lie algebra of the Lorentz group. The matrices $(R_a)^i_j$ are the representation matrices for the Lie-algebra generators T_a in the representation provided by the fermionic generators.

This result is the most general one for a theory with massive particles. In a massless theory, another set of fermionic charges may be present.

In the $N = 1$ case, the superalgebra is reduced to the so-called *Wess–Zumino algebra* (Wess and Zumino (1974))

$$\left\{Q_{\alpha}, \bar{Q}^{\dot{\alpha}}\right\} = 2(\sigma^{\mu})_{\alpha}{}^{\dot{\alpha}}P_{\mu}$$
$$\left\{Q_{\alpha}, Q_{\beta}\right\} = 0 \tag{11.1.10}$$
$$\left\{\bar{Q}^{\dot{\alpha}}, \bar{Q}^{\dot{\beta}}\right\} = 0$$

to the commutation relations of the fermionic generators with those of the
Poincaré subalgebra

$$[Q_\alpha, M_{\mu\nu}] = \tfrac{1}{2}(\sigma_{\mu\nu})_\alpha{}^\beta Q_\beta \qquad [\bar{Q}^{\dot\alpha}, M_{\mu\nu}] = -\tfrac{1}{2}(\bar{\sigma}_{\mu\nu})^{\dot\alpha}_{\dot\beta}\bar{Q}^{\dot\beta}$$

$$[Q_\alpha, P_\mu] = 0 \qquad [\bar{Q}^{\dot\alpha}, P_\mu] = 0$$

(11.1.11)

and to

$$[Q_\alpha, R] = -Q_\alpha \qquad [\bar{Q}^{\dot\alpha}, R] = \bar{Q}^{\dot\alpha}. \qquad (11.1.12)$$

Here, R is the infinitesimal generator of an Abelian group which is the trace of
the internal symmetry group G.

Being already familiar with the notion of tensor operators, the reader can
easily realize that the operators Q_α and $\bar{Q}^\alpha$ change the spin of a quantum
mechanical state by one-half due to their transformation properties (11.1.5)
and (11.1.6) with respect to the Lorentz subgroup. Simultaneously, because of
the *anticommutation* relations (11.1.3) and (11.1.4), they change the statistical
properties of states, i.e. transform bosonic states to fermionic and vice versa.
Despite the beauty of all these unified theories, there has so far been no
experimental evidence of SUSY being realized in particle physics. One of the
important predictions of unbroken SUSY theories is the existence of SUSY
partners of all known elementary particles which have the same masses as their
SUSY counterparts. The fact that no such particles have been seen implies that
SUSY must be broken. One hopes that the scale of this breaking is in the range
of 100 GeV to 1 TeV in order that it can explain the hierarchy problem of
mass differences. This leads to a conceptual problem since the natural scale of
symmetry breaking is the gravitational or Planck scale, which is of the order of
10^{19} GeV. Various schemes have been invented to try to resolve the hierarchy
problem, including the idea of non-perturbative breaking of SUSY. It was in
the context of this question that SUSY was first studied in the simplest case of
SUSY quantum mechanics (SUSY QM) by Witten (1981) (see also Cooper *et
al* (1995) and references therein). Thus, in the early days, SUSY was studied
in quantum mechanics as a testing ground for the non-perturbative methods
of investigating SUSY breaking in field theory , but it was soon realized that
this field was interesting in its own right, not just as a model for testing field
theoretical methods.

In particular, there is now a much deeper understanding of why certain
potentials are analytically solvable and an array of powerful new approximation
methods for handling potentials which are not exactly solvable.

We will use SUSY QM as a simple realization of a superalgebra involving
the fermionic and the bosonic operators, to give the reader the main ideas and an
introduction to the subject, but let us remark that non-relativistic SUSY quantum
mechanics not only has pedagogical meaning and provides us with the method
of solution of the Schrödinger equation for certain classes of potentials, but
also has practical applications and has stimulated new approaches to different

branches of physics, such as nuclear, atomic, condensed matter and statistical physics.

11.2 SUSY quantum mechanics

A quantum mechanical system characterized by a self-adjoint Hamiltonian H, acting on some Hilbert space $\mathcal{H}$, is called *supersymmetric* if there exists a *supercharge* operator Q obeying the following anticommutation relations:

$$\{Q, Q\} = 0 = \{Q^\dagger, Q^\dagger\} \qquad \{Q, Q^\dagger\} = H. \tag{11.2.1}$$

One can easily recognize in these relations the non-relativistic analogue of the Wess–Zumino algebra (11.1.10).

An immediate consequence of these relations is the conservation of the supercharge and the non-negativity of the Hamiltonian:

$$[H, Q] = 0 = [H, Q^\dagger] \qquad H \geq 0. \tag{11.2.2}$$

In 1981 Witten introduced a simple model of supersymmetric quantum mechanics (Witten (1981)). It is defined in the Hilbert space $\mathcal{H} = L^2(\mathbb{R}) \otimes \mathbb{C}^2$, that is, it characterizes a spin-$\frac{1}{2}$-like particle (with mass $m > 0$) moving along the one-dimensional line. In constructing a supersymmetric Hamiltonian on $\mathcal{H}$, let us first introduce bosonic operators A, $A^\dagger$ and fermionic operators f, $f^\dagger$:

$$A = \frac{1}{\sqrt{2m}} \frac{\mathrm{d}}{\mathrm{d}x} + W(x) \qquad A^\dagger = \frac{-1}{\sqrt{2m}} \frac{\mathrm{d}}{\mathrm{d}x} + W(x)$$

$$f = \sigma_+ = \begin{pmatrix} 0 & 1 \\ 0 & 0 \end{pmatrix} \qquad f^\dagger = \sigma_- = \begin{pmatrix} 0 & 0 \\ 1 & 0 \end{pmatrix} \tag{11.2.3}$$

where the *superpotential* W is assumed to be continuously differentiable. Obviously, these operators obey the commutation and anticommutation relations

$$[A, A^\dagger] = \frac{\sqrt{2}}{\sqrt{m}} W'(x) \qquad \{f, f^\dagger\} = 1 \tag{11.2.4}$$

and allow us to define suitable supercharges

$$Q = A \otimes f^\dagger = \begin{pmatrix} 0 & 0 \\ A & 0 \end{pmatrix} \qquad Q^\dagger = A^\dagger \otimes f = \begin{pmatrix} 0 & A^\dagger \\ 0 & 0 \end{pmatrix} \tag{11.2.5}$$

which obey the required relations $\{Q, Q\} = 0 = \{Q^\dagger, Q^\dagger\}$. Note that Q is a combination of a generalized bosonic annihilation operator and a fermionic creation operator. Finally, we may construct a supersymmetric quantum system by defining the Hamiltonian in such a way that the second relation in (11.2.1) also holds

$$H = \{Q, Q^\dagger\} = \begin{pmatrix} A^\dagger A & 0 \\ 0 & AA^\dagger \end{pmatrix} = \begin{pmatrix} H_1 & 0 \\ 0 & H_2 \end{pmatrix} \tag{11.2.6}$$

with

$$H_1 = -\frac{1}{2m}\frac{d^2}{dx^2} + W^2(x) - \frac{1}{\sqrt{2m}}W'(x) \qquad (11.2.7)$$

$$H_2 = -\frac{1}{2m}\frac{d^2}{dx^2} + W^2(x) + \frac{1}{\sqrt{2m}}W'(x) \qquad (11.2.8)$$

being the standard Schrödinger operators acting on $L^2(\mathbb{R})$.

Example: SUSY harmonic oscillator

As we have discussed in chapter 7 on the Jordan–Schwinger construction, for the usual quantum mechanical harmonic oscillator, one can introduce a Fock space of bosonic occupation numbers and the creation and annihilation operators a and $a^\dagger$, which after a suitable normalization obey the commutation relations (cf (7.1.5) and (7.1.9))

$$[a, a^\dagger] = 1 \qquad [N, a] = -a \qquad [N, a^\dagger] = a^\dagger$$
$$N = a^\dagger a \qquad H = N + \tfrac{1}{2}. \qquad (11.2.9)$$

For the case of the SUSY harmonic oscillator, one can rewrite the operators Q ($Q^\dagger$) as a product of the bosonic operator a and the fermionic operator f. Namely, we write $Q = af^\dagger$ and $Q^\dagger = a^\dagger f$, where the matrix fermionic creation and annihilation operators are defined in (11.2.3) and obey the usual algebra of the fermionic creation and annihilation operators, namely

$$\{f^\dagger, f\} = 1 \qquad \{f^\dagger, f^\dagger\} = \{f, f\} = 0 \qquad (11.2.10)$$

as well as obeying the commutation relation

$$[f, f^\dagger] = \sigma_3 = \begin{pmatrix} 1 & 0 \\ 0 & -1 \end{pmatrix}. \qquad (11.2.11)$$

The SUSY Hamiltonian can be rewritten in the form

$$H = QQ^\dagger + Q^\dagger Q = \left(-\frac{d^2}{dx^2} + \frac{x^2}{4}\right)1 - \tfrac{1}{2}[f, f^\dagger]. \qquad (11.2.12)$$

The effect of the last term is to remove the zero-point energy.

The state vector can be thought of as a matrix in the Schrödinger picture or as the state $|n_b, n_f\rangle$ in the Fock space picture. Since the fermionic creation and annihilation operators obey anti-commutation relations, the fermion number is either zero or one. We will choose the ground state of H_1 to have zero fermion number. Then we can introduce the fermion number operator

$$n_F = \frac{1 - \sigma_3}{2} = \frac{1 - [f, f^\dagger]}{2}. \qquad (11.2.13)$$

The action of the operators $a, a^\dagger, f, f^\dagger$ in this Fock space is then

$$a|n_b, n_f\rangle = |n_b - 1, n_f\rangle \qquad f|n_b, n_f\rangle = |n_b, n_f - 1\rangle \qquad (11.2.14)$$

$$a^\dagger|n_b, n_f\rangle = |n_b + 1, n_f\rangle \qquad f^\dagger|n_b, n_f\rangle = |n_b, n_f + 1\rangle. \qquad (11.2.15)$$

Of course, n_f can have only the values zero and unity. Now one can see that the operator $Q^\dagger = -iaf^\dagger$ has the property of changing a boson into a fermion without changing the energy of the state. This is the boson–fermion degeneracy, characteristic of all SUSY theories.

As is seen from (11.2.3), for the general case of SUSY QM, the operators a and $a^\dagger$ are replaced by A and $A^\dagger$ in the definition of Q and $Q^\dagger$, i.e. one writes $Q = Af^\dagger$ and $Q^\dagger = A^\dagger f$. The effect of Q and $Q^\dagger$ is now to relate the wave functions of H_1 and H_2 which have fermion number zero and one respectively, but now there is no simple Fock space description in the bosonic sector because the interactions are non-linear. Thus in the general case, we can rewrite the SUSY Hamiltonian in the form

$$H = \left(-\frac{d^2}{dx^2} + W^2\right)1 - [f, f^\dagger]W'. \qquad (11.2.16)$$

11.3 Factorization and the hierarchy of Hamiltonians

Let us now reverse our point of view and consider how SUSY can help in finding exactly the spectrum of one-dimensional Hamiltonians. It is generally difficult to solve exactly the eigenvalue problem of a (time-independent) Hamiltonian in quantum mechanics. Among the various methods developed for this purpose, a remarkably simple but powerful one is the ladder operator technique, a typical example of which is the simple harmonic oscillator. If a Hamiltonian has a discrete eigenvalue spectrum bounded from below, the energy eigenstates should be labelled by integers and formal raising and lowering operators can be written in this basis. However, it does not provide a way to find the ladder operators explicitly for a given Hamiltonian. A practical method of obtaining ladder operators has been studied using ideas of supersymmetric quantum mechanics and a concept of shape-invariant potentials (see, e.g., Cooper *et al* (1995) and references therein). This approach has (re)produced many exactly solvable potentials.

One of the key ideas of the approach is the connection between the bound state wave functions and the potential. Let us choose the ground state energy for the moment to be zero. Then one has from the Schrödinger equation that the ground state wave function $\psi_0(x)$ obeys

$$H_1\psi_0(x) = -\frac{1}{2m}\frac{d^2\psi_0}{dx^2} + V_1(x)\psi_0(x) = 0 \qquad (11.3.1)$$

so that

$$V_1(x) = \frac{1}{2m}\frac{\psi_0''(x)}{\psi_0(x)}. \qquad (11.3.2)$$

This allows a global reconstruction of the potential $V_1(x)$ from a knowledge of its ground state wave function. Once we realize this, it is now very simple to factorize the Hamiltonian using the following *ansatz*:

$$H_1 = A^\dagger A \qquad (11.3.3)$$

where A and $A^\dagger$ are given by (11.2.3). This allows us to identify

$$V_1(x) = W^2(x) - \frac{1}{\sqrt{2m}} W'(x) \qquad (11.3.4)$$

as in (11.2.7). This equation is the well known Riccati equation. The solution for $W(x)$ in terms of the ground state wave function is

$$W(x) = -\frac{1}{\sqrt{2m}} \frac{\psi_0'(x)}{\psi_0(x)}. \qquad (11.3.5)$$

This solution is obtained by using the fact that if $A\psi_0 = 0$, one automatically has

$$H_1 \psi_0 = A^\dagger A \psi_0 = 0.$$

The supersymmetric partner of H_1 is the operator $H_2 = AA^\dagger$ obtained by reversing the order of A and $A^\dagger$ (cf (11.2.8)):

$$H_2 = -\frac{1}{2m} \frac{d^2}{dx^2} + V_2(x) \qquad V_2(x) = W^2(x) + \frac{1}{\sqrt{2m}} W'(x). \qquad (11.3.6)$$

The potentials $V_1(x)$ and $V_2(x)$ are known as supersymmetric partner potentials.

For $n > 0$, the Schrödinger equation for H_1

$$H_1 \psi_n^{(1)} = A^\dagger A \psi_n^{(1)} = E_n^{(1)} \psi_n^{(1)} \qquad (11.3.7)$$

implies

$$H_2(A\psi_n^{(1)}) = AA^\dagger A \psi_n^{(1)} = E_n^{(1)}(A\psi_n^{(1)}). \qquad (11.3.8)$$

Similarly, the Schrödinger equation for H_2

$$H_2 \psi_n^{(2)} = AA^\dagger \psi_n^{(2)} = E_n^{(2)} \psi_n^{(2)} \qquad (11.3.9)$$

implies

$$H_1(A^\dagger \psi_n^{(2)}) = A^\dagger AA^\dagger \psi_n^{(2)} = E_n^{(2)}(A^\dagger \psi_n^{(2)}). \qquad (11.3.10)$$

From (11.3.7)–(11.3.10) and the fact that $E_0^{(1)} = 0$, it is clear that the eigenvalues and eigenfunctions of the two Hamiltonians H_1 and H_2 are related by

$$E_n^{(2)} = E_{n+1}^{(1)} \qquad E_0^{(1)} = 0 \qquad (11.3.11)$$

$$\psi_n^{(2)} = [E_{n+1}^{(1)}]^{-1/2} A \psi_{n+1}^{(1)} \qquad (11.3.12)$$

$$\psi_{n+1}^{(1)} = [E_n^{(2)}]^{-1/2} A^\dagger \psi_n^{(2)} \qquad (11.3.13)$$

$(n = 0, 1, 2, \ldots)$. Notice that if $\psi_{n+1}^{(1)}$ $(\psi_n^{(2)})$ of H_1 (H_2) is normalized, then the wave function $\psi_n^{(2)}$ $(\psi_{n+1}^{(1)})$ in (11.3.12) and (11.3.13) is also normalized. Thus, the operator A $(A^\dagger)$ converts an eigenfunction of H_1 (H_2) into an eigenfunction of H_2 (H_1) with the same energy. Since the ground state wave function of H_1 is annihilated by the operator A, this state has no SUSY partner. Thus the picture we obtain is that knowing all the eigenfunctions of H_1 we can determine the eigenfunctions of H_2 using the operator A, and vice versa; using $A^\dagger$ we can reconstruct all the eigenfunctions of H_1 from those of H_2 except for the ground state.

It is the commutativity of the supercharges Q and $Q^\dagger$ with the Hamiltonian H:

$$[H, Q] = [H, Q^\dagger] = 0$$

that is responsible for the degeneracy.

Let us look at one more well known potential, namely the infinite square well, and determine its SUSY partner potential. Consider a particle of mass m in an infinite square well potential of width L:

$$
\begin{aligned}
V(x) &= 0 && 0 \le x \le L \\
&= \infty && -\infty < x < 0 \qquad x > L.
\end{aligned}
\qquad (11.3.14)
$$

The ground state wave function is known to be

$$\psi_0^{(1)} = (2/L)^{1/2} \sin(\pi x/L) \qquad 0 \le x \le L \qquad (11.3.15)$$

and the ground state energy is $E_0 = \pi^2/(2mL^2)$.

Subtracting the ground state energy so that we can factorize the Hamiltonian, we have for $H_1 = H - E_0$ that the energy eigenvalues are

$$E_n^{(1)} = \frac{n(n+2)}{2mL^2}\pi^2 \qquad (11.3.16)$$

and the eigenfunctions are

$$\psi_n^{(1)} = (2/L)^{1/2} \sin\frac{(n+1)\pi x}{L} \qquad 0 \le x \le L. \qquad (11.3.17)$$

The superpotential for this problem is readily obtained using (11.3.5):

$$W(x) = -\frac{1}{\sqrt{2m}}\frac{\pi}{L}\cot(\pi x/L) \qquad (11.3.18)$$

and hence the supersymmetric partner potential V_2 is

$$V_2(x) = \frac{\pi^2}{2mL^2}[2\cos^{-2}(\pi x/L) - 1]. \qquad (11.3.19)$$

The wave functions for H_2 are obtained by applying the operator A to the wave functions of H_1. In particular, one finds that

$$\psi_0^{(2)} \propto \sin^2(\pi x/L) \qquad \psi_1^{(2)} \propto \sin(\pi x/L)\sin(2\pi x/L). \tag{11.3.20}$$

Thus these two rather different potentials corresponding to H_1 and H_2 have exactly the same spectrum, except for the fact that H_2 has one bound state less.

Thus, once we know the ground state wave function corresponding to a Hamiltonian H_1, we can find the superpotential $W_1(x)$ from (11.3.5). The resulting operators A_1 and $A_1^{\dagger}$ obtained from (11.2.3) can be used to factorize the Hamiltonian H_1. The ground state wave function of the partner Hamiltonian H_2 is determined from the first excited state of H_1 via the application of the operator A_1. This allows a refactorization of the second Hamiltonian in terms of W_2. The partner of this refactorization is now another Hamiltonian H_3. Each of the new Hamiltonians has one fewer bound states, so that this process can be continued until the number of bound states is exhausted and one can solve for the energy eigenvalues and wave functions for the entire hierarchy of Hamiltonians created by repeated refactorizations. Conversely, if we know the ground state wave functions for all the Hamiltonians in this hierarchy, we can reconstruct the solutions of the original problem.

So far we have discussed SUSY QM on the full line ($-\infty < x < \infty$). Many of these results have analogues for the n-dimensional potentials with spherical symmetry. For example, in three dimensions after a partial wave expansion

$$\psi_{lm}(r,\theta,\phi) = \frac{1}{r}R_l(r)Y_{lm}(\theta,\phi)$$

the reduced radial wave function R_l satisfies the one-dimensional Schrödinger equation ($0 < r < \infty$)

$$-\frac{1}{2m}\frac{d^2 R_l(r)}{dr^2} + \left[V(r) + \frac{l(l+1)}{2mr^2}\right]R_l(r) = E R_l(r) \tag{11.3.21}$$

with the original potential plus an angular momentum barrier.

A very interesting example of a supersymmetric system is given by the Pauli equation in three dimensions (Crombrugghe and Rittenberg (1983)). It is amusing that as a result of the existence of SUSY, the gyromagnetic ratio is equal to two. Consider the Hermitian SUSY generator of the form

$$Q = \frac{1}{\sqrt{2}}\left[\phi(r) + (p + A(r))\sigma)\right] \tag{11.3.22}$$

where $\phi(r)$ and $A(r)$ are external fields and p is the momentum operator of the particle. The relations (11.2.1) in this case reduce to just one:

$$Q^2 = H. \tag{11.3.23}$$

This equation is not in conflict with the basic anticommutation relations since the anticommutation relations $\{Q, Q\} = 0$ and $\{Q^\dagger, Q^\dagger\} = 0$ are not assumed now. Combining (11.3.22) and (11.3.23), one obtains

$$H = \frac{1}{2}\left[(p + A)^2 + \phi^2 + \{\phi, p\}\sigma + 2\phi A\sigma + (\nabla \times A)\sigma\right].$$

If A is identified with the magnetic potential, the gyromagnetic ratio is equal to two. This is a consequence of the supersymmetry.

11.4 Broken supersymmetry

In the quantum theory with an exact symmetry, the ground state (in field theory this is the vacuum state, i.e. the state without particles) must be invariant with respect to the group transformations (see, e.g., Chaichian and Nelipa (1984)). This means, in turn, that the ground state must be annihilated by the generators of the symmetry group. In the case of SUSY this gives

$$Q\psi_0 = Q^\dagger\psi_0 = 0.$$

As the Hamiltonian in supersymmetric theory is expressed in terms of the supercharges

$$H = \{Q, Q^\dagger\}$$

the supersymmetry (11.2.1) of a quantum system is said to be an unbroken symmetry (exact SUSY) if the ground state energy of H vanishes. Otherwise, SUSY is said to be broken. For unbroken SUSY the ground state of H belongs either to H_1 or to H_2 and is given by

$$\psi_0^\pm(x) = \psi_0^\pm(0) \exp\left\{\pm \int_0^x dz\, W(z)\right\}. \tag{11.4.1}$$

Obviously, depending on the asymptotic behaviour of the SUSY potential, one of the two functions $\psi_0^\pm$ will be normalizable (exact SUSY) or neither will be normalizable (broken SUSY). To be more explicit, let us introduce the *Witten index* (Witten (1982)), which (according to the Atiyah–Singer index theorem) depends only on the asymptotic values of Φ:

$$\Delta \equiv \text{ind } A = \dim \ker H_1 - \dim \ker H_2 = \tfrac{1}{2}[\text{sgn } W(+\infty) - \text{sgn } W(-\infty)] \tag{11.4.2}$$

where $\dim \ker H_i$ ($i = 1, 2$) are the numbers of eigenstates of the Hamiltonians H_i with zero eigenvalue. Hence, for the exact SUSY we have $\Delta = 1$ with the ground state belonging to H_1 and $\Delta = -1$ with the ground state belonging to H_2. For broken SUSY we have $\Delta = 0$. The spectral properties of $H_{1,2}$ are summarized as

$$\Delta = +1: \qquad E_n^{(2)} = E_{n+1}^{(1)} > 0 \qquad E_0^{(1)} = 0$$

$$\Delta = -1: \qquad E_n^{(1)} = E_{n+1}^{(2)} > 0 \qquad E_0^{(2)} = 0 \qquad (11.4.3)$$

$$\Delta = 0: \qquad E_n^{(1)} = E_n^{(2)} > 0$$

where $E_n^{(i)}$, $n = 0, 1, 2, \ldots$, denotes the ordered set of eigenvalues of H_i with $E_n^{(i)} < E_{n+1}^{(i)}$. For simplicity, we have assumed purely discrete spectra.

The class of potentials for which the SUSY-generalized operator method quickly yields all the bound state energy eigenvalues and eigenfunctions as well as the scattering matrix (in analogy with solution of the harmonic oscillator problem by the method of creation and annihilation operators) is called the class of shape-invariant potentials and includes all the popular, analytically solvable potentials.

The meaning of shape invariance is the following. If the pair of SUSY partner potentials $V_{1,2}(x)$ are similar in shape and differ only in the parameters that appear in them, then they are said to be shape invariant. More precisely, if the partner potentials $V_{1,2}(x; a_1)$ satisfy the condition

$$V_2(x; a_1) = V_1(x; a_2) + R(a_1) \qquad (11.4.4)$$

where a_1 is a set of parameters, $a_2 = f(a_1)$ is a function of a_1 and the remainder $R(a_1)$ is independent of x, then $V_1(x; a_1)$ and $V_2(x; a_1)$ are said to be shape invariant. Using this condition and the hierarchy of Hamiltonians, one can obtain the energy eigenvalues and eigenfunctions of any shape-invariant potential when SUSY is unbroken.

As can be seen even from our brief discussion, the idea of supersymmetry has found many fruitful applications both in the area of high-energy particle physics and in non-relativistic quantum mechanics.

APPENDIX A

Remarks on symmetric and self-adjoint operators

(i) We shall assume that A is a linear operator in the Hilbert space $\mathcal{H}$ defined on a set D_A dense in $\mathcal{H}$: $\bar{D}_A = \mathcal{H}$. The *adjoint operator* $A^\dagger$ to the *densely* defined operator A is defined by the relation

$$\langle A^\dagger \eta | \varphi \rangle = \langle \eta | A\varphi \rangle \tag{A.1}$$

on a domain

$$D_{A^\dagger} = \{\eta \in \mathcal{H}; \ |\langle \eta | A\varphi \rangle| < C\|\varphi\| \text{ for all } \varphi \in D_A\}.$$

This means that $\langle \eta | A\varphi \rangle$ as a function of φ is a linear bounded functional on D_A, which is dense in $\mathcal{H}$. Then by the Riesz theorem there exists an element in $\mathcal{H}$, denoted as $A^\dagger \eta \in \mathcal{H}$, such that (A.1) holds, i.e. should hold for any $\eta \in D_A^\dagger$ and $\varphi \in D_A$.

(ii) Operator A is *symmetric* if

$$\langle \psi | A\varphi \rangle = \langle A\psi | \varphi \rangle \qquad \text{for } \psi, \varphi \in D_A. \tag{A.2}$$

Comparing (A.1) and (A.2) we see that A is symmetric iff

$$A \subset A^\dagger \tag{A.3}$$

i.e. $A = A^\dagger$ on D_A, and $D_A \subset D_{A^\dagger}$. An operator A is *self-adjoint* iff

$$A = A^\dagger \tag{A.4}$$

i.e. A is symmetric *and* $D_A = D_{A^\dagger}$.

Note 1. The notions of symmetric and self-adjoint operator are equivalent if A is a bounded operator, since then one can take $D_A = \mathcal{H}$, and of course for a symmetric operator $D_{A^\dagger} = \mathcal{H}$.

Note 2. In quantum mechanics many important operators corresponding to observables are unbounded. The relevant correspondence is by *self-adjoint* operators. The *symmetry* guarantees only that the mean values $\langle \varphi | A\varphi \rangle$ are *real* for $\varphi \in D_A$. The self-adjointness guarantees much more, namely that for the self-adjoint operator there exists a *complete set* of generalized eigenstates

$$A\psi_\lambda = \lambda \psi_\lambda \qquad \lambda \text{ real.} \tag{A.5}$$

Here $\psi_\lambda \in \mathcal{H}$ for a discrete point $\lambda = \lambda_n$ of the spectrum and for the continuous spectrum ψ_λ are distributions on D_A. The completeness means that any $\varphi = \mathcal{H}$ can be expanded as

$$\varphi = \sum_n c_n \psi_{\lambda_n} + \int d\lambda \, c(\lambda) \psi_\lambda. \qquad (A.6)$$

This is *necessary* for the interpretation of quantum mechanics. Mathematically more suitable objects as generalized eigenstates are spectral projectors acting on φ as

$$E_\lambda \varphi = \int_{-\infty}^\lambda d\lambda \, (\psi_\lambda^*, \varphi) \psi_\lambda = \int_{-\infty}^\lambda d\lambda \, c(\lambda) \psi_\lambda \qquad (A.7)$$

existing for any self-adjoint operator.

We stress that if one finds a complete set of (generalized) eigenfunctions (A.5) of some symmetric operator this means that A is self-adjoint.

Example

$$A\varphi = (1/i)\partial_x \varphi$$
$$D_A = \{\varphi \in \mathcal{L}^2(0, 1), \ \varphi \text{ absolutely continuous}, \ \varphi(0) = \varphi(1) = 0 \}.$$

Note. φ is absolutely continuous if

$$\sum_{i=1}^N |\varphi(\beta_i) - \varphi(\beta_i)| < \epsilon \qquad \text{for any } 0 < \alpha_i < \beta_i < 1$$

such that

$$\sum_{i=1}^N (\beta_i - \alpha_i) < \delta.$$

This guarantees that $\partial_x \varphi$ exists almost everywhere on $(0, 1)$.

Property 1. A is symmetric on $(0, 1)$:

$$\langle \varphi | A\psi \rangle = \int_0^1 dx \, \varphi^*(x) \frac{1}{i} \partial_x \psi(x)$$
$$= \frac{1}{i}[\varphi^*(1)\psi(1) - \varphi^*(0)\psi(0)] + \int_0^1 dx \left(\frac{1}{i}\partial_x \varphi(x)\right)^* \psi(x)$$
$$= \langle A\varphi | \psi \rangle.$$

Property 2. A is not self-adjoint:

$$A^\dagger \eta = \frac{1}{i} \partial_x \eta \qquad D_{A^\dagger} = \{\eta \in \mathcal{L}^2(0, 1), \ \varphi \text{ absolutely continuous}\}$$

$$\langle \eta | A \psi \rangle = \int_0^1 \mathrm{d}x \, \eta^*(x) \frac{1}{\mathrm{i}} \partial_x \psi(x) = \frac{1}{\mathrm{i}} [\eta^*(1)\psi(1) - \eta^*(0)\psi(0)]$$
$$+ \int_0^1 \mathrm{d}x \left(\frac{1}{\mathrm{i}} \partial_x \eta(x) \right)^* \psi(x) = \langle A^\dagger \eta | \psi \rangle$$

since $\eta^*(1)\psi(1) - \eta^*(0)\psi(0) = 0$ is satisfied for *any* $\eta \in D_{A^\dagger}$ provided that $\psi \in D_A$, i.e. $\psi(1) = \psi(0) = 0$.

Note 3. If B is a symmetric ($B \subset B^\dagger$) extension of a symmetry operator A ($A \subset A^\dagger$), then

$$A \subset B \subset B^\dagger \subset B.$$

von Neumann (1955) formulated the conditions under which a self-adjoint symmetric extension exists, namely

$$A \subset B = B^\dagger \subset B$$

and presented a method for their construction.

Let $n_\pm$ be the number of linearly independent solutions in $\mathcal{H}$ of the equations

$$A^\dagger \varphi = \pm \mathrm{i} \varphi \qquad \varphi \in \mathcal{H}. \tag{A.8}$$

The self-adjoint extension exists $\Leftrightarrow n_+ = n_-$.

Example. Denote by A_α the operator

$$A_\alpha \varphi = \frac{1}{\mathrm{i}} \partial_x \varphi$$
$$D_\alpha = \left\{ \varphi \in \mathcal{L}^2(0, 1), \ \varphi \text{ absolutely continuous}, \ \varphi(0) = \mathrm{e}^{\mathrm{i}\alpha} \varphi(1) \right\}.$$

Then

$$\langle \eta | A_\alpha \psi \rangle = \int_0^1 \mathrm{d}x \, \eta^*(x) \frac{1}{\mathrm{i}} \partial_x \psi(x) = \frac{1}{\mathrm{i}} [\eta^*(1)\psi(1) - \eta^*(0)\psi(0)]$$
$$+ \int_0^1 \mathrm{d}x \left(\frac{1}{\mathrm{i}} \partial_x \eta(x) \right)^* \psi(x).$$

$D_{A_\alpha^\dagger}$ is determined by the condition

$$0 = \eta^*(1)\varphi(1) - \eta^*(0)\varphi(0) \qquad \text{for all } \varphi \in D_A$$
$$= (\eta^*(1) - \mathrm{e}^{\mathrm{i}\alpha} \eta^*(0))\varphi(0)$$
$$= (\eta^*(1) - \mathrm{e}^{\mathrm{i}\alpha} \eta^*(0))\varphi(0).$$

This is guaranteed for all $\varphi(x)$ provided that

$$\eta(0) = \mathrm{e}^{\mathrm{i}\alpha} \eta(1)$$

as is the case for φ; this means that $D_{A_\alpha^\dagger} = D_{A_\alpha}$ and since $A_\alpha^\dagger \subset A_\alpha$, we see that $A_\alpha^\dagger = A_\alpha$, i.e. we find a 1-parametric family $B = A_\alpha$, $\alpha = (0, 2\pi)$ for the self-adjoint extension of A.

Note 4. A systematic method for the construction of all self-adjoint extensions of symmetric operators with $n_+ = n_- = n < \infty$ was invented by von Neumann (1955).

Denote by $\kappa_\pm$ the n-dimensional subspace generated by the solutions $\varphi_\pm^{(i)}$), $i = 1, \ldots, n$ of (A.8). Then the extensions A_U of A are indexed by the unitary operator $U : \kappa_+ \to \kappa_-$ and

$$DA_u = \{\psi = \varphi + \varphi_+ + U\varphi_+; \ \varphi \in D_A, \varphi_+ \in \kappa_+\}$$
$$A_u \psi = A\varphi + i\varphi_+ - iu\varphi_+.$$

Note. Our explicit example on construction of the left-adjoint extension fits into this general framework.

APPENDIX B

The distinction between finite and infinite numbers of degrees of freedom in quantum mechanics

Let us assume the case of N degrees of freedom, i.e. the states are elements of the Hilbert space

$$\mathcal{H} = \mathcal{L}^2(R^N, \mathrm{d}^N x) = \bigotimes_{k=1}^{N} (R, \mathrm{d}x_k) = \bigotimes_{k=1}^{N} \mathcal{H}_k$$

i.e. the space of square integrable functions $\phi = \phi(x_1, \ldots, x_N)$ in R^N. This space is separable as it has a countable basis

$$\Phi_{n_1, \ldots, n_N} = \varphi_{n_1}(x_1) \cdots \varphi_{n_N}(x_n)$$

where $\{\varphi_n(x), n = 0, 1, \ldots\}$ is a basis in $\mathcal{L}^2(R^N, \mathrm{d}^N x)$ (e.g. formed by the eigenfunctions of the harmonic oscillator).

Under certain domain assumptions von Neumann (1955) proved that the representation of canonical commutation relations (CCRs)

$$[X_i, X_j] = [P_i, P_j] = 0 \qquad [X_i, P_j] = \mathrm{i}\delta_{ij}$$

in a separable Hilbert space, $\mathcal{H}$, is unique up to unitary equivalence. This means that if we have another set of operators satisfying

$$[X_i', X_j'] = [P_i', P_j'] = 0 \qquad [X_i', P_j'] = \mathrm{i}\delta_{ij}$$

then there exists a unitary operator U, such that

$$X_i' = U X_i U^{-1} \qquad P_i' = U P_i U^{-1}.$$

Note 1. This is a very important theorem which tells us that any representation of CCRs for N degrees of freedom is unitarily equivalent to the standard one

$$X_i \Phi = x_i \Phi \qquad P_i \Phi = \frac{1}{\mathrm{i}} \partial_i \Phi$$

i.e. there is basically only one QM of N degrees of freedom. There is a nice example from the history of QM. At the very beginning of QM there were two independent formulations of QM, one by Heisenberg and the other by

Schrödinger. Later it was shown by Jordan and Pauli that they are equivalent and this was a motivation for von Neumann to prove his theorem.

Note 2. In the case of an infinite number of degrees of freedom, the Hilbert space in question is

$$\mathcal{H} = \bigotimes_{k=1}^{\infty} \mathcal{H}_k.$$

However, this space is non-separable as its base

$$\Phi_{n_1 n_2 \ldots} = \prod_{k=1}^{\infty} \varphi_{n_k}$$

is non-countable, even in the case where $\mathcal{H}_k$ itself is finite dimensional. For example, for an infinite chain of spins any $\mathcal{H}_2$ is two dimensional, having the base formed by φ_1 (= spin up) and φ_0 (= spin down). To any base element $\Phi_{n_1 n_2 \ldots}$ we can assign the real number

$$O_{n_1 n_2 \ldots} = \sum_{k=1}^{\infty} \frac{n_k}{2^k}$$

from the interval $(0, 1)$ (remember that $n_k = 0$ or 1), but the set $(0,1)$ is non-countable. Thus, our Hilbert space is non-separable. The von Neumann theorem is not applicable, and usually there are many inequivalent representations of CCRs (von Neumann (1938), Thirring (1983)). This situation, still mathematically not completely understood, is typical, e.g. for infinite spin systems in solid state physics, or in quantum field theory. Fortunately, we are dealing here with quantum mechanics of finite numbers of degrees of freedom and can use such nice results as, for example, given in the von Neumann theorem.

BIBLIOGRAPHY

Abramowitz M and Stegun I A 1970 *Handbook of Mathematical Functions* (New York: Dover)

Aharonov Y and Bohm D 1959 *Phys. Rev.* **115** 485

Bargmann V 1961 *Commun. Pure Appl. Math.* **14** 187

Bargmann V 1962 *Rev. Mod. Phys.* **34** 829

Bargmann V 1964 *J. Math. Phys.* **5** 862

Barut A O and Raczka R 1977 *Theory of Group Representations and Applications* (Warsaw: Polish Scientific Publishers)

Bethe H A and Salpeter E E 1957 *Quantum Mechanics of One- and Two-electron Atoms* (Berlin: Springer)

Biedenharn L C 1953 *J. Math. Phys.* **31** 289

Biedenharn L C, Blatt J M and Rose M E 1952 *Rev. Mod. Phys.* **24** 249

Biedenharn L C and van Dam H 1956 *Theory of Angular Momentum (Reprint Collection)* (New York: Academic)

Blatt J M and V F Weisskopf 1959 *Theoretical Nuclear Physics* (New York: Wiley)

Boerner H 1963 *Representation of Groups* (Amsterdam: North-Holland)

Bohm D 1989 *Quantum Theory* (New York: Dover)

Brink D M and Satchler G R 1968 *Angular Momentum* (Oxford: Clarendon)

Chaichian M and Nelipa N F 1984 *Introduction to Gauge Field Theories* (Berlin: Springer)

Chaichian M and Demichev A 1996 *Introduction to Quantum Groups* (Singapore: World Scientific)

Coleman S and Mandula J 1967 *Phys. Rev.* **159** 1251

Condon E U and Shortley G H 1953 *Theory of Atomic Spectra* (Cambridge: Cambridge University Press)

Cooper F, Khare A and Sukhatme U 1995 *Phys. Rep.* **251** 267

de Crombrugghe M and Rittenberg V 1983 *Ann. Phys.* **151** 99

de-Shalit A and Talmi I 1963 *Nuclear Shell Theory* (New York: Academic)

Dirac P A M 1981 *The Principles of Quantum Mechanics* (Oxford: Oxford University Press)

Edmonds A R 1957 *Angular Momentum in Quantum Mechanics* Princeton, NJ: Princeton University Press)

Elliot J P 1953 *Proc. R. Soc.* A **218** 345

Elliot J P and Dawber P G 1985 *Symmetries in Physics* vol 1 (London: Macmillan)

Fano U and Racah C 1959 *Irreducible Tensorial Sets* (New York: Academic)

Flügge S (ed) 1957 *Handbuch der Physik* (Berlin: Springer)

Forte S 1992 *Rev. Mod. Phys.* **64** 193

Galindo A and Pascual P 1990–91 *Quantum Mechanics 1–2* 2nd edn (Transl. from Spanish by J D Garcia and L Alvarez-Gaumé) (Berlin: Springer)

Gasiorowicz S 1996 *Quantum Physics* (New York: Wiley)

Gates S J, Grisaru M T, Roček M and Siegel W 1983 *Superspace: or One Thousand and One Lessons in Supersymmetry* (London: Benjamin–Cummings)

Gelfand I M, Minlos R A and Shapiro Z Ya 1963 *Representations of Rotation Group and Lorentz Group and their Applications* (New York: Pergamon)

Goldhaber A S and Mackenzie R 1988 *Phys. Lett.* **214B** 471

Goldin G A, Menikoff R and Sharp D H 1980 *J. Math. Phys.* **21** 650

Goldin G A, Menikoff R and Sharp D H 1981 *J. Math. Phys.* **22** 1664

Golfand Y A and Likhtman E S 1971 *JETP Lett.* **13** 323

Gradshteyn I S and Ryzhik I M 1984 *Tables of Integrals, Series and Products* (New York: Academic)

Green M B, Schwartz J H and Witten E 1987 *Superstring Theory* vols 1 and 2 (Cambridge: Cambridge University Press)

Haag R, Łopuszański J and Sohnius M 1975 *Nucl. Phys.* B **88** 257

Hagedorn R 1963 *Relativistic Kinematics: a Guide to the Problems of High-Energy Physics* (New York: Benjamin)

Hamilton J 1959 *The Theory of Elementary Particles* (Oxford: Clarendon)

Hilgerved J and Wouthuysen S A 1963 *Nucl. Phys.* **40** 1

Hill E L 1957 *Rev. Mod. Phys.* **23** 253

Humphrey J E 1972 *Introduction to Lie Algebras and Representation Theory* (New York: Springer)

Iengo R and Lechner K 1992 *Phys. Rep.* **213** 179

Ishidzu T, Horie H, Obi S, Sato M, Tanabe J and Yanagawa S 1960 *Tables of the Racah Coefficients* (Tokyo: Pan-Pacific)

Jackiw R and Redlich A N 1983 *Phys. Rev. Lett.* **50** 555

Jacob M and Wick G C 1959 *Ann. Phys., NY* **7** 404

Jacobson N 1961 *Lie Algebras* (New York: Wiley)

Jahn H A 1951 *Proc. R. Soc.* A **205** 192

Jahnke-Emde 1948 *Tables of Higher Functions* (Stuttgart: Teubner)

Jones H F 1990 *Groups, Representations and Physics* (Bristol: Institute of Physics)

Jordan P 1935 *Z. Phys.* **94** 531

Källen G 1964 *Elementary Particle Physics* (Reading, MA: Addison-Wesley)

Klauder J R and Skagerstam B-S 1985 *Coherent States, Applications in Physics and Mathematical Physics* (Singapore: World Scientific)

Landau L D and Lifshitz E M 1981 *Quantum Mechanics* (New York: Pergamon)

Laughlin R B 1983 *Phys. Rev.* B **23** 3383

Leinaas J M and Myrheim J 1977 *Nuovo Cimento* B **37** 1

Lerda A 1992 *Anyons (Lecture Notes in Physics 14)* (Berlin: Springer)

Madelung E 1950 *Die Matematischen Hilfsmittel des Physikers* (Berlin: Springer)

Magnus W and Oberhettinger F 1949 *Formulas and Theorems for the Special Functions of Mathematical Physics* (New York: Chelsea)

Merzbacher E 1961 *Quantum Mechanics* (New York: Wiley)

Messiah A 1970 *Quantum Mechanics* vols 1 and 2 (Amsterdam: North-Holland)

Morse P M and Fesbach H 1953 *Methods of Theoretical Physics* (New York: McGraw-Hill)

Naimark M A 1964 *Linear Representations of the Lorentz Group* (Oxford: Pergamon)

Nikiforov A F, Uvarov V B and Levitan Yu L 1965 *Tables of Racah Coefficients* (Transl. from the original Russian edition (Computing Center of the Academy of Sciences of the USSR, Moscow 1962) by P Basu) (New York: Pergamon)

Noether E 1918 *Nachr. d. Kgl. Ges. d. Wiss. Göttingen* p 235

Particle Data Group 1996 *Review of Particle Properties Phys. Rev.* D **54** 172

Prange R E and Girvin S M (eds) 1990 *The Quantum Hall Effect* (Berlin: Springer)

Racah G 1942 *Phys. Rev.* **62** 438

Regge T 1958 *Nuovo Cimento* **10** 544

Regge T 1959 *Nuovo Cimento* **11** 116

Rose M E 1957 *Elementary Theory of Angular Momentum* (New York: Wiley)

Ross G R 1984 *Grand Unified Theories* (London: Benjamin–Cummings)

Rotenberg M, Bivins R, Metropolis N and Wooten J K Jr 1959 *The 3j- and 6j-symbols* (Cambridge, MA: MIT Press) (table and formulae)

Schiff L I 1955 *Quantum Mechanics* (New York: McGraw-Hill)

Schweber S S 1961 *An Introduction to Relativistic Quantum Field Theory* (New York: Row and Peterson)

Schwinger J 1952 *US Atomic Energy Commission* NYO-3071; reprinted in Biedenharn and van Dam (1956) pp 229–79

Thirring W 1983 *Quantum Mechanics of Large Systems (A Course in Mathematical Physics 4)* (New York: Springer)

Tung Wu-Ki 1985 *Group Theory in Physics* (Singapore: World Scientific)

van der Waerden B L 1974 *Group Theory and Quantum Mechanics* (Berlin: Springer)

Varshalovich D A, Moskalev A N and Kersonshii V K 1988 *Quantum Theory of Angular Momentum* (Singapore: World Scientific)

Vilenkin N Ya 1968 *Special Functions and the Theory of Group Representations (Amer. Math. Soc. Transl. of Math. Monographs 22)*

Volkov D V and Akulov V P 1973 *Phys. Lett.* **46B** 109

von Neumann J 1938 *Compositio Math.* **6** 1–77

von Neumann J 1955 *Mathematical Foundations of Quantum Mechanics* (Princeton, NJ: Princeton University Press)

Weinberg S 1995 *Quantum Theory of Fields* vol 1 (New York: Cambridge University Press)

Wess J and Zumino B 1974 *Nucl. Phys.* B **70** 39

Wess J and Zumino B 1974 *Nucl. Phys.* B **78** 1

Wess J and Bagger J 1983 *Supersymmetry and Supergravity* (Princeton, NJ: Princeton University Press)

West P 1987 *Introduction to Supersymmetry and Supergravity* (Singapore: World Scientific)

Weyl H 1931 *Theory of Groups and Quantum Mechanics* (Princeton, NJ: Princeton University Press)

Wigner E P 1931 *Gruppentheorie und ihre Anwendung auf die Quantenmechanik der Atomspektren* (Braunschweig: Vieweg) (the following reference is an English translation of this book)

Wigner E P 1959 *Group Theory and Its Application to the Quantum Mechanics of Atomic Spectra* (New York: Academic)

Wilczek F 1982 *Phys. Rev. Lett.* **48** 114

Wilczek F 1982 *Phys. Rev. Lett.* **49** 957

Wilczek F 1990 *Fractional Statistics and Anyon Superconductivity* ed F Wilczek
   (Singapore: World Scientific) pp 3–101 (and articles therein)
Witten E 1981 *Nucl. Phys.* B **188** 513
Witten E 1982 *Nucl. Phys.* B **202** 253
Wybourn B G 1974 *Classical Groups for Physicists* (New York: Wiley–Interscience)
Zhelobenko D P 1973 *Compact Lie Groups and Their Applications (Amer. Math. Soc.*
   Transl. of Math. Monographs 40)

INDEX

For Product Safety Concerns and Information please contact our EU
representative GPSR@taylorandfrancis.com Taylor & Francis Verlag GmbH,
Kaufingerstraße 24, 80331 München, Germany

Printed and bound by CPI Group (UK) Ltd, Croydon, CR0 4YY
01/05/2025
01858547-0001